Sustainable Plastics

Sustainable Plastics

Environmental Assessments of Biobased, Biodegradable, and Recycled Plastics

Second Edition

Joseph P. Greene
California State University
Chico, California

WILEY

This edition first published 2023

Registered Office
John Wiley & Sons, Inc., 111 River Street, Hoboken, NJ 07030, USA

Editorial Office
111 River Street, Hoboken, NJ 07030, USA

For details of our global editorial offices, customer services, and more information about Wiley products visit us at www.wiley.com.

Wiley also publishes its books in a variety of electronic formats and by print-on-demand. Some content that appears in standard print versions of this book may not be available in other formats.

Library of Congress Cataloging-in-Publication Data applied for:
Hardback ISBN: 9781119882060

Cover Image: @Japril17/Shutterstock
Cover Design by Wiley

Set in 9.5/12.5pt STIXTwoText by Straive, Pondicherry, India

Contents

Acknowledgements

I would like to acknowledge Ms. Joann Greene for her support and love on the writing of this book.

1

Introduction to Sustainability

1.1 Sustainability Definition

Sustainability has many definitions. The most common definition of sustainability has its roots in a 1987 United Nations conference, where sustainability was defined as "meeting the needs of the current generation without compromising the ability of future generations to meet their needs" (WCED 1987). Sustainable materials, processes, and systems must meet this definition and not compromise the ability of future generations to provide for their needs while providing for the needs of the current generation. Thus, for plastics manufacturing, materials and processes used today should not deplete resources for future generations to produce plastic materials.

Sustainability can be measured by the outcomes of using a material, process, or system on the environment, society, and economy. The three components of sustainability have economic, social, and environmental aspects and are related with each other as shown in Figure 1.1.

Materials, processes, and systems can have environmental, economic, and societal impact. Sustainable materials, processes, and systems have all three impacts. For example, the development of materials will have environmental impacts of using raw materials, energy sources, and transportation that come from natural resources, which can create air, land, and/or water pollution; economic impacts are creating commerce, jobs, and industries; and societal impacts are creating roles for jobs and services. Some new materials for clothing were evaluated for sustainability (Provin et al. 2020). They found that relating new materials to the sustainable development goals of 2030 Agenda from the United Nations is necessary due to the important issues presenting challenges at the global level, in relation to the economic, social, and environmental pillars of sustainability.

Sustainable Plastics: Environmental Assessments of Biobased, Biodegradable, and Recycled Plastics, Second Edition. Joseph P. Greene.

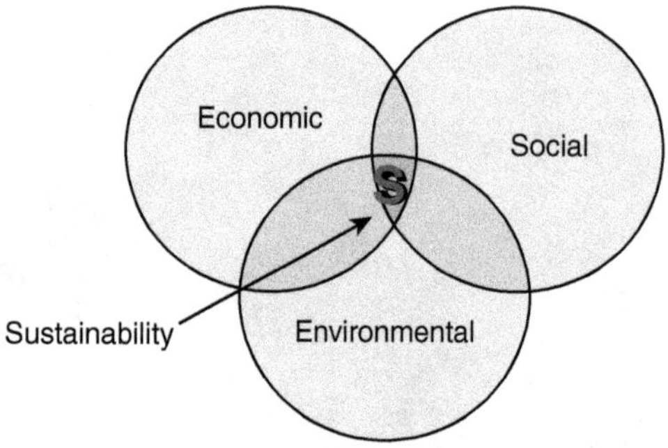

Figure 1.1 Sustainability definition.

Organizations are often analyzed with a "Triple Bottom Line" approach to evaluate the social, economic, and environmental performances of a company (Vanclay, F., 2004). This approach is the key to creating a sustainable organization. The "Triple Bottom Line" was used for biofuels as an excellent example of a sustainable fuel (Sala 2020). Biofuels meet the new economic paradigms that are related to the green economy, the bio-economy, and the circular economy. They developed an essential approach to apply life cycle thinking to production and consumption patterns to evaluate environmental and socioeconomic burdens and benefits, in an integrated manner. Lastly, they found that for socioeconomic sustainability, the assessment is going beyond the profit- and finance-oriented perspectives, including instead externalities associated to the activities under evaluation.

Examples of sustainability measures were developed for using a holistic approach from sustainability measurements of technology use in the marine environment (Basurko and Mesbahi 2012). The environmental effects of ballast water were measured with an integrated quantitative approach of sustainable assessment. The systematic approach can provide environmental, economic, and social sustainability for marine technologies.

The sustainable tool allows for the inclusion of sustainability principles to the design and operations of marine products. Sustainability can be effectively incorporated into the design phase of products and services and create reduced environmental, social, and economic impacts. The sustainable tool was created with LabView® software with SimPro® life cycle assessment (LCA) program to provide an integrated approach with a single indicator to reduce the environmental, social, and economic impacts of ballast water effects on the ocean quality. Another approach for the triple bottom line in sustainability was found for biomass ethanol production in China (Wang et al. 2020). They used a multi-regional input output (MRIO)-based hybrid LCA to estimate the sustainability of ethanol production in China. Employment, economic stimulus, and energy use were assessed. Bioethanol was found to be more effective on energy savings and economic stimulus than regular gasoline. Second-generation bioethanol had the highest energy return on the investment. Lastly, supply chain sectors make up the majority of social and environmental impacts.

1.1.1 Societal Impacts of Sustainability

The first aspect of sustainability can measure the impacts of products and processes on the society. The societal impact of using a material and manufacturing process can be measured by the effects on the population and the roles of the workers in the community. Sustainable manufacturing processes are defined as providing proper wages for the workers and a clean and safe work environment. The method and environment of producing a manufactured product can result in impacts on a person, group, and community. In another research work, sustainability was evaluated with supply chain management concepts (Vermeulen and Seuring, 2009). The authors propose three-generation approaches. Single firm approaches are the first generation. Joint product sector approaches are the second generation. And cross-sectoral approaches are the third generation. They propose that the various forms of sustainable supply chain governance clarify two aspects that have hardly been addressed in the general analysis of value chains: first, these varying forms of interaction, cooperation, and compelling rules in the value chain are an instrument of competition, partly based on specific quality assets of the products (namely the environmental and socio-ethical performance of value chain partners), and second, these forms of interaction and cooperation include other types of societal actor – apart from newly created nonprofit governance institutions and their (for profit) auditing and control bodies, consumer NGOs, development NGOs, and environmental NGOs also play diverse roles. They recommend that sustainable supply chain management and governance are important to companies, consumers, NGOs, and even governmental agencies. They propose that the challenges of climate change, energy provision, and creating wealth for an increasing world population will broaden the need for sustainability management and sustainable supply chain management and governance in the near future.

The wages, benefits, hours per week, safety, and other human resources provided to an individual worker contribute to the quality of the product or process and the ability of that product or process to maintain its presence in the marketplace. A workplace that produces a product or process without wages and benefits that are appropriate to the workers in the region can lead to high turnover rates of workers, poor worker morale, and loss of personal buy-in for workers. The product or service will not be sustainable since it may not last if few workers are available or the environment may suffer tragic losses due to health or safety concerns. Poor working conditions and poor wage structures may benefit the economics of the current company but may lead to poor working environments for future workers and thus is not sustainable.

Sustainable workplaces feature the maintaining of welfare levels in the future (WCED 1987). Welfare can be defined as a subjective measure of the sum of all individuals' utilities generated from the consumption of goods, products, and services (Perman et al. 2003).

1.1.2 Economic Impacts of Sustainability

The second aspect of sustainability can measure the economic impacts of using a material and manufacturing process to produce products. Sustainable manufacturing processes are defined as providing proper wages for the workers and clean and safe work environments. Economic impacts of sustainability can be measured with a capital approach that can be defined as maintaining economic, environmental, human, and social capital over time for future generations (Kulig et al. 2010). The capital approach can be proposed as a theoretical basis for sustainable development indicators (Atkinson and Hamilton 2003; World Bank 2006; UNECE 2014). The capital approach provides a theoretical approach by measuring all capital stocks in their own units. The capital approach can provide consistent, theoretically sound, and policy-relevant comparisons between countries (Kulig et al. 2010).

The economic benefits of sustainability were developed for water reuse. Sgroi et al. (2018) found that economics are the major barriers to actual development of water reuse. A holistic approach is needed to evaluate the sustainability of water reuse. Circular economy may lead to a "paradigm shift" to enhance resource recovery. Segregation at source may be a starting point for sustainable on-site resource recovery. The economic impact of using a material can be measured by the effects on the creation of jobs and industry for communities. The creation of jobs can lead to creation of taxable bases and tangible property. In addition, the use of sustainable materials and processes can lead to reduced energy, transportation, waste disposal, and utility costs for manufacturing operations. Sustainable enterprises can be defined as "Lean and Green," where manufacturing costs are minimized, and manufactured materials are made with reduced environmental impacts.

Recycling of metals, plastics, glass, paper, wood, waste inks and concentrates, waste oils, and industrial fluids can reduce the amount of trash that is sent to landfills and hazardous disposal sites and reduce the waste disposal costs. Use of recycled or bio-based plastics can reduce the manufacturing costs of some plastics. Use of lower energy pumps, motors, and lighting can reduce energy costs for plastics manufacturing.

The incorporation of sustainability into the business plan can lead to a design for sustainability paradigm where an eco-design approach can lead to integrating social, economic, environmental, and institutional aspects into the supply chain of an eco-friendly product line. This can lead to healthy organizations providing good jobs to healthy employees and contributing to the social network of the organization and community.

1.1.3 Environmental Impacts of Sustainability

The third aspect of sustainability can measure the environmental impacts of producing a product or system in terms of usage of natural resources for raw materials, energy, and real estate land. The production of plastic products can generate greenhouse gases (GHGs), solid and liquid waste, air pollution, water pollution, and toxic chemicals. Environmental aspects are measured with the life cycle process are explained later.

Strategic environmental assessment can be used to provide a basis for establishing sustainability for products and services (White and Noble 2013). Strategic environmental assessment can help ensure that policies, plans, and programs are developed in a more environmentally sensitive way. Strategic environmental assessment can support sustainability by providing a framework for decision-making, setting sustainability objectives, ensuring consideration of other sustainable of using strategic environmental assessment of sustainability. This can include some of the following:

- Providing a decision support framework for sustainability.
- Being adaptive to the decision-making process.
- Incorporating sustainability objectives and principles.
- Considering relevant sustainability issues early on.
- Adopting sustainability criteria.
- Identifying and evaluating other sustainable alternatives.
- Trickling-down sustainability.
- Capturing large-scale and cumulative effects.
- Enabling institutional change and transformational learning.

Environmental aspects of sustainability can be measured by monitoring resource depletion and pollution generation during the production of products or services. Resource depletion can include land use, energy usage, water usage, fossil fuel usage, among others. The pollution emissions can include GHGs, water pollution, air pollution, climate change, toxic chemical released, human toxicity, carcinogens released, summer smog creation, acidification, eutrophication, among others.

An important environmental concern is the increased amount of GHGs in the atmosphere. GHGs are gases in the atmosphere that absorb and emit thermal radiation within the thermal infrared range causing the planet to increase in temperature. During plastic manufacturing, GHGs are produced by the energy sources needed to mine the raw materials, processing the raw materials into pellets, conversion of the pellet into finished products, and transportation. GHGs comprise of gases that contribute to global warming by creating a layer of insulating gases that insulate the planet. These gases absorb and emit radiation within the thermal infrared range. GHGs include methane, carbon dioxide, water vapor, fluorocarbons, nitrous oxide, and ozone.

Carbon dioxide is the largest contributor to global warming due to its volume. Methane has a global warming rate of 22 times the rate for carbon dioxide. Typically, the production of these gases is listed in LCAs as CO2 equivalent (CO2eq). Thus, the formation of GHGs is listed as CO2eq. Reductions in GHGs can be done with lowering energy usage for products and services.

1.2 Green Chemistry Definitions

The American Chemistry Institute established green chemistry principles. The green chemistry engineering principles provide a framework for scientists and engineers to design and build products, processes, materials, and systems with lower environmental impacts. Green chemistry principles can be used to develop chemical products and processes that reduce or eliminate the use and generation of hazardous or toxic chemicals. The 12 principles of green chemistry are as follows (Anastas and Warner 1998):

1) Prevention
2) Atom economy
3) Less hazardous chemical synthesis
4) Designing safer chemicals
5) Safer solvents and auxiliaries
6) Design for energy efficiency
7) Uses of renewable feedstock
8) Reduce derivatives
9) Use of catalytic reagents
10) Design for degradation
11) Real-time analysis for pollution prevention
12) Inherent safer chemistry

Prevention of waste generation during the manufacturing of the chemicals can help reduce environmental impacts of chemical production. Atom economy guides developers in incorporating all materials in the creation of chemicals. Synthetic chemicals should be created with little or no toxicity to the human health and the environment. Solvents, separation agents, and other auxiliary substances should be used sparingly or not at all. Energy usage should be minimized in the creation of chemical substances. Renewable feedstock should be the material source of the chemical substances rather than fossil fuel-based sources.

Creation of unnecessary intermediates or derivatives should be minimized or avoided if possible to reduce chemical waste. Catalytic reagents should be used rather than stoichiometric reagents. Chemical products should be designed to biodegrade in a disposal environment rather than be a persistent pollutant. Real-time, in-process monitoring and control of hazardous substances should use analytical methodologies. Chemical substances and processes should minimize the potentials for accidental chemical spills, explosions, and fires.

The 12 green chemistry definitions can be grouped into three areas: reduction in energy usage, reduction in waste, and reduction in pollution. The reduction in energy area includes design for energy efficiency, use of renewable feedstock, and reducing derivatives principles. The reduction in waste area includes prevention, atom recovery, and use of catalytic reagents principles. The reduction in pollution includes less hazardous chemical synthesis, reducing derivatives, designing safer chemicals, designing safer solvents and auxiliaries, design for degradation, pollution prevention, and inherent safer chemistry principles. These three areas are used to define sustainable manufacturing.

Recently, a chapter in a book on green chemistry studied the importance of sustainability to business with the connections of green chemistry and green engineering (Coish et al., 2018). The value creation of green initiatives is further supported by the results of a survey in 2009 of more than 1500 worldwide executives and managers about their perspectives on business sustainability (Berns et al., 2009). The study provided evidence that sustainability has the potential to affect multiple value creation levers over both the short- and long-term goals for the company. They found that six aspects are common to the notions of eco-green, sustainable, and environmental innovation. Those aspects are as follows:

1) Innovation object: Product, process, service, method.
2) Market orientation: Satisfy needs/be competitive on the market.
3) Environmental aspect: Reduce negative impact (optimum: zero impact).
4) Phase: Full life cycle must be considered (for material flow reduction).
5) Impulse: Intention for reduction may be economical or ecological.
6) Level: Setting a new innovation/green standard to the firm.

Company leaders with a vision that incorporates sustainability play a key role in ensuring that sustainability is part of the organization's creative process. Leaders and managers can create the value system of a company that can impact the type of innovation (e.g. incremental vs. radical) and the extent of innovation that is pursued by the company. The value-adding connections between the management of eco-innovation and the principles of green chemistry and green engineering are essential in creating a sustainable company. Sustainability and use of green chemistry should be of interest to business leaders, managers across various departments, innovation experts, sustainability officers, directors of research and development, product designers, engineers, and other people from across all disciplines who have an interest in business sustainability.

1.3 Green Engineering Definitions

Green engineering can be defined as a process to develop products, processes, or systems with minimal environmental impacts. The full product life cycle is developed when evaluating the environmental sustainability of the product, process, or system. Green engineering can be a foundation of sustainability. The 12 principles of green engineering are as follows (McDonough et al. 2003; Anastas and Zimmerman 2003).

1) Inherent rather than circumstantial
2) Prevention instead of treatment
3) Design for separation
4) Maximize efficiency
5) Output-pulled versus input-pushed
6) Conserve complexity
7) Durability rather than immortality
8) Meet need, minimize excess
9) Minimize material diversity
10) Integrate material and energy flows
11) Design for commercial "afterlife"
12) Renewable rather than depleting resources

Sustainable engineering is based on maximizing product throughput, quality, efficiency, productivity, space utilization, and reducing costs. Products are designed with inherently nonhazardous methods and nontoxic materials. Waste should be reduced at its source and not discarded after production. Production operations should be designed to minimize energy consumption and material use. Energy and materials should require a product requirement rather than a material input. Material and energy inputs should be based on renewable sources rather than from fossil fuel sources.

End-of-life options for the product should be designed at the beginning of a product life rather than at the end of it. The design goal should be product-targeted durability rather than product immortality.

Universal functionality should not be a design goal. Multicomponent products should be designed to promote disassembly and value retention. Integration and interconnectivity with available energy and material flows should be designed into products, processes, and systems.

Environmental engineering covers a wide range of applications of applying science and engineering principles to improve the natural environment of air, water, and land resources. Environmental engineering provides healthy water, air, and land for human use that is compatible with other organisms. It provides platform for balance between humans and other species on the planet. It also remediates environmental pollution issues. Recently, biological solutions to problems in environmental engineering often involve engineers integrating apparently disjointed biological knowledge, and tailoring this knowledge to address specific engineering challenges. The emerging discipline of environmental biotechnology contributes to the field of environmental engineering. Biological solutions help in assessing the risk to human health and determining the effectiveness of environmental engineering design decisions to reduce this risk to an acceptable level. Environmental engineers are helping waste water treatment plants to become more sustainable with the use of green engineering principles. Green engineering provides a biological solution to the problem of highly concentrated organic pollution in waste water treatment plants. In such waste water treatment plants, the processes of microbial degradation of organic waste with biomass production followed by sedimentation are encouraged to occur in a highly controlled environment. Alternative technologies for total nitrogen removal have been developed by environmental engineers that avoid some of the inefficiency of nitrification followed by denitrification (Oerther, 2005).

1.4 Sustainability Definitions for Manufacturing

Environmental aspects of product manufacturing include production of liquid and solid wastes, air pollution, water pollution, and GHG emissions. Discharges from manufacturing facilities can lead to pollution of the sewers, water treatment plants, and neighborhoods. Pollution prevention in communities with manufacturing operations can be achieved with regional sustainability programs that provide to small- and medium-sized manufacturing companies pollution prevention technical assistance and financial incentives to reduce pollution at the manufacturing sources rather than at the waste water and solid waste disposal sites (Granek and Hassanali 2006). Pollution often includes heavy metals, particulates, sulfates, phosphates, petroleum-based oils, solid wastes, oil-based inks and concentrates, and other contaminants. Sustainable practices can reduce the pollutants by installing filters, using water-based inks, bio-based oils, and recovery units for waste water effluent.

Sustainability can be defined in many ways for production companies to reduce GHGs and reduce pollution. Often missing from sustainability analysis, though, is waste generation. Products or services that are sustainable must also not produce significant amounts of solid or liquid waste. Sustainable products and practices should encourage the use of recycled materials during the production of products and processes and encourage the recycling of waste materials during the production of products and processes.

The essential components of sustainable products and services are ones with reduced GHG emission, reduced pollution, and reduced waste generation. Sustainable products, processes, and systems minimize the generation of GHGs, waste, and pollution. Thus, sustainable manufacturing incorporates producing products and processes with:

1) reduced GHGs emissions,
2) reduced solid waste, and
3) reduced pollution.

This definition will be used in subsequent chapters in the book. The first component of sustainable manufacturing processes is the reduction in GHGs. Reductions in GHGs can be done with lowering energy usage, which has direct cost reduction implications. Sustainable materials and processes minimize the generation of CO2eq gases. The second component of sustainable manufacturing is the reduction in waste generation. This can be listed for plastics manufacturing as the solid wastes are generated during the extraction of raw materials, production of the plastic pellets, and conversion of the pellet into plastic products. The listing of waste generation is listed as *kilogram of solid waste*. California in the United States has a law that requires state agencies and schools to achieve greater than 50% diversion rate of solid wastes (California Assembly Bill 939 2014), wherein over 50% of the trash that could be sent to landfill is sent to recycling, composting, or reuse. Reductions in waste generation can reduce the cost for manufacturing operations. Sustainable materials and processes minimize the generation of solid waste.

The third component of sustainable manufacturing processes is the reduction in pollution of air, land, and water. The pollution component can be defined in LCAs as creation of chemicals that cause eutrophication, acidification, and human health concerns.

Eutrophication can be defined as the addition of nitrates and phosphates to the land through the use of fertilizers and soil conditioners. Eutrophication is a very common pollutant from fertilizers in farming or from natural causes. Eutrophication can deplete oxygen in ocean and freshwater lakes causing algae and phytoplankton blooms in the water.

Acidification can occur in ocean and freshwater, as well as in soil when the pH is reduced due to the presence of sulfur and nitrous oxides. The presence of sulfur and nitrous oxides in the atmosphere can be released into the soil and waterways during rain storms. Sulfur and nitrous oxides are released during the combustion of fossil fuels at energy plants, burning of plastics as fuel, and during the combustion of fuels.

Toxic chemical pollution is caused by the presence of toxins that can cause human health problems, including cancer, blindness, sterility, and other health concerns. Combustion of fuels can lead to release of carcinogenic materials into the environment. Sustainable materials and processes reduce the release of pollution in the land, air, and water.

1.5 Life Cycle Assessment (LCA)

LCAs are an essential component of sustainability and can be used to scientifically determine the environmental effects of products, processes, and systems. LCA can be used to calculate the energy and raw materials consumed and the resulting carbon footprint, waste, and pollution generated in the production of a product or process. LCA is needed to establish the sustainability of products and processes because it follows a worldwide thorough approach to establishing measurable environmental outcomes of products and processes. LCA will be more fully explained in later chapters.

1.6 Lean and Green Manufacturing

Sustainability is an essential component of manufacturing today. Plastics manufacturing can lead the way in producing products with lower carbon footprint, lower waste, and lower pollution through the use of recycled and bio-based materials. Lean and Green are essential components of the manufacturing industry. Lean and Green manufacturing for plastics can be a unique feature of plastics manufacturers and can provide sustainable products for a promising marketplace.

1.7 Summary

Sustainable materials, processes, and systems must not compromise the ability of future generations to provide for their needs while providing for the needs of the current generation. The three components of sustainability have economic, social, and environmental aspects. Organizations are often analyzed with a "Triple Bottom Line" approach to evaluate the social, economic, and environmental performances of a company. The first aspect of sustainability can measure the impacts of products and processes on the society. The societal impact of using a material and manufacturing process can be measured by the effects on the population and the roles of the workers in the community.

The second aspect of sustainability can measure the economic impacts of using a material and manufacturing process to produce products. Sustainable manufacturing processes are defined as providing proper wages for the workers and clean and safe work environments.

The third aspect of sustainability can measure the environmental impacts of producing a product or system in terms of usage of natural resources for raw materials, energy, and real estate land. The production of plastic products can generate GHGs, solid and liquid wastes, air pollution, water pollution, and toxic chemicals.

Green chemistry principles can be used to develop chemical products and processes that reduce waste generation, energy, and production of toxic chemicals during the creation of chemicals. Green engineering principles are based on maximizing product throughput, quality, efficiency, productivity, and space utilization, as well as reducing hazards, pollution, and costs. Sustainable products, processes, and systems minimize the generation of GHGs, waste, and pollution. LCAs are an essential component of sustainability and can be used to scientifically determine the environmental effects of products, processes, and systems.

References

Anastas, P. and Warner, J. (1998). *Green Chemistry: Theory and Practice*, 30. New York: Oxford University Press.

Anastas, P.T. and Zimmerman, J.B. (2003). Design through the 12 principles of green engineering. *Environ. Sci. & Technol.* 37 (5): 94A–101A.

Atkinson, G. and Hamilton, K. (2003). Savings, growth, and the resource cure hypothesis. *World Dev.* 31 (11): 1893–1807.

Basurko, O.C. and Mesbahi, E. (2012) "Methodology for the sustainability assessment of marine technologies." *J. Clean. Prod.* doi:https://doi.org/10.1016/j.jclepro.2012.01.022.

Berns, M., Townsend, A., Khayat, Z. et al. (2009). Sustainability and competitive advantage. *MIT Sloan Manage. Rev.* 51: 19e26.

California Assembly Bill 939 (2022) History of California Solid Waste Law, 1985–1989, Cal Recycle. https://leginfo.legislature.ca.gov/faces/billTextClient.xhtml?bill_id=198919900AB939 (accessed January 2022).

Coish, P., McGovern, E., Zimmerman, J.B., and Anastas, P.T. (2022). Chapter 3. The value-adding connections between the management of eco-innovation and the principles of green chemistry and green engineering, Green Chemistry. http://dx.doi.org/10.1016/B978-0-12-809270-5.00033-9 (accessed January 2022).

Granek, F. and Hassanali, M. (2006). The Toronto Region Sustainability Program: insights on the adoption of pollution prevention practices by small to medium-sized manufacturers in the Greater Toronto Area (GTA). *J. Clean. Prod.* 14: 572–579.

Green Chemistry (2013) The twelve principles of green chemistry, American Chemical Society Green Chemistry Institute. http://www.acs.org/content/acs/en/greenchemistry/about/principles/12-principles-of-green-chemistry.html (accessed January 2022).

Kulig, A., Kolfoort, H., and Hoekstra, R. (2010). The case for the hybrid capital approach for the measurement of the welfare and sustainability. *Ecol. Indic.* 10 (2): 118–128.

McDonough, W., Braungart, M., Anastas, P.T., and Zimmerman, J.B. (2003). Applying the principles of green engineering to cradle-to-cradle design. *Environ. Sci. Technol.* 37 (23): 434A–441A.

Oerther, D. (2005). Chapter 6. Biological solutions. In: *Environmental Solutions* (ed. F. Agardy and N. Nemerow). 127–141 Elsevier, Inc. https://www.researchgate.net/publication/261215148_Chapter_6_Biological_Solutions (accessed January 2022).

Perman, R., Ma, Y., McGilvray, J., and Common, M. (2003). *Natural Resource and Environmental Economics*. Upper Saddle River, NJ: Pearson Higher Education.

Provin, A., de Aguiar Dutra, A.R., Machado, M.M., and Cubas, A.L.V. (2020). New materials for clothing: rethinking possibilities through a sustainable approach. *J. Clean. Prod.* 282: 124444.

Sala, S. (2020). Chapter 2 – Triple bottom line, sustainability and sustainability assessment, an overview. In: *Biofuels for a More Sustainable Future: Life Cycle Sustainability Assessment and Multi-Criteria Decision Making*, 47–72.

Sgroi, M., Vagliasindi, F.G.A.; Roccar, P. (2022) Feasibility, sustainability and circular economy concepts in water reuse. https://www.researchgate.net/publication/322876353_Feasibility_sustainability_and_circular_economy_concepts_in_water_reuse (accessed January 2022).

The World Bank (2013) Washington, DC. WealthofNationsconferenceFINAL.pdf (accessed June 2013).

UNECE (United Nations Economic Commission for Europe). (2014) Measuring sustainable development. Prepared in cooperation with the OECD and Eurostat, United Nations Economic Commission for Europe, United Nations, New York and Geneva. chrome-extension://efaidnbmnnnibpcajpcglclefindmkaj/viewer.html?pdfurl=https%3A%2F%2Fwww.oecd.org%2Fgreengrowth%2F41414440.pdf&clen=772427&chunk=true (accessed January 2022).

Vanclay, F., (2004). The triple bottom line and impact assessment: how do TBL, EIA, SIA, SEA and EMS relate to each other? *Journal of Environmental Assessment Policy and Management*, 6 (3), 265–288.

Vermeulen, W. and Seuring, S. (2009). Sustainability through the market – the impacts of sustainable supply chain management: introduction. *Sustain. Dev.* 17: 269–273.

Wang, C. Malik, A., Wang, Y., Change, Y., Lenzen, and M., Zhou, D. (2020). The social, economic, and environmental implications of biomass ethanol production in China: a multi-regional input–output-based hybrid LCA model. *J. Clean. Prod.* 249 (10): 119326.

WCED (World Commission on Environment Development) (1987) *Our Common Future, World Commission on Environment and Development*, p. 10. Oxford University Press, Oxford, UK.

White, L. and Noble, B. (2013). Strategic environmental assessment for sustainability: a review of a decade of academic research. *Environ. Impact Assess. Rev.* 42 (2013): 60–66.

World Bank (2006) Where is the wealth of nations? Measuring capital for the XXI century. https://econpapers.repec.org/bookchap/wbkwbpubs/7505.htm (accessed January 2022).

2

Environmental Issues

2.1 The Planet Is Warming

According to the World Meteorological Organization, the average temperatures of the planet in 2011 were the second highest strong "La Niña" temperatures in recorded history. Measurements on land and sea from over 10 000 locations in the world found that an average global temperature was $14.41 \pm 0.11\,°C$ (World Meteorological 2012). In 2010, the National Oceanic and Atmospheric Administration's (NOAA) National Climatic Data Center (NCDC) published a report, which found that 2010 was the warmest of the average yearly temperature on record for the planet (State of the Climate 2012).

In the Northern Hemisphere, the snow cover from December to February melted in May to the lowest area covered as recorded by satellites. The mountain glaciers on the Earth lost size and mass from the previous year. This trend of glacier reductions has continued since 1990 (State of the Climate 2012).

The air temperatures over the land of the Earth, especially in the Artic, have increased in 2010 from the previous year. The NOAA report found that the average temperature in the United States in 30-year timespan from 1981 to 2010 was a half degree Fahrenheit warmer than from the previous 30-year timespan from 1971 to 2000 (State of the Climate 2012). As shown in Figure 2.1, greenhouse gases (GHGs) can cause the temperatures of the land and sea to increase.

Figure 2.2 shows the levels of carbon dioxide concentrations in the last 800 000 years. The levels of CO_2 concentrations have remained relatively flat until the last 70 years.

In recent years since 1980, the levels of CO_2 in the atmosphere are steadily increasing every year as shown in Figure 2.3.

Sustainable Plastics: Environmental Assessments of Biobased, Biodegradable, and Recycled Plastics, Second Edition. Joseph P. Greene.

Figure 2.1 Graphic of planet warming with a blanket of greenhouse gases. Source: Artwork courtesy of Ms. Vanessa Vaquera of Chico, CA.

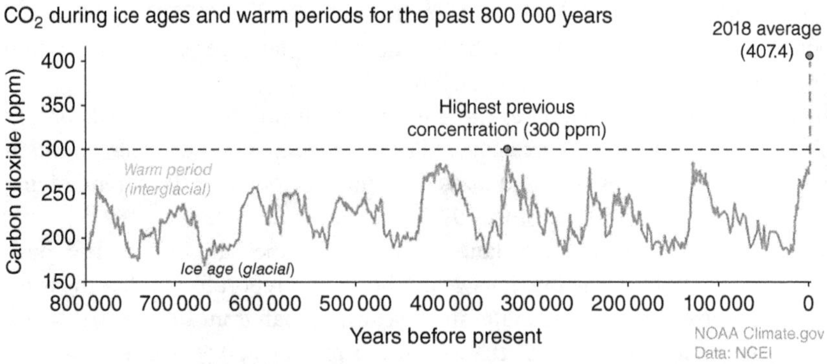

Figure 2.2 Carbon dioxide levels over 800 000 years. Source: CO_2 Levels (2020).

The CO_2 concentrations are shown in Figure 2.3 from 1975 to 2020. The resulting temperature from 1860 to 2020 is shown in Figure 2.4.

Figure 2.4 shows that the planet's average surface temperature has risen about 2.05 °F (1.14 °C) since the late nineteenth century, a change driven largely by increased carbon dioxide and other human-made emissions into the atmosphere. Most of the warming occurred in the past 40 years, with the six warmest years on record taking place since 2014. Not only was 2016 the warmest year on record, but eight months out of that year – from January through September, with the

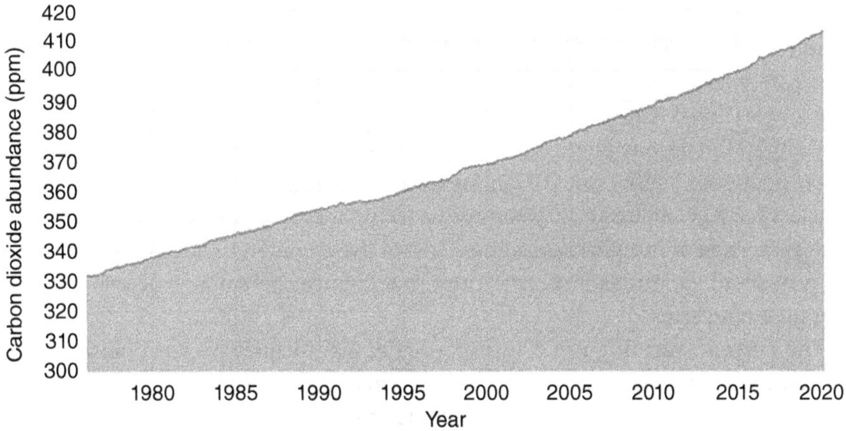

Figure 2.3 Levels of CO_2 in the atmosphere since 1980. Source: CO_2 Levels (2020).

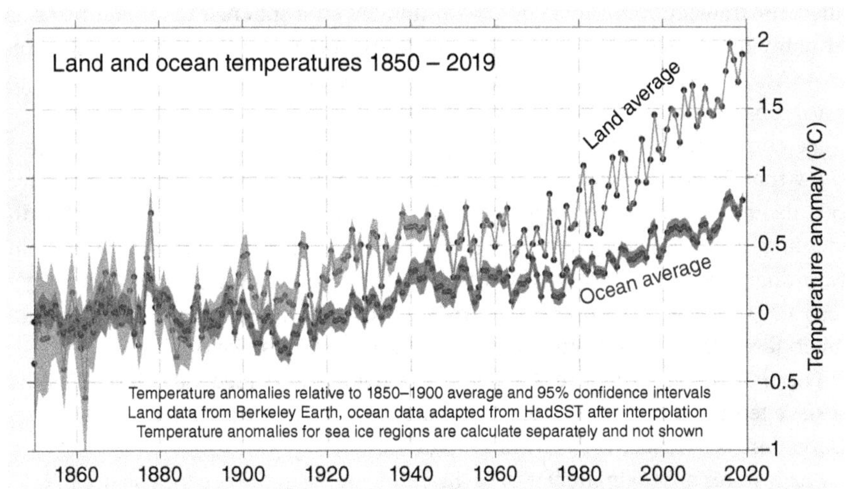

Figure 2.4 Global average temperature of the world from 1850 to 2019. Source: World temperature.

exception of June – were the warmest on record for those respective months. Figure 2.4 shows that since 1976 the difference between the average temperature and the twentieth century average was greater than 0. Figure 2.3 shows that since 1976 the yearly average temperature was warmer than the twentieth century average of 13.9 °C. Figure 2.4 clearly shows that the planet is warming. The surface temperature is warming at a faster rate than the water temperature. In 2011, the global average land surface temperature was 0.8 °C (1.49 °F) above the twentieth century average of 8.5 °C (47.3 °F). In 2011, the global average ocean and sea temperature was 0.40 °C (0.72 °F) above the twentieth century average of 16.1 °C (60.9 °F) (Global Analysis 2013).

Earth's climate has changed throughout history. Just in the last 650 000 years there have been seven cycles of glacial advance and retreat, with the abrupt end of the last Ice Age about 11 700 years ago marking the beginning of the modern climate era – and of human civilization. Most of these climate changes are attributed to very small variations in Earth's orbit that change the amount of solar energy our planet receives.

The current warming trend is of particular significance because most of it is extremely likely to be due to the result of human activity since the mid-twentieth century and proceeding at an unprecedented rate.

Earth-orbiting satellites and other technological advances have enabled scientists to study the climate-changing phenomena. The heat-trapping nature of carbon dioxide and other gases was demonstrated in the 1800s. Their ability to affect the transfer of infrared energy through the atmosphere is the scientific basis of many instruments flown by National Aeronautics and Space Administration (NASA). There is no question that increased levels of GHGs must cause Earth to warm up in response.

Ice cores drawn from Greenland, Antarctica, and tropical mountain glaciers show that Earth's climate responds to changes in GHG levels. Ancient evidence can also be found in tree rings, ocean sediments, coral reefs, and layers of sedimentary rocks. This ancient, or paleoclimate, evidence reveals that current warming is occurring roughly 10 times faster than the average rate of Ice-Age-recovery warming. Carbon dioxide from human activity is increasing more than much faster than it did from natural sources after the last Ice Age.

The time series in Figure 2.5 shows the five-year average variation of global surface temperature from 2019. Lighter areas indicates areas cooler than average. Darker areas indicates areas warmer than average.

Earth stores approximately 90% of the extra energy in the ocean. The ocean covers about 70% of Earth's surface. The ocean plays a large part in Earth's environment. One of its largest roles is to soak up energy (heat) and distribute it more evenly throughout the Earth. The ocean also helps to absorb Earth's CO_2. But our world is a water world. The ocean covers 70% of Earth's surface. The average depth of the ocean is about 2.7 miles. The ocean contains about 97% of all the water on Earth. The ocean plays a significant role in the climate change in the environment. The ocean soaks up energy in the form of heat and distributes it more evenly around the Earth. The ocean also soaks up CO_2.

The ocean does an excellent job of absorbing excess heat from the atmosphere. The top few meters of the ocean store as much heat as Earth's entire atmosphere.

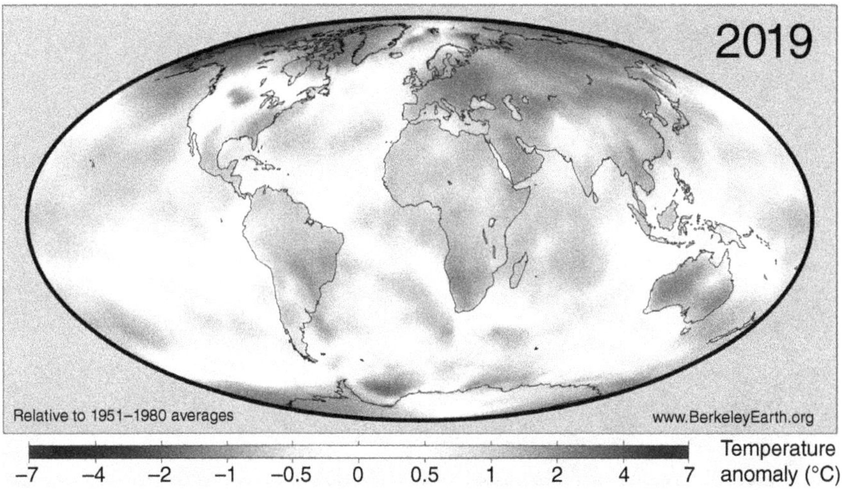

Figure 2.5 Global average temperature of the world from 2019. Source: Global Analysis (2013).

So, as the planet warms, it is the ocean that gets most of the extra energy. But if the ocean gets too warm, then the plants and animals that live in it must adapt or perish. All of the creatures in the ocean depend upon the supply of planktons, which are tiny plants and animals. These small creatures are at the bottom of the food chain.

Planktons include many different kinds of tiny animals, plants, or bacteria that just float and drift in the ocean. Other tiny animals such as krill (sort of like little shrimp) eat the plankton. Fish and even whales and seals feed on the krill. In some parts of the ocean, krill populations have dropped significantly in recent years.

Coral is another ocean creature in trouble due to global warming. Coral is a very fragile animal that builds a shell around itself and lives in very well with a certain kind of colorful algae. The algae make food from sunlight with a process called photosynthesis. The coral and algae share the food. Fish love to swim around the coral and algae and hide from predators. As the ocean warms, it increases in acidification with a lowering of the pH levels. This reduces the amount of food that the algae create and the coral die and turn white in a process called bleaching (Coral bleaching 2020).

2.2 Melting of Glaciers

One consequence of the warmer surface temperatures is the melting of glaciers worldwide, as shown in Figure 2.6. Glaciers are found throughout the Northern and Southern Hemispheres, including Africa, North and South America, Asia,

Figure 2.6 Melting glaciers illustration. Source: Artwork courtesy of Ms. Vanessa Vaquera of Chico, CA.

Europe, New Zealand, and Antarctica. Over 130000 glaciers are listed in the World Glacier Index (WGI; World Glacier Inventory 2012). The WGI includes information about the glaciers including area, elevation, geographic location, length, and classification. In Norway, information on 1627 glaciers was published in 1988. The average temperature in Norway is expected to rise by2.3 °C by 2100. The climate change scenario can result in approximately 98% of the Norwegian glaciers being disappeared and approximately 34% of the glacier area being reduced by 2100 (Nesje et al. 2008). In Asia, glaciers in the Everest region have lost mass in recent years. Between 1999 and 2004, satellite measurements found that the rate of ice loss is twice as much as from 1977 and 1999. Good agreement was found between satellite measurements and field measurements between 2002 and 2004 (Berthier et al. 2007). Based on the analysis of entrapped air from ice cores extracted from permanent glaciers from various regions around the globe, it has been demonstrated that **global warming began 18,000 years ago**, accompanied by a steady rise in atmospheric **carbon dioxide**. Humans are quite likely the cause of a large portion of the 80 ppm rise in CO_2 since the dawn of the Industrial Revolution, and from a distance, it looks possible that increasing CO_2 may cause atmospheric temperatures rise. (Global warming 2022).

In South America, 72 glaciers in Northern and Southern Patagonia were studied for glacier retreat. Twenty of the 72 glaciers were found to retreat from 3 to 37.9%. Glacier retreat is attributed to average air temperature, basin geometry, glacier dynamics, and response time (Lopez et al. 2010).

In North America, Mendenhall Glacier in Alaska is undergoing substantial recession and thinning. The glacier has retreated 3 km during the twentieth century. The glacier has thinned 5.5 km^3 in the last 50 years. The shrinking of the glacier is attributed to surface melting and to lake calving. Lake calving can be defined as when an edge of an iceberg or glacier separates from the iceberg or

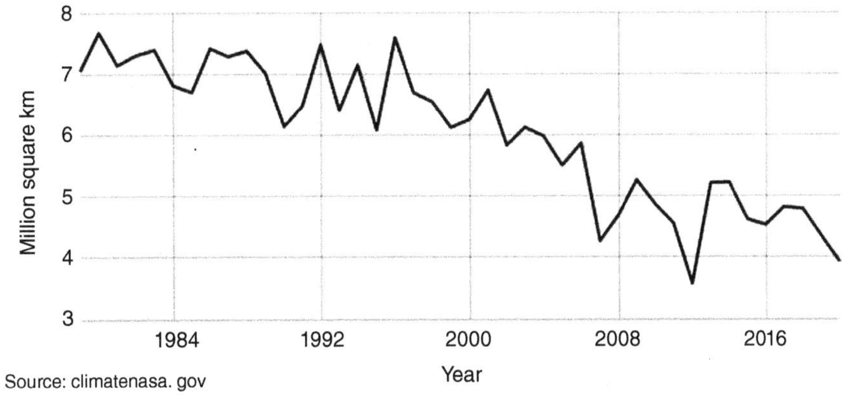

Figure 2.7 Decline in the Arctic sea ice.

glacier. Mean temperatures in Juneau, Alaska, decreased slightly from 1947 to 1976 and then increased since then. The average temperature in Juneau increased 1.4 °C since 1943. Glacier melting in Alaska flows into the Mendenhall River. Approximately 50% of the total river discharge in the summer is due to glacier melting (Motyka et al. 2002).

An important feature of the Arctic Ocean is its floating sea-ice cover that has traditionally ranged from 16 million km^2 in March to a minimum coverage of 7 million km^2 at the end of summer melt season in September. Based on regression analysis during the period between 1979 and 2006, the ice extent (fractional ice cover) has declined every month. The decline is the largest during September with a retreat of $8.6 \pm 2.9\%$/decade or approximately 10 000 km^2/year (Serreze et al. 2007).

Both the extent and thickness of Arctic sea ice have declined rapidly over the last several decades. According to the National Snow and Ice Date Center, average September sea-ice extent has been reduced by 32% from 1979 to 2017 (Artic Ice Decline 2020). Figure 2.7 shows the decline in the Arctic sea ice. The decline in the average September area is 13.1%/decade.

Ice, which covers approximately 10% off the Earth's surface, is disappearing rapidly. The ice can be in the form of glaciers in Greenland, Iceland, the Artic, and the Antarctica (Ibid).

Approximately, 400 billion tons of ice disappeared due to total glacier loss since 1994 (Ibid). The Greenland and Antarctic ice sheets have decreased in mass. Data from of ice loss per year between 1993 and 2019, while Antarctica lost about 148 billion tons of ice/year (NASA Gravity Recovery 2021).

2.3 Rising Seas

The levels of the oceans are rising. This can be attributed to thermal expansion of the water, land water storage change, and melting of glaciers and icebergs. Sea levels are measured with satellite altimetry or with in situ and remote sensing equipment. Recent ice mass loss and melting glaciers have increased the rate of sea-level rise to 3.1 mm/year (Cazenave et al. 2008). Between the years of 1993 and 2010, melting glaciers and polar ice caps can account for approximately 30% of the sea-level rise (Cogley 2009).

Sea-level rise in the tropical Pacific and Indian Oceans has been studied in recent years. The sea level is influenced by thermal expansion of the oceans from the warmer temperatures, melting polar ice caps and glaciers, and the atmospheric changes caused by El Niño. Large variability during El Niño years and the shortness of many of the individual tide-gauge records contribute to uncertainty in historical rates of sea-level rise.

From 1993 to 2001, the western Pacific and eastern Indian Oceans experienced large rates of sea-level rise that approached 30 mm/year, whereas the eastern Pacific and western Indian Oceans experienced a sea-level fall approaching −10 mm/year (Church et al. 2006). From 1950 to 2001, the average sea-level rise (relative to land) from the six longest tide-gauge records is 1.4 mm/year. After correcting for glacial isostatic adjustment and atmospheric pressure effects, the rate of sea-level rise is 2.0 mm/year, which is close to the estimates of the global averages. The relative sea-level rise at Funafuti, Tuvalu, is 2 ± 1 mm/year over the period 1950–2001. The analysis clearly indicates that sea levels in this region are rising (Church et al. 2006). The rising of the sea levels can cause flooding in low-lying areas in Los Angeles, New Orleans, New York City, among others. Figure 2.8 displays the consequence of flooding for homes.

In addition to global warming and melting of the ice caps, increase in GHG concentrations can also alter the planet's hydrologic cycle (Mitchell et al. 1987;

Figure 2.8 Rising sea waters causing flooding of a home. Source: Artwork courtesy of Ms. Vanessa Vaquera of Chico, CA.

Held and Soden 2006). If changes in rainfall intensity and spatial distribution are substantial, then significant amounts of rainfall can occur and pose a serious risk caused by climate change (Wentz et al. 2007).

The number of record high-temperature events in the United States has been increasing, while the number of record low-temperature events has been decreasing, since 1950. The United States has also witnessed increasing numbers of intense rainfall events (Intense rainfall, USGCRP 2017).

2.4 Causes of Global Warming

In its Fifth Assessment Report, the Intergovernmental Panel on Climate Change, a group of 1300 independent scientific experts from countries all over the world under the auspices of the United Nations, concluded there is a more than 95% probability that human activities over the past 50 years have warmed our planet (Climate change causes 2020).

The industrial activities that our modern civilization depends upon have raised atmospheric carbon dioxide levels from 280 to 414 parts per million in the last 150 years. The panel also concluded there is a better than 95% probability that human-produced GHGs such as carbon dioxide, methane, and nitrous oxide have caused much of the observed increase in Earth's temperatures over the past 50 years.

2.4.1 Increased Greenhouse Gases

The National Climate Data Center report from the NOAA found that the levels of GHGs in the atmosphere increased in 2010 from the previous year (NOAA 2012). In fact, the rate that the GHG increased from 2009 to 2010 was the largest yearly rate of increase since 1980. Many scientists have concluded that carbon dioxide and other GHGs can trap heat in the Earth's lower atmosphere (Majorowicz and Skinner 1997; Golovanova et al. 2001; Beltrami and Bourlon 2004; Viola et al. 2010).

2.4.2 Sources of CO_2eq Emissions

The sources of CO_2eq emissions are from burning of fossil fuels. "Carbon dioxide equivalents" refer to the amount of carbon dioxide that would give the same warming effect as the effect of the GHGs being emitted. According to the United States Environmental Protection Agency (EPA) from 2006 as shown in Figure 2.8, 40% of the CO_2 emissions are from burning fossil fuels for electricity, 31% from transportation, 14% from industrial sources, 10% from residential or commercial sources, and 5% from other sources (EPA Inventory of Greenhouse Gas Emissions

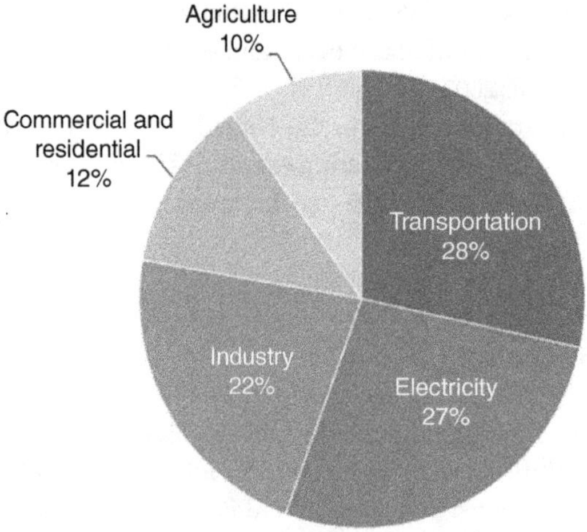

Figure 2.9 Sources of CO_2eq emissions in 2018 for the United States per market segment. Source: EPA emissions (2020).

and Sinks 2013). In the industrial sector, steel and concrete industries have the highest CO_2eq emissions. Petrochemical industry, including plastics, has the tenth largest contribution. Figure 2.9 shows the sources of CO_2eq emissions from the US market segment (EPA emissions 2020).

Figure 2.10 shows the types of gases that account for the majority of the emissions.

Figure 2.11 shows the sectors of CO_2eq emission from 1990 to 2015 (Ibid). Figure 2.11 shows that the economic sectors remain consistent from year to year. Transportation and electricity generation have been the largest contributors to the carbon dioxide emissions from 1990 to 2015. Residential use has been the lowest contributor during the time period.

Figure 2.12 shows the global CO_2eq emission by sector in 2018. From the figure, transportation is the largest contributor to the carbon dioxide emissions in 2018. Industry accounts for only 32% of the emissions. Building materials and construction had the lowest contribution during 2018 (Global emission by sector 2020).

Figure 2.13 shows the breakdown of the global GHG emissions by industry (GHG emission by industry 2020). The iron and steel industry is the largest contributor to GHG emissions, with 25% of the emissions. Concrete industry is the second largest contributor, with approximately 19% of the emissions. The plastics industry is a much lower contributor, with 5% of the emissions. Aluminum industry is the lowest contributor and only releases 3% of the GHG emissions. The total global GHG emission was 9.9 $GtCO_2eq$.

Scientists attribute the heating of the planet to human activity causing a "greenhouse effect" as shown in Figure 2.1. The "greenhouse effect" is caused when the atmosphere traps heat radiating from Earth toward space. Certain gases in the

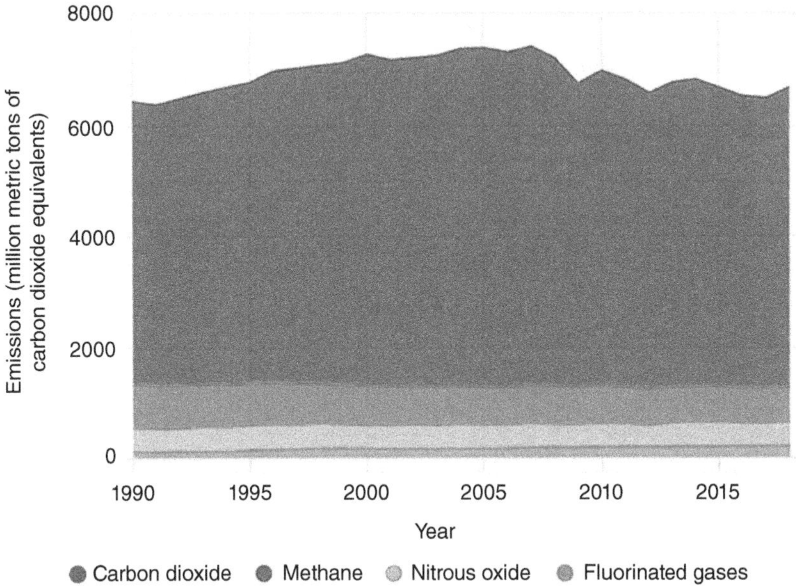

Figure 2.10 Types of gases in the CO_2eq emissions in 2018 for the United States per market segment. Source: EPA emissions (2020).

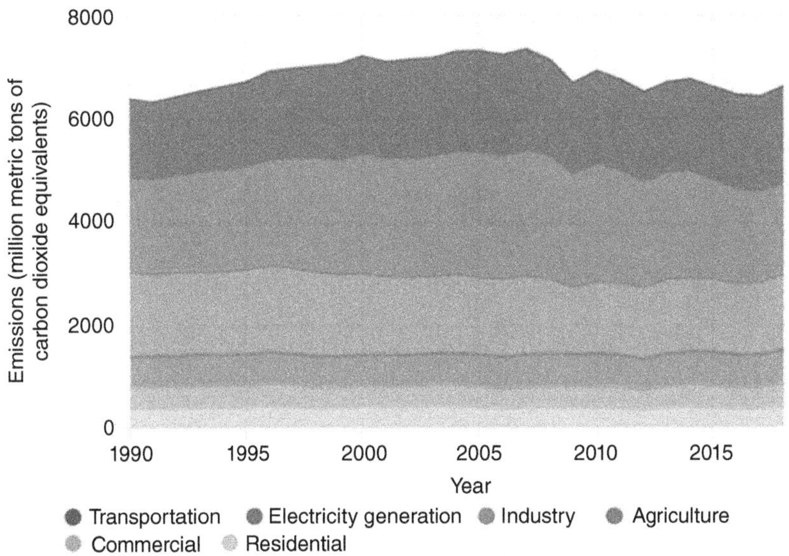

Figure 2.11 US greenhouse CO_2eq emissions from 1990 to 2015. Source: EPA emissions (2020). NOAA/Marine debris from Wikpedia 2020/Public Domain.

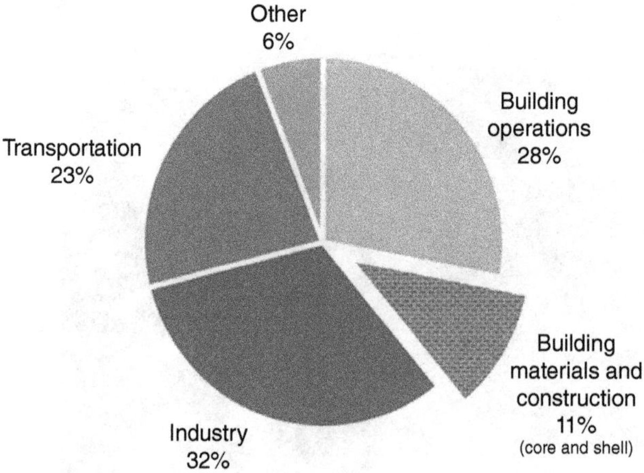

Figure 2.12 Global GHG emissions by sector in 2018. Source: Global emission by sector (2020).

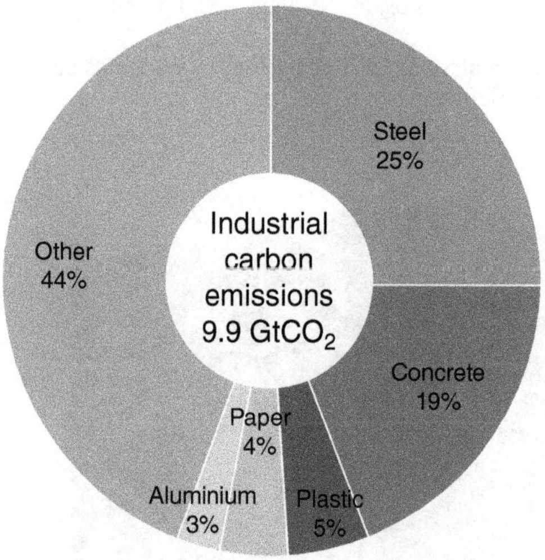

Figure 2.13 Global GHG emissions by sector in 2018. Source: Global emission by sector (2020).

atmosphere block heat from escaping. There are five significant "greenhouse gases" (GHG gases 2020) that contribute to the greenhouse effect, which include:

1) *Water vapor:* The most abundant GHG, but, importantly, it acts as a feedback to the climate. Water vapor increases as the Earth's atmosphere warms, but so does the possibility of clouds and precipitation, making these some of the most important feedback mechanisms to the greenhouse effect.

2) *Carbon dioxide (CO_2)*: A minor but very important component of the atmosphere, carbon dioxide is released through natural processes such as respiration and volcano eruptions and through human activities such as deforestation, land use changes, and burning fossil fuels. Humans have increased atmospheric CO_2 concentration by 47% since the Industrial Revolution began. This is the most important long-lived "forcing" of climate change.
3) *Methane:* A hydrocarbon gas produced both through natural sources and human activities, including the decomposition of wastes in landfills, agriculture, and especially rice cultivation, as well as ruminant digestion and manure management associated with domestic livestock. On a molecule-for-molecule basis, methane is a far more active GHG than carbon dioxide, but also one that is much less abundant in the atmosphere.
4) *Nitrous oxide:* A powerful GHG produced by soil cultivation practices, especially the use of commercial and organic fertilizers, fossil fuel combustion, nitric acid production, and biomass burning.
5) *Chlorofluorocarbons (CFCs):* Synthetic compounds entirely of industrial origin used in a number of applications, but now largely regulated in production and release to the atmosphere by international agreement for their ability to contribute to destruction of the ozone layer.

Figure 2.14 shows that carbon dioxide is the most prevalent GHG, methane is second, and fluorinated gases are the last in concentrations.

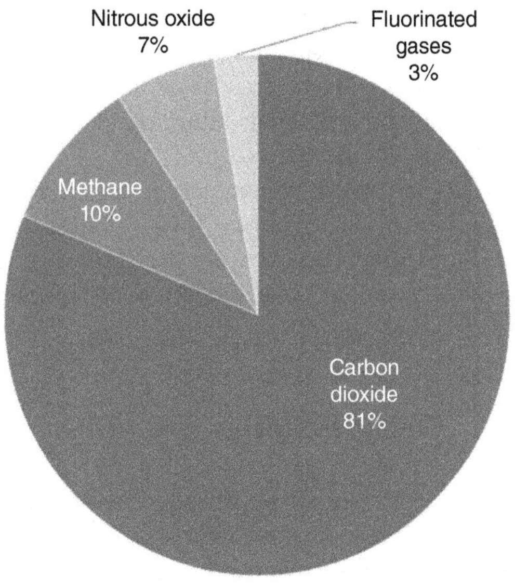

Figure 2.14 GHG concentrations. Source: GHG gas concentrations. https://www.epa.gov/ghgemissions/overview-greenhouse-gases, (December 2020).

The importance of reducing CO_2 emissions is a significant feature of sustainable products for industry.

2.4.3 Anti-Warming Theory

Theories are established to refute the claims of GHG as causes for climate change and the climate change as a balance of nature. The warming of the planet can be caused by Milankovitch climate oscillations. Periodical changes in the orbit of the Earth cause climatic changes termed "Milankovitch oscillations," leading to large changes in the size and location of species' geographical distributions (Dynesius and Jansson 2000). Climate has fluctuated widely during the history of the Earth. Climatic variability increases in amplitude toward longer time scales, but have a marked peak on the time scale of 10 000–100 000 years caused by Milankovitch oscillations (Berger 1989). The oscillations of the Earth axis can cause temperature variations on the planet that is beyond seasonal variations.

The increased temperatures for the planet may be part of repeating cycle over the last 500 000 years. (Earth warming 2022). We may be part of global warming that began over 18 000 years ago. The increased carbon dioxide may also be part of a repeating cycle over the last 400 000 years.

Current global warming cannot be explained by changes in energy from the Sun as shown as follows. Since 1750, the average amount of energy coming from the Sun either remained constant or increased slightly (NASA climate 2020). If the warming were caused by a more active Sun, then scientists would expect to see warmer temperatures in all layers of the atmosphere. Instead, they have observed a cooling in the upper atmosphere, and a warming at the surface and in the lower parts of the atmosphere. That is because GHGs are trapping heat in the lower atmosphere. Climate models that include solar irradiance changes cannot reproduce the observed temperature trend over the past century or more without including a rise in GHGs.

GHG warming can cause an increase in the temperature in the middle of the troposphere, about 5 km up. However, from 1999 to 2002, the temperature of the mid-troposphere has actually decreased lightly and surface temperatures have ceased warming, even as CO_2 concentrations have continued to increase (Tropospheric 2012).

2.5 Ocean Pollution and Marine Debris

Marine debris or ocean litter is a worldwide problem. One study found that 60–80% of ocean litter is comprised of plastics (Derraik 2002). Plastic ocean debris is most often litter that was brought to the beach by beach visitors or from storm

Figure 2.15 Marine debris.

drains that collect litter on the ground and then empty into the oceans. Beach litter is comprised of cigarette butts, plastic packaging, bottles, caps, or bags that are discarded after use. Ocean plastic litter can include, as well, pre-production plastic pellets that flow to the ocean from unintended release in ocean container ships or from plastic manufacturing facilities. Plastic pellets were found on New Zealand Beaches. (Gregory 2010). Marine debris is shown in Figure 2.15.

Recent reports have identified plastics as a significant source of ocean litter throughout the world and the physical danger that the plastic litter represents to sea life. It has been estimated that more than 300 million tons are produced each year, yet only 10% of plastic products get recycled (Marine plastic pollution 2020). It is estimated that as much as 7 million tons of plastic are added to the oceans and freshwater lakes every year (Ibid). The fish in the North Pacific can ingest as much as 48 million pounds of plastic debris annually (Ibid).

Plastics never decompose on their own in the ocean. It breaks into smaller pieces, which can be dangerous to wildlife when ingested. In addition, microplastics that are less than 6 mm form microparticles that come from larger pieces that have broken down over time, or pre-produced particles used in manufacturing. This section describes the plastic debris problem in more detail and identifies the types of plastics that are found on beaches throughout the world.

Debris can be attached to a natural food source, such as fish eggs, and be ingested accidentally. It can also be mistaken for a food source like jellyfish or plankton. When animals ingest plastic, it can cause irritation to the animal's digestive system. Plastic pieces can remain in the stomach rather than passing through and cause the animal to feel full and avoid eating. This can lead to starvation or malnutrition. The plastic pieces themselves can lead to internal injuries as it is digested, intestinal blockages, and other potentially fatal complications.

2.5.1 Plastic Marine Debris

Oceanic gyres are circulating water zones in the world composed of large-scale ocean currents. The size of the 11 continental-scale gyres ranges from 1000 nautical miles along the major axis in the Arctic Ocean to 5000 nautical miles along the major axis in the South Pacific. The gyres are designated by traditional oceanographic designations, for example, Beaufort, North Pacific Subtropical, Pacific Subarctic, South Pacific Subtropical, North Atlantic Subtropical, Atlantic Subarctic, South Atlantic Subtropical, Antarctic Circumpolar, and Indian Subtropical gyres. Two gyres in the Arctic Ocean do not have designations (Ebbesmeyer and Scigliano 2009).

Five gyres around the world, as shown in Figure 2.16, are known to collect excessive marine debris. The five gyres include the North and South Pacific gyres, North and South Atlantic gyres, and the Indian Ocean gyre.

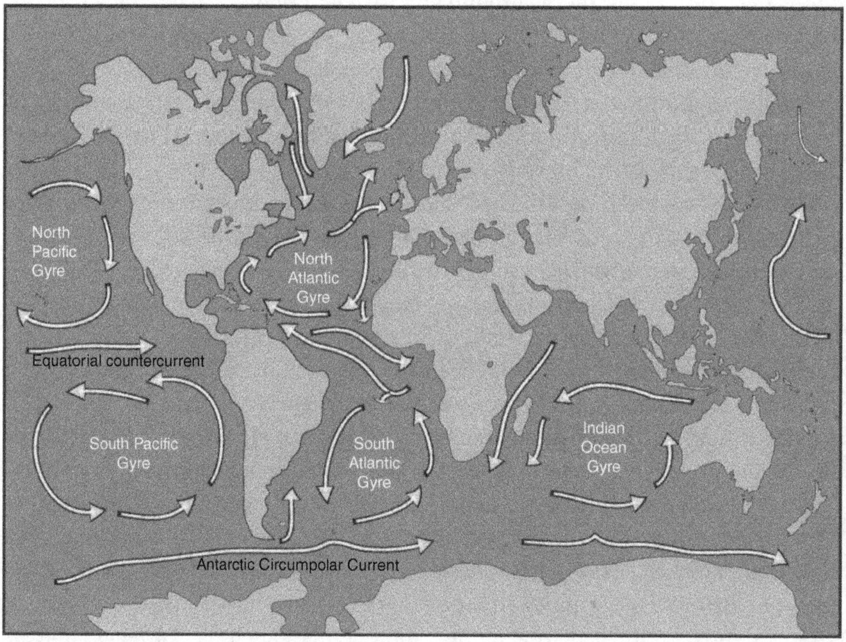

Figure 2.16 Worldwide gyres. Source: Artwork courtesy of Ms. Vanessa Vaquera of Chico, CA.

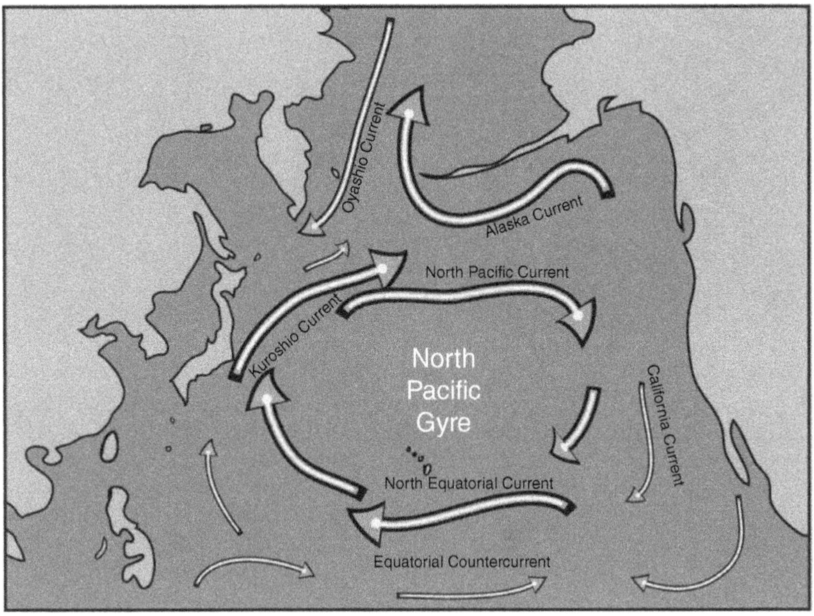

Figure 2.17 North Pacific gyre. Source: Artwork courtesy of Ms. Vanessa Vaquera of Chico, CA.

The North Pacific gyre, shown in Figure 2.17, is the most studied of the five gyres. It also can be described as the "Great Pacific Garbage Patch," the "Eastern Garbage Patch," or the "Pacific Trash Vortex." The North Pacific gyre is a gyre of marine litter in the central North Pacific Ocean located approximately between 135–155 °W and 35–42 °N. The gyre is characterized by exceptionally high concentrations of floating plastic and other debris that have been trapped by the currents of the North Pacific gyre. In the gyre, plastic pieces are usually broken into fragments due to wave action and UV exposure. The plastic debris is not continuous throughout the gyre. Instead, a series of rings with high concentrations of floating debris comprises the gyre. The rings of floating debris expand and contract based on atmospheric conditions. In the Southern and Pacific Ocean off Japan, the amount of plastic debris has increased significantly in the last 20 years (Plastic Debris from Rivers to Sea 2009). The Southern California Coastal Water Research Project (SCCWRP) and the Algalita Marine Research Foundation (AMRF) in recent years have conducted studies to identify and quantify ocean litter in the four marine habits: beach, ocean bottom, ocean water column, and ocean surface. Plastics can accumulate in each of these four areas. The ocean bottom collects larger plastic materials, such as fishing gear

Figure 2.18 Debris found on a lake in California in 2012.

and heavier objects. The water column collects plastic fragments that are suspended by ocean currents. The ocean surface collects plastic fragments, floating plastic trash, and plastic pre-production pellets. The beach environment collects several plastic materials that come from beach litter and ocean debris that can interrupt the beach experience through encounters with plastic litter on the beaches (Figures 2.17 and 2.18).

Plastic debris can cause pollution for humans and injury to fish, seabirds, and marine mammals. Ocean litter is known to have affected at least 267 species worldwide, including 86% of all sea turtle species, 44% of all seabird species, and 43% of all marine mammal species (Plastic Debris from Rivers to Sea 2009). Plastic fragments and floating debris can cause fatalities to marine life as a result of ingestion, starvation, suffocation, infection, drowning, and entanglement. Seabirds and short-tailed shearwaters that feed on the ocean surface are especially prone to ingesting plastic debris that floats.

The best way to reduce the plastic pellet pollution on the beaches is to have better pellet control at plastic manufacturing operations. The plastics industry is promoting ways to reduce amounts of plastic pellets released into the environment with the Operation Clean Sweep (OCS) program. The goal of OCS is to assist plastic manufacturing operations with good housekeeping and pellet containment practices to achieve zero pellet loss in the plastic manufacturing plant (Operation Clean Sweep 2007). The OCS program can help reduce the amount of plastic pellet litter in the oceans.

2.5.1.1 Persistent Organic Pollutants

Persistent organic pollutants (POPs) can be attracted to plastic pollution in the ocean. The organic pollutants that float in the oceans are absorbed onto the plastic. POPs are organic compounds that reside in the oceans and are resistant to environmental degradation through chemical, biological, and photolytic processes (Jones and de Voogt 2009).

POPs have been observed to persist in the environment and spread out over long distances. POPs can accumulate in human and animal tissue and become part of the human food chain. POPs have potential significant impacts on human health and the environment. POPs in the marine environment attach to plastic debris, which then are consumed by marine animals, many of which are in the human food chain. The plastic pellets and fragments can transport toxic substances in the marine environment to animals and ultimately to humans (Rios et al. 2007). Researchers in Morocco found polychlorinated biphenyls (PCBs) in sediments of coastal communities (Piazza et al. 2009).

Twelve different chemicals are commonly referred to as POPs, including aldrin, chlordane, dichlorodiphenyl-trichloro-ethene (DDT), dieldrin, endrin, heptachlor, hexachlorobenzene, mirex, PCBs, polychlorinated dibenzo-*p*-dioxins, polychlorinated dibenzofurans, and toxaphene. Other POP chemicals are considered carcinogenic and include polycyclic aromatic hydrocarbons (PAHs), brominated flame retardants, as well as some organometallic compounds such as tributyltin.

Many POPs have an origin as pesticides. Some POPs are produced from natural sources. Other POPs are released from electrical equipment or released in the production of solvents, polyvinyl chloride, and pharmaceuticals (Rios et al. 2007). Some POPs are highly toxic, carcinogenic, and can cause chronic health effects, including endocrine disruption, mutagenicity, and carcinogenicity. The toxic POPs can be collected by plastics floating in the oceans causing potential danger to sea life that ingest the plastic. PCBs and DDTs were absorbed on polypropylene and polyethylene in high concentrations in a simulated marine environment (Mata et al. 2001).

Dioxins are present in the ocean waters due to the burning of organic materials, and can be a persistent marine pollutant. It is possible for dioxins to get absorbed onto plastic particles in the ocean. In 2009, research on POPs in the world oceans was undertaken at an international workshop on the occurrence, effects, and fate of microplastic debris (University of Washington 2009). The workshop focused on small plastic debris in the oceans, including the amount, location, and environmental impacts of small plastics. Participants presented current research on microplastic occurrence and movement in the oceans, direct effects of microplastics on marine organisms, interactions of POPs with plastics and the potential of plastics to adsorb and desorb these pollutants in the marine environment and to organisms, and the effect plastics could have on the cycling of POPs. Small plastic pieces in the oceans can have high concentrations of toxic chemicals that are floating in the waters. The workshop illustrated the research, which concluded that plastic debris can absorb and transport organic contaminants, such as PCBs, up to 105–106 times the ambient seawater concentrations.

Ocean pollution can also be caused from agricultural, municipal, and industrial runoffs. The pesticides, herbicides, and insecticides used on commercial, agricultural, and residential properties can flow from the ground and into storm sewers during the rainy season. The amount of pesticides, insecticides, and herbicides can be significant and cause significant environmental harm. These pollutants can be a source of POPs. Pesticides can also be a source of environmental pollution that might migrate toward plastics in the oceans. Pesticides released from storm runoff in Australia were found to damage the Great Barrier Reef Marine Park by creating a hazy cloud in the water over the reef and blocking out sunlight and reducing photosynthesis (Barrier Reef 2022). The rural land use has changed in the last 100 years causing more pesticide use and subsequent release into the marine environment. The pesticides include organochloride, organophosphate, phenoxy, triazine, urea, mercurial, and azole group pesticides. Urban and industrial water discharges significantly pollute the marine environment. A wide variety of pesticides, including PAHs, terbutylazine, and PCBs, were found in ocean waters of Spain with a new extraction technique that will allow measurements of semi-volatile substances (Perez-Carrera et al. 2007). The new technique will allow other researchers to study the effects of pesticides on marine life. Chlorohydrocarbon pesticides were found in Australian sediments throughout the coastal regions in the country. DDT, in particular, was found throughout the region (Connell et al. 2002). Pesticide, herbicide, and insecticide runoffs from farms and urbane dwellings can also contribute to the POPs. In addition, plastic debris can contain pollutants such as PCBs far in excess of what is found naturally in seawater. Though banned in the United States in 1979, and on the international front in 2001, they are still found in waterways today.

2.5.2 Worldwide Coastal Cleanup

The persistence of ocean debris can be determined from the analysis of litter collected in beach cleanups around the world. The Ocean Conservancy's International Coastal Cleanup program encourages people around the world to remove trash and debris from the world's beaches and waterways and identify the sources of debris. Trash can migrate to the ocean from waterways and storm drains hundreds of miles inland. In 1985, the Ocean Conservancy conducted a study on plastic marine garbage for the EPA (Highlights in the Fight against Marine Debris 2010). In 1986, the first beach cleanup was conducted along the Texas coast. Volunteers collected 124 tons of trash from 122 miles of coastline. Since then the number of countries participating in the cleanup has expanded from 35 countries in 1992 to 104 countries in 2008. The Ocean Conservancy published reports every year on the collection efforts of worldwide volunteers. In 2012, the Ocean Conservancy published "Trash Free Seas. International Coastal Cleanup 2012 Data Release" (Ocean Conservancy 2012). It represented 598 076 volunteers who covered 20 776 miles and collected 9 184 428 pounds of trash on the beaches throughout the world (Ocean Conservancy 2012).

In 2009, the Ocean Conservancy published "A Rising Tide of Ocean Debris and What We Can Do About It." The list of countries is shown in Table 2.1.

The report presented ocean debris data recorded by nearly 400 000 volunteers in 104 countries and 42 US states (Ocean Conservancy's International Coastal Cleanup 2010). In 2008, approximately 73% of cleanups were performed on ocean beaches and 27% were completed on inland waterways and lakes.

The marine report features a *Marine Debris Index* that includes a breakdown of the amount and type of trash in the ocean and waterways collected in one day. The report also emphasizes environmental impacts of ocean trash on sea life. Volunteers collected more than 11 million pieces of trash from cigarette butts to

Table 2.1 Amount of debris collected and miles covered by nations in the 2009 "Ocean Cleanup Program."

Country	People	Pounds	Miles
Australia	139	6839	4.7
Brazil	9197	60 772	1260.4
Canada	34 220	317 156	1857.4
China	383	656	0.8
France	56	417	5.1
Germany	79	439	2.1
Greece	2686	30 521	56.6
India	26 038	170 300	330.2
Italy	204	5582	13.7
Japan	8802	39 547	65.7
Mexico	16 426	236 931	281.1
Philippines	114 418	1 069 443	149.4
Spain	703	18 457	18.3
United Kingdom	2729	22 409	1805.8
United States	200 190	3 861 630	9573.8

grocery bags and fast food wrappers. These were collected from beaches and the ocean floor along 17 000 miles of coastline. Beach cleanup results from 15 countries are listed in Table 2.1. For most countries, the majority of litter was found on land and not underwater. In the United States, 99% of the litter collected was found on the land versus underwater. Figure 2.18 shows debris that was collected at a beach on a Northern California freshwater lake. The items are similar to items collected along beaches throughout the world. The debris included aluminum cans, cigarette butts, plastic bottles, glass bottles, plastic fragments, clothing, plastic packaging, and plastic cups (Figure 2.18).

Table 2.2 illustrates the combination of beach and underwater litter on beaches throughout the world. The beach cleanup includes marine and freshwater lakes on land and underwater. The results show that the United States had the most volunteers collecting land and underwater litter, the greatest amount of debris collected, and the most beach distance covered. China collected the least amount of debris and covered the least amount of beach area. Table 2.1 illustrates that Canada, the Philippines, and Puerto Rico also collected large amounts of litter on the beaches and underwater.

The results from the beach cleanup can be normalized to visualize the concentrations of litter per person and per mile of beach. Thus, the amount of beach

Table 2.2 Litter ratios of total debris collection for worldwide coastal cleanup in 2011.

Country	Pounds collected per person	Pounds collected per mile
Australia	49	1455
Brazil	7	48
Canada	9	171
China	2	820
France	7	82
Germany	6	209
Greece	11	539
India	7	516
Italy	27	407
Japan	4	602
Mexico	14	843
Philippines	9	7158
Spain	26	1009
United Kingdom	8	12
United States	19	403

trash over the beach area can be better understood. It is preferable to have the least amount of marine litter per mile.

Table 2.2 illustrates that the Philippines, Australia, and Spain have the highest concentrations of beach litter per mile on land and underwater. The Australian concentration is high because of lower number of participants covering fewer miles than in 2008, where 16 pounds of debris were collected per person and 363 pounds of debris were collected per mile. These countries are followed by China, Mexico, and Japan. The United States has significantly less beach and underwater litter density than the top six countries. Brazil and the United Kingdom had the lowest litter density and can be considered the cleanest of the group per mile of beach.

In 2020, the Ocean Conservancy published a report entitled, "Together We Are Team Ocean" (Ocean Conservancy 2020). In that report, the team sponsored over 7000 cleanup events across 116 countries with 943 195 volunteers over 869 miles. They collected 20 772 394 pounds of trash in 1 067 735 items. Many of the items were plastic products. They found over 7 million pieces of plastic, over 2.3 million pieces of plastic foam, and over 400 000 pieces of glass that measured less than 2.5 cm. The researchers found the following in 2019: (Ibid)

1) Microplastics were found in ice cores taken from Antarctica.
2) Microplastics were found in the deep oceans.
3) Microplastics were found in food and beverage products.
4) Rubber dust is a large source of microplastics in the United States.

As many as 61 000 hermit crabs are estimated to become entrapped and die each year on just one remote island due to plastic marine debris. Scientists are concerned that their discovery may have serious implications for crab species globally (Lavers et al. 2020).

Table 2.3 shows the ocean cleanup summaries for 15 countries, in 2019.

In 2019, volunteers collected over 7.7 million pounds of items on the shoreline. China and Brazil had the most polluted beaches. Brazil also had the most ocean debris collected on the beaches. Brazil also had the most amount of debris collected per person.

The beach collection data can be analyzed by the number of items gathered on the beaches rather than the weight of the items. In 2011, the top 10 participating countries were:

1) The United States
2) The Philippines
3) Canada
4) Hong Kong
5) Dominican Republic
6) Mexico

Table 2.3 Ocean cleanup ratios for 15 countries.

Country	Pounds collected per person	Pounds collected per mile
Australia	4.62	88.82
Brazil	100.96	3777.42
Canada	4.48	124.10
China	11.30	3015.40
France	23.60	90.63
Germany	2.91	110.09
Greece	6.21	295.66
India	14.54	551.07
Italy	5.27	168.52
Japan	2.97	0.65
Mexico	3.86	448.36
Philippines	4.87	370.63
Spain	1.68	286.05
United Kingdom	2.28	449.94
United States	17.19	361.58

Source: Ocean Conservancy (2020).

7) India
8) Peru
9) Ecuador
10) Puerto Rico

In 2011, volunteers collected over 10.5 million items of litter on beaches throughout the world, as compared to over 11.4 million items of litter on beaches in the world that were collected in 2008. The top 10 marine debris items that were collected in 2011 are listed in Table 2.4 and include plastic, aluminum, glass, and paper. The top 10 items accounted for 81% of the debris in 2011 versus 73% in 2008. Cigarettes and cigarette filters are the most common litter items and are usually made with cellulose acetate plastic. Cigarette debris is followed by plastic debris including food wrappers, beverage bottles, trash bags, caps, lids, bottles, cups, plates, spoons, and straws. Glass bottles, aluminum cans, and paper bags round out the top 10 items. Table 2.4 also lists the common plastic materials that can be used for the debris item. Note that PS represents polystyrene, PP represents polypropylene, HDPE represents high density polyethylene, LDPE represents low

Table 2.4 Top 10 worldwide marine debris items in 2011.

Products	Materials: Plastic and other	Number of items	Percentage of total US debris
Cigarette/cigarette filters	Cellulose acetate	1 025 044	28%
Food wrappers and containers	PS, PET, LDPE, HDPE, PVC	401 800	11%
Caps and lids	HDPE, PP, aluminum	370 252	10%
Plastic bags	HDPE and LDPE	245 773	7%
Plastic beverage bottles	PET	227 046	6%
Cups, plates, knives, spoons	PS, PP, PS foam, PP foam, paper	168 478	5%
Straw/stirrers	HDPE, PP	166 601	5%
Beverage cans	Aluminum	158 796	4%
Beverage bottles	Glass	150 344	4%
Paper bags	Paper	73 928	2%
	Total	2 988 062	82%

Source: Ocean Conservancy (2012).

density polyethylene, PET represents polyethylene terephthalate, and PVC represents polyvinyl chloride. Plastics can be found in 7 of the top 10 marine debris items collected throughout the world.

The following table lists the items that were found in 2019 as a comparison to those found in 2011. Table 2.5 lists items found on ocean shores in 2019 (Ocean Conservancy 2020). In 2019, more plastic items in food wrappers were found on beaches and waterways than cigarette products. These food wrappers are made mostly from polystyrene, polyester terephthalate, polypropylene, PVC, and polyethylene.

Table 2.6 lists the sources of marine debris found in the world in 2008. The source of these items is mostly related to human recreational activities and smoking. The world averages for the sources of marine debris demonstrate that 61% of the items that were collected were related to shoreline and recreational activities, 31% of the items were related to smoking, 5% related to ocean and waterway activities, 2% to dumping activities, and 1% to medical and personal hygiene activities. In North America, the sources of ocean debris were similar to the world averages. The sources of marine debris by region are listed in Table 2.6.

This can indicate littering as the primary source of ocean debris in North America. Smoking usually occurs by people who are on or near the beaches around the world. Very little ocean debris is caused by dumping from ocean

Table 2.5 Items found on beaches and fresh waterways in 2019.

	Plastic item	Quantity	%
1	Food wrappers	4 771 602	28.57
2	Cigarette butts	4 211 962	25.22
3	Plastic bottles	1 885 833	11.29
4	Plastic bottle caps	1 500 523	8.98
5	Straws and stirrers	942 992	5.65
6	Plastic cups and plates	754 969	4.52
7	Plastic grocery bags	740 290	4.43
8	Plastic takeout containers	678 312	4.06
9	Other plastic bags	611 100	3.66
10	Plastic lids	605 778	3.63
	Total	16 703 361	100

Source: Ocean Conservancy (2020).

vessels. Rather, sources of ocean debris can be from trash collection facilities near streams or coasts, river and flood waters, industrial outfalls, storm drain discharge, untreated municipal sewage, and recreational use of beaches.

A report in 2009 from Ocean Conservancy found that the source of the debris items is mostly related to human recreational activities and smoking (Ocean Conservancy's International Coastal Cleanup 2010).

Table 2.6 lists the marine debris around the world. Littering from shoreline and recreational activities is the primary source of ocean debris in the world that

Table 2.6 Sources of marine debris by regions of the world.

Source	North America	Central America	South America	Caribbean	South East Asia	Worldwide
Shoreline and recreational	53.1%	23.7%	71.2%	81.5%	79.6%	61%
Smoking	35.2%	74%	19.1%	9.8%	11.1%	31%
Ocean and waterway	4.9%	1.6%	5.9%	5.3%	7.3%	5%
Dumping	2.1%	0.3%	2.4%	2.2%	1.1%	2%
Medical and personal hygiene	4.7%	0.4%	1.4%	1.2%	0.9%	1%

accounts for approximately 61% of the worldwide marine debris items. Smoking activities contribute to approximately 31% of the worldwide marine debris items. The source of marine debris from smoking can be attributed to smokers who are on or near the beaches around the world. Very little ocean debris is caused by dumping from ocean vessels or of medical and personal hygiene products.

Shoreline and recreational activities are the most significant sources of marine debris in North America, South America, the Caribbean islands, and South East Asia. Smoking-related activities are the most significant source of marine debris in South America.

2.5.3 US Coastal Cleanup

The persistence of marine debris in the United States is similar to the persistence of worldwide debris. The United States has a lot of beaches that are used for recreational activities, as do countries in the Caribbean, South America, and South East Asia. Increased trash collection on beaches and restricted smoking areas can reduce the amount of debris that ends up in the oceans.

Table 2.7 lists cigarettes and cigarette filters as the most common litter item in the United States, which is followed by plastic debris including food wrappers,

Table 2.7 Top 10 US marine debris items in 2011.

Products	Materials: Plastic and other	Number of items	Percentage of total US debris
Cigarette/cigarette filters	Cellulose acetate	1 362 741	35%
Food wrappers and containers	PS, PET, LDPE, HDPE, PVC, PP	397 231	10%
Caps and lids	HDPE, PP, aluminum, steel, tin	379 589	10%
Plastic bags	LDPE and HDPE	229 758	6%
Plastic beverage bottles	PET	210 568	5%
Cups, plates, knives, spoons	PS, PP, PS foam, PP foam, paper	167 108	4%
Beverage bottles: glass	Glass	168 849	4%
Straw/stirrers	HDPE, PP	155 659	4%
Beverage cans	Aluminum	163 441	4%
Cigar tips	Cellulose	74 399	2%
	Total	3 309 343	84%

Source: Ocean Conservancy (2012).

Table 2.8 Top 10 US marine debris Items in 2019.

	Plastic item	Quantity	%
1	Cigarette butts	900178	37
2	Food wrappers	387025	16
3	Plastic bottle caps	383434	16
4	Plastic bottles	223088	9
5	Straws and stirrers	182481	7
6	Plastic grocery bags	103710	4
7	Plastic lids	74839	3
8	Other plastic bags	74162	3
9	Plastic takeout containers	68912	3
10	Plastic cups and plates	57676	2
	Total	2455505	100

Source: Ocean Conservancy (2020).

containers, caps, lids, plastic bags, bottles, cups, plates, spoons, and straws. Glass bottles, aluminum cans, and paper bags round out the top 10 items. Plastics can be found in seven of the top 10 marine debris items collected in the United States.

Other beach litter items were classified as dangerous to aquatic life. These items can cause death or injury due to entanglement, suffocation, injury, or starvation. Plastic pollution and litter are problems with plastics and need to be addressed with sustainable materials to reduce the waste and pollution.

Table 2.8 lists the US marine debris items in 2019.

The United States collected over six million items in 2019. Cigarette butts were the most prolific item in the US beaches accounting for 37% of the debris as shown in Table 2.8

2.6 Chemical Pollution from Plastics

Pollution can be created during the production of plastic products. The most common areas of pollution-related concern include ozone layer depletion, atmospheric emissions, smog generation, aquatic eutrophication, terrestrial eutrophication, aquatic acidification, toxic chemical generation, and carcinogenic material generation.

Ozone depletion and atmospheric emissions can come from byproducts generated during the creation of plastics. The EPA provides a list of two classes of ozone-depleting chemicals (EPA Ozone Protection 2013). The chemicals are primarily chlorinated and brominated compounds, for example, CFCs. Ozone-depleting chemicals can be used and released while producing plastic pellets and products. Atmospheric emissions include chemicals classified as pollutants. Atmospheric emissions primarily are associated with the combustion of fossil fuels for energy and transportation requirements in manufacturing of products, including plastics. Atmospheric emissions can include carbon dioxide, carbon monoxide, hydrocarbons, nitrous oxides, sulfur oxides, and particulates. Smog can be formed from many chemicals including carbon monoxide, particulates, and hydrocarbons. Life cycle assessment (LCA) software can provide emission data on the production of ozone-depleting chemicals and atmospheric emissions (LCA Nature Works™ 2009).

Chemical pollution can include production of polluting substances that can lead to aquatic eutrophication, terrestrial eutrophication, aquatic acidification, toxic chemicals, and carcinogenic substances. Eutrophication is a bloom of vegetation in aquatic or terrestrial environments. It is caused by the addition of artificial nitrates and phosphates in the ecosystem. Acidification is a reduction in the pH of the aquatic environment. It is caused by the uptake of carbon dioxide in the atmosphere. As the levels of carbon dioxide increase, more of it can dissolve in ocean water and freshwater areas. The carbon dioxide can react with the water and form carbonic acid resulting in a lower pH levels. Toxic chemicals can be formed during the production of energy for manufacturing facilities. The toxic chemicals can include benzene, dioxins, toluene, xylene, ethyl benzene, styrene, cyanide, etc. The toxic chemicals can also be carcinogenic. LCA software can provide generation data on the production of toxic and carcinogenic chemicals during the creation of plastic products (LCI Summary for PLA and PET 12-Ounce Water Bottles 2013).

2.7 Landfill Trash

Most of the debris collected on land is placed in landfill operations. Modern landfills are well designed and engineered facilities that are operated and monitored to meet federal regulations. The landfills are designed to accept solid debris and prevent contamination of the land and groundwater from the solid debris. Landfills are, typically, designed according to the "cap-and-seal" strategy of the EPA, as shown is Figure 2.19, to prevent the leachate from contaminating groundwater near the landfill. The area below the landfill is covered in geosynthetic clay to provide a barrier for the landfill. American Society for Testing and Materials (ASTM) standards are developed to design the liner systems using geosynthetic clay liners (Koerner and Narejo 1995).

In the United States, 1908 municipal solid waste (MSW) landfills were in operation during 2008. The operations at landfills in 2020 is shown in Figure 2.20.

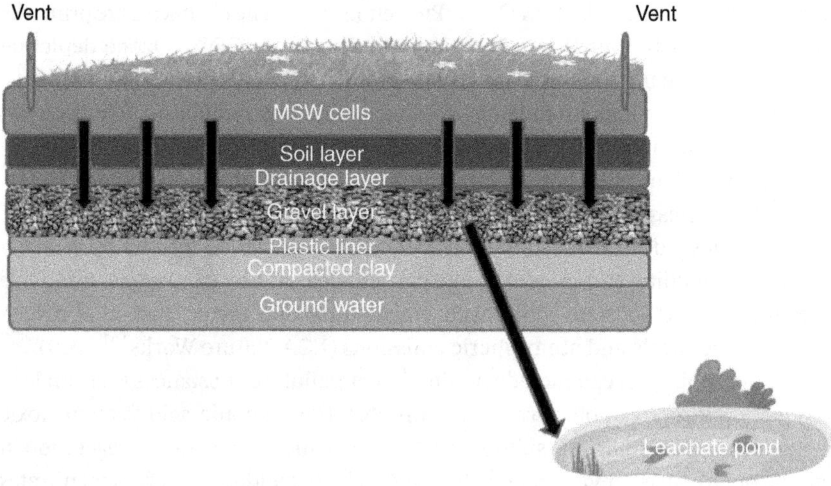

Figure 2.19 Cross-sectional view of cap-and-seal landfill.

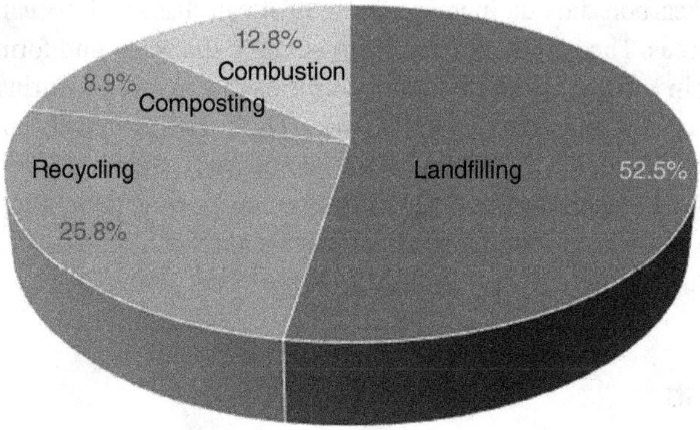

Figure 2.20 Operations at a landfill in 2020. Source: Landfill operations (2020).

A total of 289.5 million tons of solid debris were collected and buried in landfill, recycled, or composted. Approximately, 69% of the solid debris was buried in landfill, 24% was recycled and composted, and 7% was combusted via waste to energy. Table 2.9 lists the regional disposal of MSW at landfills during 2008 (van Haaren et al. 2010).

The landfills are vented to release the trapped landfill gases that are made from biogas, methane, and carbon dioxide. The landfill gases can be collected and used for energy purposes or burned. Methane can be recovered from active landfill sites called bioreactors where biodegradation is enhanced with the addition of liquids, air, and microbial processes. The bioreactor can be of three designs that include

Table 2.9 Diversion rates of MSW in landfills during 2008.

Region	Landfill (%)	Recycling/composting (%)	Waste to energy (%)
Mid-Atlantic	59	27	14
Midwest	78	22	1
Rocky Mountains	88	11	1
South	79	13	8
West	52	46	2
United States Total	69	24	7

Source: MSW Landfill EPA (2008).

anaerobic, aerobic, and hybrid. With anaerobic designs, moisture is added in an anaerobic environment to generate landfill gas. With aerobic designs, leachate is removed from the bottom of the reactor and recirculated with air back into the landfill. Landfill gas is removed from the landfill. With hybrid bioreactor designs, waste biodegradation is accelerated through alternating aerobic and anaerobic treatments that rapidly biodegrade organics in the landfill and collect landfill gases at the bottom of the reactor (EPA Landfill 2013).

Landfills are capped with ground cover and grasses to provide an enclosed "tomb-like" structure for the solid debris. Very little biodegradation occurs in the landfills due to the lack of oxygen, cool temperatures, dry environment, and limited microorganisms. A study from the University of Arizona found that newspapers did not biodegrade in the landfill and could be read after 100 years from when it was first printed. Food scraps and organics found in the landfill were still recognizable after 30 years in the landfill (Rathje and Murphy 2001). At the end of the service life, plastic products can be either collected for recycling or thrown away with the trash. Waste disposal companies usually collect the plastics with other recycled products. Plastics, metals, and glass are sorted from the refuse and sent to recyclers. The solid waste can be recycled or sent to an incinerator or landfill.

On average in the United States, the waste generation sent to landfills is approximately 700 000 tons/day. With plastics accounting for approximately 13% of the refuse, the average plastic disposal is 91 000 tons/day. There are approximately 2600 landfills in the United States in 2020. A map of the landfills in the United States in 2020 is provided in Figure 2.21 (Landfills USA 2020).

California has the most landfills in the United States. Rhode Island has the fewest landfills in the United States.

The items in the landfill are mostly paper, food waste, and yard trimmings. These are separated and then composted. Most landfill facilities have compost and recycling operations. The items in the landfill are shown in Figure 2.22.

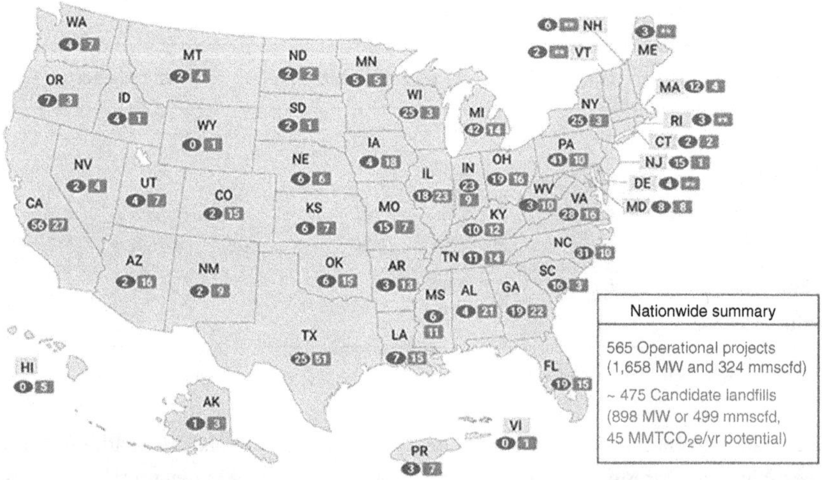

Figure 2.21 Map of landfills in the United States in 2020. Source: Landfills USA (2020).

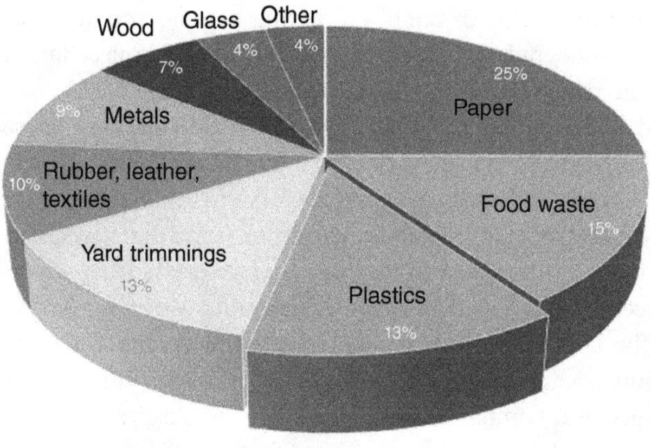

Figure 2.22 Contents of the landfills in the United States in 2018.

The items in a landfill on average, in 2018, are as follows:

1) Paper, 25%
2) Food waste, 15.2%
3) Plastics, 13.2%
4) Yard trimmings, 13.1%
5) Rubber, leather, and textiles, 9.7%
6) Metals, 9.4%
7) Wood, 6.7%
8) Glass, 4.2%
9) Other, 3.5%

Plastics historically were present at about 10% by weight. Thus, in 2018, the amount of plastics has increased in the landfills. By volume, the percentage is higher due to the low density of plastics. EPA refers to trash, or MSW, as various items consumers throw away after they are used. These items include bottles and corrugated boxes, food, grass clippings, sofas, computers, tires, and refrigerators. Food waste comprised the fourth largest material category, estimated at 63.1 million tons or 21.6% of total generation in 2018. Yard trimmings comprised the next largest material category, estimated at 35.4 million tons, or 12.1% of total generation, in 2018. This compares to 35 million tons (16.8% of total generation) in 1990 (Ibid). In 2018, plastic products generation was 35.7 million tons, or 13.2% of generation. This was an increase of 4.3 million tons from 2010 to 2018, and it came from durable goods and the containers and packaging categories. Plastics generation has grown from 8.2% in 1990 to 12.2% in 2018. Plastics generation as a percent of total generation has varied from 12.2 to 13.2% over the past eight years (Ibid).

In 2018, 2.7 million tons of selected consumer electronics were generated, representing less than 1% of MSW generation. Selected consumer electronics include products such as TVs, VCRs, DVD players, video cameras, stereo systems, telephones, and computer equipment.

The total MSW recycled was 69 million tons, with paper and paperboard accounting for approximately 67% of that amount. Metals comprised about 13%, while glass, plastic, and wood made up between 4 and 5%. Total MSW recycling by material, in 2018, was 69.0 million tons: paper and paperboard, 66.60%; glass, 4.43%; metals, 12.63%; plastics, 4.38%; wood, 4.49%; rubber, leather, and textiles, 6.06%; and other, 1.41% (Ibid).

The total generation of MSW in 2018 was 292.4 million tons (US short tons, unless specified) or 4.9 pounds/person/day (MSW EPA 2019).

Of the MSW generated, approximately 69 million tons were recycled and 25 million tons were composted. Together, 93.9 million tons of MSW were recycled and composted, equivalent to a 32.1% recycling and composting rate. An additional 17.7 million tons of food were managed by other methods. Other food management includes the following management pathways: animal feed, bio-based materials/biochemical processing, co-digestion/anaerobic digestion, donation, land application, and sewer/wastewater treatment. In addition, nearly 35 million tons of MSW (11.8%) were combusted with energy recovery and more than 146 million tons of MSW (50%) were landfilled (Ibid).

For 2011 in the United States, the total waste disposed of in landfills was approximately 250 million tons. Of the MSW, approximately 53.6% was buried in a landfill, 34.7% was recycled or composted, and 11.7% was converted to energy through combustion. As shown in Table 2.10, plastics accounted for approximately 12.7% by weight of the solid debris (EPA Waste 2013). For 2008, as reported in a California statewide waste characterization study, approximately 40 million tons of solid waste was disposed of in landfills as MSW. Plastics accounted for

Table 2.10 Contents of the landfill in California, 2008.

1.	Organics	32.40%
2.	Inert and other	29.10%
3.	Paper	17%
4.	Plastic	9.60%
5.	Metal	4.60%
6.	Special waste	3.90%
7.	Glass	1.40%
8.	Other	1.70%
9.	CA total	100.00%

Source: CA Landfill (2020).

Table 2.11 Contents of the landfill in California, 2018.

1.	Organics	34.00%
2.	Paper	17.00%
3.	Inert and other	14.00%
4.	Plastic	12.00%
5.	Other	9.00%
6.	Special waste	7.00%
7.	Metal	5.00%
8.	Glass	2.00%
9.	CA Total	100.00%

Source: CA Landfill (2020).

approximately 9.6% of the waste by weight. Plastic trash bags comprised 1% and plastic film comprised 1.7% of the waste stream. The commercial sector generated approximately 50% of the solid waste, the residential sector generated approximately 30% of the solid waste, and the self-hauled sector generated approximately 20% of the solid waste. For 2008, plastics contributed to 12% by weight of the waste stream for the commercial waste, 11.3% of the waste from residential waste, and 5.8% of the waste stream in self-hauled waste (California 2008).

Table 2.11 lists the top 10 items found in a landfill on average in California. For the landfills in the United States, approximately 56% of the items are organic in nature. Of the landfills in California, approximately 50% of the items are organic in nature. Plastics accounted for 12.7 weight% of the waste in the United States and 9.6% of the solid waste in California.

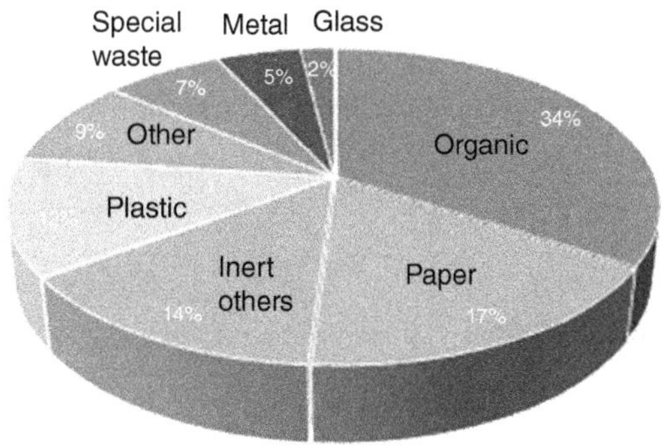

Figure 2.23 Contents of landfill in California in 2018. Source: CA Landfill (2020).

In 2018, California had the following contents in the landfill as shown in Table 2.11 and Figure 2.23.

As shown in the figure, the amount of plastic in the landfill increased from 9.6 to 12%. The amount of organics also increased from 32.4 to 34%. Inert and others also increased from 1 to 14%.

2.8 Summary

The land and ocean temperatures are rising over the last 100 years. The average temperatures of the planet in 2011 were the second highest "strong La Niña" temperatures in recorded history. Glaciers throughout the world are melting due to the increased surface temperature of the planet. The levels of the oceans are rising. This can be attributed to thermal expansion of the water, land water storage change, and melting of glaciers and icebergs. Carbon dioxide and other GHGs have increased dramatically since 1960 in the world. The increase in the levels of GHGs in the atmosphere is the likely cause of the temperature rise on the planet. The primary sources of carbon dioxide emissions are from energy generation and transportation.

The predominant sources of marine debris around the world are from smoking activities and littering during shoreline and recreational activities. Very little ocean debris is caused by dumping from ocean vessels or of medical and personal hygiene products.

Plastic debris accounts for 60–80% of marine pollution. The top debris items are cigarette products, plastic bags and bottles, plastic packaging and disposal cutlery, glass bottles, aluminum can, and paper bags. Cigarette and cigar products comprise the majority of solid debris items found throughout the world. Common plastic debris items are plastics bags, containers, bottles, and disposable dinnerware.

Landfills in the United States are designed according to the "Cap and Seal" strategy of the EPA, where the landfill has a liner of clay and is sealed on top with grass and vegetation. The landfills in the United States are similar to a "dry tomb" structure with very little moisture, oxygen, and microorganisms. Very little biodegradation occurs in the typical landfill.

References

Artic Ice Decline (2020). https://mynasadata.larc.nasa.gov/phenomenon-changes-snow-and-ice-extent#:~:text=Declining%20Arctic%20sea%20ice,the%20period%20of%201979%2D2017. (accessed December 2020).

Barrier Reef (2022). https://www.barrierreef.org/the-reef/threats/poor-water-quality (accessed February 2022).

Beltrami, H. and Bourlon, E. (2004). Ground warming patterns in the Northern Hemisphere during the last five centuries. *Earth Planet. Sci. Lett.* 227: 169–177.

Berger, A. (1989). The spectral characteristics of pre-Quaternary climate records, an example of the relationship between the astronomical throery and geosciences. In: *Climate and Geo-Sciences* (ed. A. Berger, A. Schneider and J.C. Duplessy), 47–76. Dordrecht: Kluwer.

Berthier, E., Arnaud, Y., Kumar, R. et al. (2007). Remote sensing estimates of glacier mass balances in the Himachal Pradesh (Western Himalaya, India). *Remote Sens. Environ.* 108: 27–338.

CA Landfill (2020). https://www2.calrecycle.ca.gov/WasteCharacterization/Study (accessed December 2020).

California (2008). Statewide waste characterization. http://www.calrecycle.ca.gov/publications/Detail.aspx?PublicationID=1346 (accessed October 2013).

Carpenter, E. and Smith, K. Jr. (1972). Plastics on the Sargasso Sea surface. Environmental Science, *Science* 175: 1240–1241.

Cazenave, A., Lombard, A., and Llovel, W. (2008). Present-day sea level rise: a synthesis. *Comptes Rendus Geosci.* 340: 761–770.

CEE Plastics in the Ocean (1987). More than a litter problem. Report prepared by the Center for Environmental Education (now called Center for Marine Conservation), Washington, DC, under Contract No. 68-02-4228, to the Environmental Protection Agency, p. 128. Washington, DC). https://www.govinfo.gov/content/pkg/CZIC-gc1085-o43-1987/html/CZIC-gc1085-o43-1987.htm (accessed May 2022).

Church, J., White, N., and Hunter, J. (2006). Sea-level rise at tropical Pacific and Indian Ocean islands. *Glob.Planet. Change* 53: 155–168.

Climate change causes (2020). https://www.ipcc.ch/site/assets/uploads/2018/02/ipcc_wg3_ar5_summary-for-policymakers.pdf (accessed December 2020).

Coral bleaching (2020). https://oceanservice.noaa.gov/facts/coralreef-climate.html#:~:text=Climate%20change%20leads%20to%3A,to%20the%20smothering%20of%20coral (accessed December 2020).

CO_2 levels (2020). https://www.climate.gov/sites/default/files/BAMS_SOTC_2019_co2_paleo_1000px.jpg. (accessed December 2020).

Cogley, J. (2009). Geodetic and direct mass balance measurements: comparison and joint analysis. *Ann. Glaciol.* 50: 96–100.

Colton, J.B. (1974). Plastics in the ocean. *Oceanus* 18: 61–64.

Connell, D., Miller, G., and Anderson, S. (2002). Chlorohydrocarbon pesticides in the Australian marine environment after banning in the period from the 1970s to 1980s. *Mar. Pollut. Bull.* 45: 78–83.

Day, R., Shaw, D., and Ignell, S. (1990) The quantitative distribution and characteristics of marine debris in the north Pacific Ocean, In: *Proceedings of the Second International Conference on Marine Debris, (2–7 April 1989), Honolulu, Hawaii. NOAA Tech. Mem. NMFS, NOAA-TM-NMFS-SWFSC-154. Department of Commerce, National Oceanic and Atmospheric Administration, National Marine Fisheries Service, Washington, DC* (eds. R.S. Shomura and M.L. Godfrey), pp. 247–266.

Derraik, J. (2002). The pollution of the marine environment by plastic debris: a review. *Mar. Pollut. Bull.* 44: 842–452.

Dynesius, M. and Jansson, R. (2000). Evolutionary consequences of changes in species' geographical distributions driven by Milankovitch climate oscillations. *Proc. Natl. Acad. Sci. USA* 97 (16): 9115–9120.

Earth warming (2022). https://www.climate.gov/news-features/climate-qa/hasnt-earth-warmed-and-cooled-naturally-throughout-history (accessed September 2022).

Ebbesmeyer, C. and Scigliano, E. (2009). *Flotsametrics and the Floating World*, 201. New York: HarperCollins Publishers.

EPA emissions (2020). https://www.epa.gov/ghgemissions/inventory-us-greenhouse-gas-emissions-and-sinks (accessed December 2020).

EPA Inventory of Greenhouse Gas Emissions and Sinks (2013). Environmental protection agency. http://www.epa.gov/climatechange/Downloads/ghgemissions/US-GHG-Inventory-2013-ES.pdf (accessed June 2013).

EPA Landfill (2013). Bioreactors. http://www.epa.gov/osw/nonhaz/municipal/landfill/bioreactors.htm (accessed October 2013).

EPA Ozone Protection (2013). Science. http://www.epa.gov/ozone/science/ods/ (accessed July 2013).

Fergala, A., AlSayed, A., Khattab, S. et al. (2018). Development of methane-utilizing mixed cultures for the production of polyhydroxyalkanoates (PHAs) from anaerobic digester sludge. *Environ. Sci. Technol.* 52 (21): 12376–12387.

Ferri, J.M., Garcia-Garcia, D., Sánchez-Nacher, L. et al. (2016). The effect of maleinized linseed oil (MLO) on mechanical performance of poly(lactic acid)-thermoplastic starch (PLA-TPS) blends. *Carbohyd. Polym.* 147: 60–68.

Franca, D.G., Bezerra, E.B., De Souza Morais, D.D. et al. (2016). Hydrolytic and thermal degradation of PCL and PCL/bentonite compounds. *Mater. Res.* 19: 618–627. https://doi.org/10.1590/1980-5373-MR-2015-0797.

GHG emission by industry (2020). https://www.researchgate.net/figure/Global-emissions-of-carbon-dioxide-a-by-major-sector-and-b-within-industry_fig1_242582149 (accessed December 2020).

GHG gases (2020). https://climate.nasa.gov/causes/#:~:text=Water%20vapor.,mechanisms%20to%20the%20greenhouse%20effect. (accessed December 2020).

GHG gas concentrations (2020). https://www.epa.gov/ghgemissions/overview-greenhouse-gases (accessed December 2020).

Global emission by sector (2020). https://architecture2030.org/buildings_problem_why/ (accessed December 2020).

Global Analysis (2013). NOAA. http://www.ncdc.noaa.gov/sotc/global/2011/13 (accessed June 2013).

Global warming (2022). https://geocraft.com/WVFossils/temp_vs_CO2.html (accessed September 2022).

Golovanova, I., Harris, R., Selezniova, G., and Stuic, P. (2001). Evidence of climatic warming in the southern Urals region derived from borehole temperatures and meteorological data. *Glob.Planet. Change* 29: 167–188.

Gregory, M. (2010). Plastic pellets on New Zealand beaches. *Mar. Pollut. Bull.* 8: 82–84.

Hays, H. and Cormons, G. (1974). Plastic particles found in tern pellets, on coastal beaches, and at factory sites. *Mar. Pollut. Bull.* 5 (3): 44–46.

Held, I. and Soden, B. (2006). An assessment of climate feedbacks in coupled ocean–atmosphere models. *J. Climate* 19: 3354–3360.

Helnrich, D., M. Raberg, P. Fricke, et. al., Synthesis gas (syngas)-derived medium-chain-length polyhydroxyalkanoate synthesis in engineered *Rhodosplrillum rubrum*, *Appl. Environ. Microbiol.*, 82 20, 6132–6140 (2016).

Highlights in the Fight against Marine Debris (2010). Take a day to leave a legacy. The International Coastal Cleanup, June 2006. http://act.oceanconservancy.org/site/DocServer/2006_ICC_Fact_Sheet.pdf?docID=1661 (accessed June 2014).

Jones, K. and de Voogt, P. (2009). Persistent organic pollutants: state of Science. *Appl. Catal. B Environ.* 89: 645–650.

Koerner, R. and Narejo, D. (1995). Bearing capacity of hydrated geosynthetic clay liners. *J. Geotechn. Eng.* 1995: 82–85.

Landfill operations (2020). https://www.google.com/url?sa=i&url=https%3A%2F%2Fwww.biocycle.net%2Fbiocycle-world-174%2F&psig=AOvVaw2--oj1ZB46T0CEnJxcL0Ht&ust=160944135993700 0&source=images&cd=vfe&ved=2ahUKEwi5v9G1svbtAhV3IDQIHQGaCSMQr4kDegUIARCyAQ (accessed December 2020).

Landfills USA (2020). https://www.epa.gov/lmop/project-and-landfill-data-state (accessed December 2020).

LCA NatureWorks™ (2009). Life cycle assessment of food packaging made of Ingeo™ biopolymer and RPET. http://www.natureworksllc.com/~/media/

The_Ingeo_Journey/Ingeo_vs_rPET/IFEU_LCA__Ingeo_Full_Report_012709_FINAL_pdf.pdf (accessed June 2013).

LCI Summary for PLA and PET 12-OunceWater Bottles (2013). http://www.containerrecycling.org/assets/pdfs/plastic/LCA-PETandPLA2007.pdf (accessed June 2013).

Lavers, J.L., Sharp, P.B., Stuckenbrock, S., and Bond, A.L. (2020). Entrapment in plastic debris endangers hermit crabs. *J. Hazard. Mater.* 387: 121703.

Liu, L., et al. (2009). Mechanical properties of poly(butylene succinate) (PBS) biocomposites reinforced with surface modified jute fibre. *Compos. Part A Appl. Sci. Manuf.* 40 (5): 669–674.

Lopez, P., Chevallier, P., Favier, V. et al. (2010). A regional view of fluctuations in glacier length in southern South America. *Glob. Planet. Change* 71: 85–108.

Majorowicz, J. and Skinner, W. (1997). Potential causes of differences between ground and surface air temperature warming across different ecozones in Alberta, Canada. *Glob. Planet. Change* 15: 79–91.

Marine plastic pollution (2020). https://marinesanctuary.org/blog/plastic-pollution-ocean-conservation-challenge/ (accessed December 2020).

Mata, Y., Isobe, T., Takada, H. et al. (2001). Plastic resin pellets as a transport medium of for toxic chemicals in the marine environment. *Environ. Sci. Technol.* 35: 318–324.

Mitchell, J., Wilson, C., and Cunnington, W. (1987). On CO_2 climate sensitivity and model dependence of results. *Q. J. R. Meteorol. Soc.* 113: 293.

Moore, S.L., Sutula, M., Bitner, T.V., Lattin, G., Schiff, K.C. (2016). Southern California Bight 2013 Regional Monitoring Program: Volume III. Trash and Marine Debris. Technical Report 928. Southern California Coastal Water Research Project Authority, Costa Mesa, CA.

Motyka, R., O'Neel, S., Connor, C., and Echelmeyer, K. (2002). Twentieth century thinning of Mendenhall Glacier, Alaska, and its relationship to climate, lake calving, and glacier run-off. *Glob. Planet. Change* 35: 93–112.

MSW EPA (2019). https://www.epa.gov/facts-and-figures-about-materials-waste-and-recycling/ (accessed December 2020).

MSW Landfill EPA (2008). https://archive.epa.gov/epawaste/nonhaz/municipal/web/html/ (accessed December 2020)

NASA climate (2020). https://climate.nasa.gov/causes/ (accessed December 2020).

NASA Gravity Recovery (2021). https://climate.nasa.gov/news/3010/study-2019-sees-record-loss-of-greenland-ice/. (accessed December 2020).

NASA Gravity Recovery (2022). https://climate.nasa.gov/vital-signs/ice-sheets/ (accessed January 2022).

Nesje, A., Bakke, J., Dahl, S. et al. (2008). Norwegian mountain glaciers in the past, present and future. *Glob. Planet. Change* 60: 10–27.

NOAA (2012). https://www.ncdc.noaa.gov/sotc/global/201013 (accessed February 2022).

Ocean Conservancy (2012). Ocean conservancy's fighting for trash free seas. International Coastal Cleanup 2012 Data Release. http://www.oceanconservancy.org/our-work/marine-debris/ (accessed June 2014).

Ocean Conservancy's International Coastal Cleanup (2010). Trash travels from our hands to the sea, around the globe, and through time, http://act.oceanconservancy.org/images/2010ICCReportRelease_pressPhotos/2010_ICC_Report.pdf (accessed June 2014).

Ocean Conservancy (2020). https://oceanconservancy.org/wp-content/uploads/2020/10/FINAL_2020ICC_Report.pdf (accessed December 2020).

Operation Clean Sweep (2007). American Chemistry Council. http://www.opcleansweep.org/overview/ (accessed June 2013).

Perez-Carrera, E., Leon, V., Parra, A., and Gonzalez-Mazo, E. (2007). Simultaneous determination of pesticides, polycyclic aromatic hydrocarbons and polychlorinated biphenyls in seawater and interstitial marine water samples, using stir bar sorptive extraction–thermal desorption–gas chromatography–mass spectrometry. *J. Chromatogr. A* 1170: 82–90.

Piazza, R., Moumni, B., Bellucci, L. et al. (2009). Polychlorinated biphenyls in sediments of selected coastal environments in Northern Morocco. *Mar. Pollut. Bull.* 58: 431–438.

Plastic Debris from Rivers to Sea (2009). Algalita Institute. http://algalita.org/research.html (accessed June 2009).

Plastic Pellets in the Aquatic Environment (2008). Sources and recommendations. chrome-extension://efaidnbmnnnibpcajpcglclefindmkaj/viewer.html?pdfurl=https%3A%2F%2Fdigitalcommons.csumb.edu%2Fcgi%2Fviewcontent.cgi%3Farticle%3D1136%26context%3Dcaps_thes&clen=335756 (accessed February 2022).

Rathje, W. and Murphy, C. (2001). *Rubbish!: The Archeology of Garbage*, 1ste. Tucson, AZ: The University of Arizona Press.

Rios, L., Moore, C., and Jones, P. (2007). Persistent organic pollutants carried by synthetic polymers in the ocean environment. *Mar. Pollut. Bull.* 54: 1230–1237.

Ryan, P. (1988). Effects of ingested plastic on seabird feeding: evidence from chickens. *Mar. Pollut. Bull.* 19(3): 125–128.

Serreze, M., Holland, M., and Stroeve, J. (2007). Perspectives on the Arctic's shrinking sea-ice cover. *Science* 315: 1533–1536.

State of the Climate in 2010 (2012). http://www1.ncdc.noaa.gov/pub/data/cmb/bamssotc/2010/bams-sotc-2010-brochure-hi-rez.pdf (accessed February 2012).

Tanaka, K., Takada, H., Yamashita, R. et al. (2013). Accumulation of plastic-derived chemicals in tissues of seabirds ingesting marine plastics. *Mar. Pollut. Bull.* 69: 219–222.

The United States Environmental Protection Agency Harbor Studies Program Survey at Honolulu, Hawaii, and Kahana Bay Beach Observations (1992). Draft Report prepared under contract to Environmental Protection Agency Office of Wetlands, Oceans, and Watersheds, Contract No. 68-C8-0105. Battelle Ocean Sciences.

chrome-extension://efaidnbmnnnibpcajpcglclefindmkaj/http://www.globalgarbage.org/13%20EPA%20Plastic%20Pellets.pdf (accessed May 2022).

The University of Washington (2009). Effects of Microplastics on Marine Environment Focus of International Workshop at UW Tacoma, National Oceanic and Atmospheric Administration (NOAA). http://www.noaanews.noaa.gov/stories2008/20080904_marinedebris.html (accessed August 2009).

Tropospheric and Stratospheric Temperature Record from Satellite Measurements (2012). The National Climatic Data Center, NOAA Satellite and Information Service. http://www.geocraft.com/WVFossils/Reference_Docs/NOAA_Satellite_Temps_to_2007.pdf (accessed April 2012).

US EPA, 1992, U.S. Environmental Protection Agency (1992). Plastic pellets in the aquatic environment: sources and recommendations. Washington, DC, EPA842-B-92-010. chrome-extension://efaidnbmnnnibpcajpcglclefindmkaj/viewer.html?pdfurl=http%3A%2F%2Fwww.globalgarbage.org%2F13%2520EPA%2520Plastic%2520Pellets.pdf&clen=1003299&chunk=true (May 2022).

USGCRP (2017). Intense rainfall. In: *Climate Science Special Report: Fourth National Climate Assessment, Volume I* (ed. D.J. Wuebbles, D.W. Fahey, K.A. Hibbard, et al.), 470. Washington, DC: U.S. Global Change Research Program doi: 10.7930/J0J964J6. (December 2020).

van Haaren, R., Themelis, N., and Goldstein, N. (2010). The state of garbage in America, 17th National Survey of MSW Management in the U.S. BioCycle. http://www.biocycle.net/images/art/1010/bc101016_s.pdf (accessed June 2014).

Viola, F., Paiva, S., and Savi, M. (2010). Analysis of global warming dynamics from temperature time series. *Ecol. Model.* 221: 1964–1978.

Wang, D., Zhao, J., Zeng, G. et al. (2015). How does poly(hydroxyalkanoate) affect methane production from the anaerobic digestion of waste-activated sludge? *Environ. Sci. Technol.* 49 (20): 12253–12262.

Wentz, F., Ricciardulli, L., Hilburn, K., and Mears, C. (2007). How much more rain will global warming bring? *Science* 317: 5835–5843.

Wikpedia (2020). Marine debris. https://en.wikipedia.org/wiki/Marine_debris (accessed December 2020).

Wong, C., Green, D., and Cretney, W. (1990). Methods to manage and control plastic wastes. https://nepis.epa.gov/Exe/ZyPURL.cgi?Dockey=100019KK.TXT (accessed February 2022).

World Glacier Inventory Documentation (2012). http://nsidc.org/data/docs/noaa/g01130_glacier_inventory/ (accessed April 2012).

World Meteorological Organization Concludes Planet is getting warmer (2012). http://www.minnpost.com/environment/2011/12/world-meteorologicalorganization-concludes-planet-getting-warmer (accessed February 2012).

3

Life Cycle Information

3.1 Life Cycle Assessment for Environmental Hazards

Life cycle assessments (LCAs) include many environmental hazards. LCA can be used to compare the environmental impacts of producing plastic products. Sustainability for manufacturing can be categorized into generation of greenhouse gases (GHGs), solid and liquid wastes, and air and water pollution.

The United Nations Environment Program and the Society for the Environmental Toxicology and Chemistry provide a Web site with information on LCA including approaches for capability development, methodologies, data, resources, and impacts. The Life Cycle Initiative includes LCA publications and manual. The goal of the initiative is to enable people around the world to use LCA more frequently and efficiently. The Life Cycle Initiative can establish a global network of LCA experts to establish and manage best practices of LCA across multiple industries and product sectors (Life Cycle Initiative 2013). The U.S. National Energy Laboratory (NREL) created a US life cycle inventory database to provide support for developers of LCA models of products and services (U.S. NREL 2014). The database provides "cradle-to-gate" and "cradle-to-grave" (life cycle inventory) LCI information to account for energy and material flows into and out of environments for products and services. The goals of the LCI database are to maintain data quality of critically reviewed LCI data for US materials, products, and process; maintain compatibility with international LCI databases; and to support US industry competitiveness.

Sustainable Plastics: Environmental Assessments of Biobased, Biodegradable, and Recycled Plastics, Second Edition. Joseph P. Greene.

3.2 Life Cycle Assessment Definitions

LCA is a methodology to assess the environmental impacts of a product, process, or service. ISO 14040:2006 is an international standard for LCA. LCA is performed with a four-step process that includes:

1) Definition of goal and scope.
2) Inventory of relevant energy, material, and transportation inputs and related liquid, gas, and solid generated outputs.
3) Evaluation of the impacts of releases on the environment associated with the emissions and waste generation by viewing the environmental impacts per functional unit.
4) Interpretation of the LCA results.

The integrated four-step LCA process is described in Figure 3.1.

3.2.1 LCA Step 1: Goal and Scope Development

The first step in the LCA process is to establish the goal and scope of the LCA. The purpose, assumptions, and functional unit are established for the LCA. The functional unit is used to compare different materials, processes, or services. The functional unit ensures that comparisons can be made between

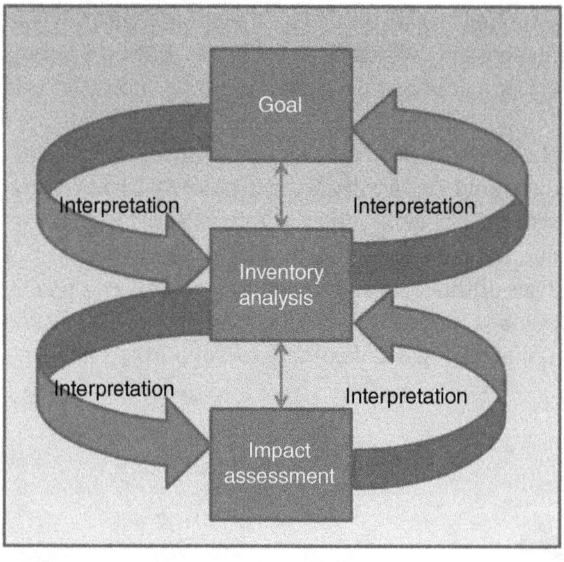

Figure 3.1 Life cycle assessment four-step definition.

different materials, processes, and services. Assumptions can include transportation by truck or rail, end-of-life is recycled, waste-to-energy, landfill, or composting process. The scope in the LCA will be determined if end-of-life environmental effects are considered or if the LCA will end after the manufacturing process. "Cradle-to-gate" LCA life cycle is one where the materials and energy required to produce a product from raw materials to final product are used in the LCA calculations. Likewise, the "cradle-to-gate" LCA will provide environmental emissions, waste generation, and pollution generated during the production of a product from raw materials to final product. An LCA that considers manufacturing, product use, and end-of-life is referred to as "cradle to grave." An LCA that includes chemical recycling or reuse can be referred to as "cradle to cradle." "Cradle-to-gate" and "cradle-to-grave" analysis will be discussed later in this chapter.

3.2.2 LCA Step 2: LCI Development

The second step in the LCA process, LCI, called life cycle inventory, catalogs all of the various materials, energy, and transportation inputs needed to produce the raw materials and manufactured products or systems. The LCI also catalogs the emissions and wastes generated in the production process or system. Figure 3.1 lists the inputs and outputs for the LCI. The U.S. Department of Energy provides a US LCI database roadmap (US Department of Energy 2009). The goals of the database project are to:

- Maintain LCI data quality and transparency.
- Provide LCI data for common industrial materials, products, and processes in the United States.
- Be compatible with international LCI databases.
- Provide LCI resources for US industry and data accessibility.

The US LCI tool can help US companies integrate LCA into business plans and environmental analysis for new products and services. The US LCI tool can provide LCI data for many US manufactured processes and an expanded database of LCI information. The US LCI tool had contributions from the U.S. Department of Energy, U.S. General Services Administration, U.S. Environmental Agency, U.S. Forest Service, U.S. Navy, U.S. Green Building Council, American Chemistry Council (ACC), Athens Institute, CORRIM, Franklin and Associates, Institute for Environmental Research and Education, Portland Cement Association, Sylvatica, and Vehicle Recycling Partnership of USCAR.

The LCI database provides LCI information for many industries including:

- Air transportation
- Chemical manufacturing

- Crop production
- Electric component and computer manufacturing
- Fabricated metal product manufacturing
- Forestry and logging
- Mining
- Nonmetallic mineral product manufacturing
- Oil and gas extraction
- Paper manufacturing
- Petroleum and coal products
- Plastics product manufacturing
- Plastics and rubber products manufacturing
- Primary metal manufacturing
- Rail transportation
- Transit and ground passenger transportation
- Truck transportation
- Utilities
- Waste management and remediation
- Wood products manufacturing
- Biomass

For plastics manufacturing, the LCI database provides LCI information for all major plastics. The database provides references of LCI data from the ACC.

3.2.3 LCA Step 3: LCA Development

The third step in the LCA process is to assess the environmental impacts from the inventory collection of Step 2. The assessment step typically normalizes the input and outputs to the LCI as per the normalized unit. Thus, the environmental effects can be viewed as per a grouping of the product, for example, 1000 grocery bags, 10000 cellular phones, and 100000 vehicles. For example, the environmental impacts can be viewed as 10000 cellular phone cases made from Acrylonitrile butadiene styrene (ABS) versus 10000 cellular phone cases made from PP. Then, the amount of energy required to produce 10000 cellular phone cases from ABS can be compared to the amount of energy required to produce 10000 cellular phones from PP. Likewise, the carbon footprint, solid waste, and pollution can be calculated as per the common grouping.

3.2.4 LCA Step 4: Interpretation of Results

The fourth step is the interpretation of the data and is involved in all of the previous three steps. The data should be analyzed in each of the three steps for consistency and accuracy. The scope and assumptions should be reviewed in each of the steps of the LCA process to ensure thoroughness. The last step interprets the results from the LCA and provides conclusions and recommendations for minimizing environmental impacts of products, processes, and services.

3.3 ISO 14040/14044 Life Cycle Assessment Standards

The International Organization for Standardization (ISO) published standards for LCA. LCA compares environmental performance of products in terms of GHG emissions, pollution generation, waste generation, energy consumption of water consumption, and other resource consumption. LCA compares these items in terms of a measurable quantity of the products.

ISO 14040: 2009 is an international standard of the LCA with a focus on principles and framework. ISO 14040 provides a summary of the LCA practice and applications for LCA. It also includes limitations of the LCA process. ISO 14040 describes the four-step process to develop an LCA that includes goal, LCI, LCA, and interpretation. ISO 14040 also provides reporting aspects of an LCA and the critical review and limitations of the LCA. ISO 14040 does not specify methodologies of the individual LCI and LCA phases of the LCA (ISO 14040:2006 2006). ISO 14044:2006 is an international standard of the LCA with a focus on requirements and methodology of conducting an LCA. ISO 14044 provides guidance on the preparation of the LCI and assessments of the LCI through the LCA. ISO 14044 provides guidance on the interpretation of the LCA results as well as sensitivity analysis of the data used in the LCA (ISO 14044:2006 2006).

LCI can be used to evaluate the material, energy, and raw materials necessary to produce a product or system, as well as the environmental impacts of the product or system. If the scope of the LCA ends with the creation of a product or system, then it is referred to as "cradle-to-gate" assessment as shown in Figure 3.2. The energy requirements, raw materials, and water consumed are tabulated to produce a product or system. Likewise, the air and water pollution, waste generation, and GHG emissions are calculated for the product or system. Cradle-to-gate LCA can be used to evaluate the environmental impacts of producing a product or system. This can be beneficial to manufacturing companies that are looking into ways to produce a product or system with lower environmental impacts.

LCA can be created with dedicated software packages, for example, GaBi Software, SimaPro, and Sustainable Minds (GaBi Software Product Sustainability 2013; SimaPro 2013; Sustainable Minds 2013). LCI can be used to evaluate the raw materials, energy, and water requirements to create a product or system that includes transportation and end-of-life options for the products.

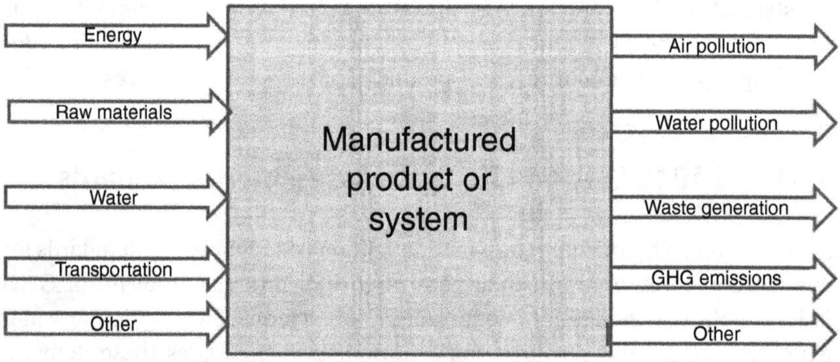

Figure 3.2 Life cycle inventory process for "cradle-to-gate" analysis.

This LCA process is referred to as "cradle to grave" and is shown in Figure 3.3. Likewise, the air and water pollution, waste generation, and GHG emissions are calculated for the product or system. Cradle-to-grave LCA can be used to evaluate the overall impacts of producing a product or system on the environment. This can be beneficial to manufacturing companies that are trying to reduce the transportation energy and environmental costs on a product. Also, companies can evaluate different end-of-life scenarios for the product for composting, recycling, waste-to-energy, or landfilling options. Cradle-to-grave LCA can be helpful for companies to establish triple bottom line accounting for the development of sustainable products and to incorporate design for sustainability principles in the design of a product or system.

3.4 Sensitivity Analysis

The results in an LCA study can be affected by many sources of uncertainty. The sources of uncertainty can be found in the choices used for assumptions, scope, boundaries, impact assessment methods, and the quality of the available data. In addition, assumptions made for the inclusion of end-of-life, transportation, and pollution can significantly affect the results in an LCA.

Key to the relevance of any LCA study is the quality of data. This can be measured with sensitivity analysis (Cellura et al. 2011). LCA studies should include a section on sensitivity analysis and identify areas in the LCA that may be unreliable or inaccurate. The uncertainty in the LCA study should require the data to be calculated with critically reviewed methods. LCA results should include a range of results that incorporates variations in the input data.

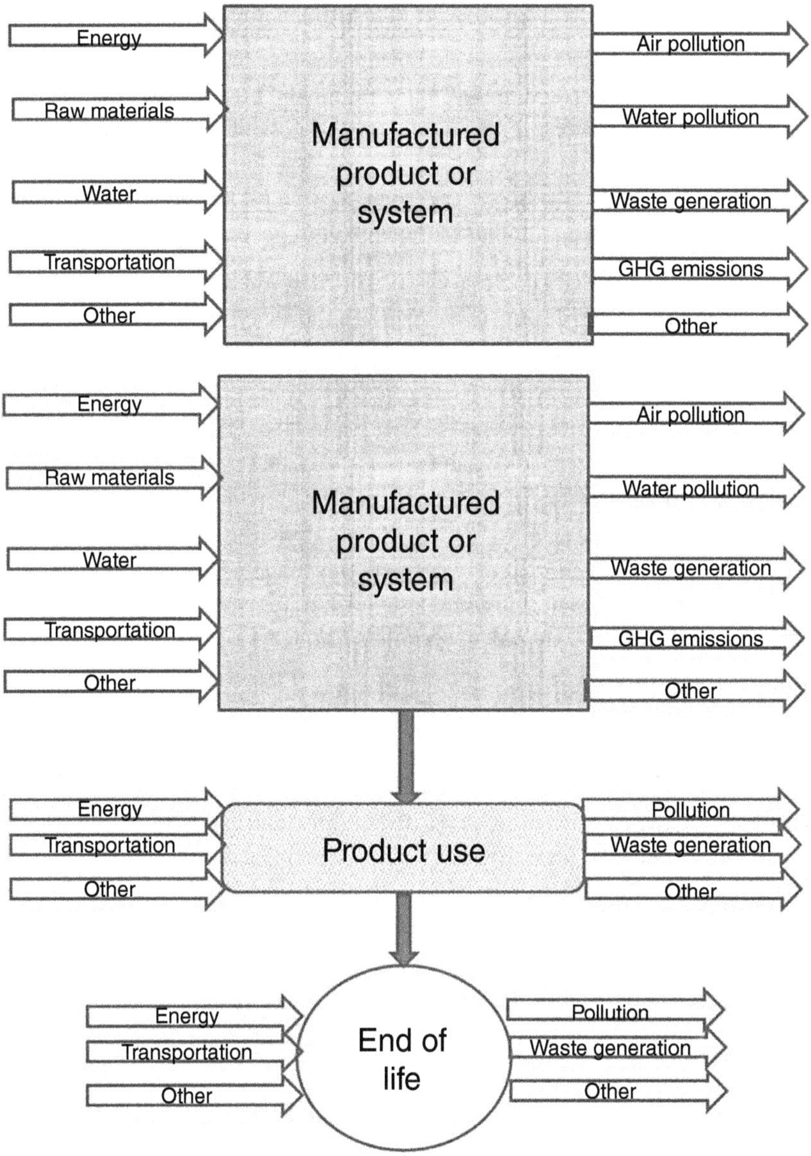

Figure 3.3 Life cycle inventory process for "cradle-to-grave" analysis.

Three procedures of analysis can be used to estimate the uncertainty in an LCA study (May and Brennan 2003):

1) Gravity analysis to determine the data with the highest contribution.
2) Uncertainty analysis to determine the range of possible results based on data uncertainty.
3) Sensitivity analysis that assesses the influence of a parameter, or independent variable, on the value of another parameter, or dependent variable.

As an example, sensitivity analysis was used to determine the uncertainty in an LCA on Italian roof tiles (Cellura et al. 2011). In the study, uncertainty was found in several sources of data in the LCA. The results revealed that in some cases significant differences in energy usage and environmental impacts can be obtained with different assumptions. The research found that uncertainty can be of the following types:

- Parameter uncertainty due to incomplete input data.
- Model uncertainty due to linear assumptions for environmental relationships.
- Methodological uncertainty due to assumptions made with LCA parameters.
- Spatial uncertainty due to short- and long-term time scales.
- Data uncertainty due to the use of secondary input data.

The Italian tile LCA case found that significant variability in the LCA was due to the use of secondary data to calculate the environmental impacts of clay tiles. Assumptions for transportation and electricity sources caused variation in the LCA data, as well as the use of subjective choices in the LCI data from different evaluators.

Standard databases and site-specific inventories can be used to reduce the inconsistencies in the LCA and provide a more reliable LCA. Primary energy, electricity, transportation, and fuel usage data should be provided by local databases and evaluated for consistency by LCA experts.

3.5 Minimal Acceptable Framework for Life Cycle Assessments

LCA can be produced from a variety of sources and provide a variety of results. Standardization of LCA inputs and outputs would benefit the usefulness of LCA. Minimal acceptable features of an LCA are that it follows ISO 14040/14044 standards and uses a four-step process indicated previously. In addition, the LCA should include several items listed in the definition of sustainability, that is, GHG emissions, waste generation, and pollution. The LCA should also include at a minimum end-of-life considerations. Water usage is an important consideration for some areas of the world but not in all areas, and thus is not included in the minimal framework. The pollution areas that are most commonly evaluated in LCAs are water eutrophication and acidification. Thus, an acceptable methodology for LCA should include the following at a minimum:

- Equivalent functional unit for the LCA.
- Energy required LCI calculation per functional unit.
- Equivalent carbon footprint LCI calculation per functional unit.
- Waste generation LCI calculation per functional unit.
- Eutrophication generation LCI calculation per functional unit.
- Acidification generation LCI calculation per functional unit.
- Transportation per functional unit from raw materials to plastics conversion operations.
- End-of-life scenarios.
- ISO 14040 and 14044 compliant.
- Uncertainty analysis.

3.6 Life Cycle Inventory for Petroleum-Based Plastics

LCI can be used to calculate the environmental impacts of producing petroleum-based plastics. The LCI for petroleum-based plastics is based on ISO 14040 and 4044 and provided by the ACC for nine plastic resins and four polyurethane procurer resins. The cradle-to-gate analysis can provide a foundation for understanding the energy requirements, GHG emissions, waste generation, and pollution with the most common petroleum-based plastics. The LCA process for polyethylene terephthalate (PET) plastic can be used as an example (ACC LCA 2011; Nine resins).

3.6.1 LCI for PET Pellets

The LCI for the production of PET pellets in a cradle-to-gate analysis was provided by Franklin and Associates in 2011. PET plastic was produced from crude oil in a five-step process outlined in Figure 3.4. The LCA provides energy and water requirements, GHG emissions, waste generation, and pollution production.

For PET plastic, the pellet production process includes the following:

1) Extracting crude oil
2) Producing naphtha
3) Producing benzene and ethylene oxide
4) Producing ethylene glycol and terephthalic acid
5) Polymerizing PET and producing pellet

The LCA process takes into account all of the energy, raw materials, water, and fossil fuels required in the production of PET pellets. The first step is the extraction

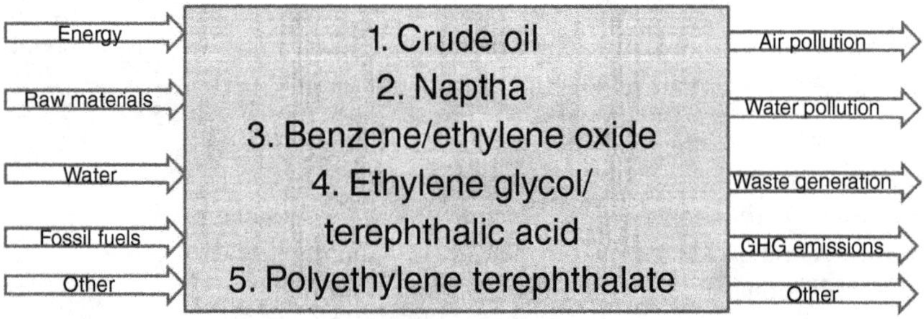

Figure 3.4 "Cradle-to-gate" process of producing PET plastic pellet.

of crude oil. The process requires electricity, fossil fuels, natural gas, water, and other materials. The second step is the production of naphtha with the use of fossil fuels, electricity, water, and other materials. The third step is the production of benzene and ethylene oxide with the use of electricity, fossil fuels, water, and other materials. The fourth step is the production of ethylene glycol and terephthalic acid with the use of fossil fuels, electricity, water, and other materials. The last step is the polymerization of PET and conversion into PET pellets with the use of electricity, fossil fuels, water, and other materials. The air and water emission, waste generations, and pollution were calculated based on data compiled by Franklin Associates from surveys from 17 resin and precursor manufacturers in North America. Other upstream information was provided by Franklin internal database. Fuels and energy databases are from the US LCI database. All calculations followed ISO 14040 and 14044 requirements (Cradle to Gate Life Cycle 2011).

The assumptions in the LCA analysis are as follows:

Assumptions

- 1000 kg of plastic resins.
- Plastic resins were produced from crude oil and natural gas.
- Fuels used to transport plastic resins are included.
- Off-gas fuel is reused in the production process.
- ISO 14040 and 14044 guidelines were used.
- Energy and fuel data included US LCI database from 2003 database.
- Data collected came from plants in the United States, Mexico, and Canada.
- Cradle-to-gate analysis was used. End-of-life options, conversion to plastic products, product use, and transportation to retail outlets are not considered.
- Water consumption was not calculated due to the lack of corresponding data for the raw materials and intermediate chemicals.
- Land use and erosion were not considered due to lack of quality data sources.

The LCA considers environmental impacts of the process, fuel, end-of- life, and energy material resource. For appropriate comparisons with other LCA studies,

Table 3.1 LCI of environmental impacts of four plastic resins.

Environmental impact	PET	GPPS	HDPE	PP
Mass (kg)	1	1	1	1
Specific gravity	1.37	1.06	0.92–0.95	0.9
Energy consumed (GJ)	0.0704	0.0952	0.0783	0.077
Carbon footprint (kg CO_2eq)	2.733	3.242	1.897	1.860
Solid waste generated (kg)	0.141	0.110	0.0771	0.085

we will consider only the environmental effects of the process of each material. The environmental categories considered are:

- GHG emission
- Energy usage
- Waste generation
- Water usage

3.6.2 LCA Sensitivity Analysis

The analysis has limitations for cradle to gate and not cradle to grave. The LCA did not consider the fuel production, energy values are for data from one particular year and not averages of 10 years.

3.6.3 LCA for PET, GPPS, HDPE, and PP Pellets

LCI can also be calculated for other plastic resins and used as a comparison. Table 3.1 presents "cradle-to-gate" analysis of the environmental impacts of four plastic resins that can be used for plastic applications (Cradle to Gate Life Cycle 2011). Table 3.1 demonstrates that PP and HDPE produce lower GHG emissions per kilogram of resin and solid waste per kilogram of resin than PET and GPPS.

Additional LCAs will be calculated for PET, PP, PS, and other plastic products in Chapter 7. The LCA of plastic products can be used based on the information in Table 3.1 to calculate the cradle to grave of plastic products made.

3.7 Life Cycle Assessment for Biobased Poly Lactic Acid

LCA can be used to calculate the environmental impacts of producing biobased PLA plastic. The LCA for the production of PLA pellets in a cradle-to-gate analysis was provided with an LCA on Ingeo® PLA. The PLA was produced with the new lactic acid production process with reduced environmental impacts for the Ingeo production system. The LCA provides energy and water requirements, GHG emissions, waste generation, and pollution production (Madival et al. 2009; Vink et al. 2010).

For PLA, the manufacturing process includes the following:

1) Harvesting corn
2) Isolating starch
3) Converting starch to dextrose
4) Fermenting glucose to lactic acid via bacteria
5) Polymerizing lactic acid to poly lactide pellets

The process of producing PLA is described in Figure 3.5. The LCA process takes into account all of the energy, raw materials, water, and fossil fuels required in the production of Ingeo pellets. The first step is the harvesting of the corn in the fields where the corn is grown, harvested, dried, and transported to the corn wet mill. The process requires fertilizers, electricity, fossil fuels, natural gas, and other materials. The second step is the production of starches and dextrose sugars with the use of fossil fuels, electricity, steam, water, and other materials. The third step is the fermentation to lactic acid with the use of electricity, fossil fuels, water, steam, and other materials. The fourth step is the production of lactide from lactic acid with the use of fossil fuels, electricity, steam, water, and other materials. The last step is the polymerization of polylactide and conversion into PLA pellets with the use of electricity, fossil fuels, water, and other materials.

Lactic acid can be produced from chemical or biotechnological methods. Chemical synthesis is based on the hydrolysis of lactonitrile by strong acids, or by base-catalyzed degradation of sugars, oxidation of propylene glycol, or by chemical reactions of acetaldehyde, carbon monoxide, and water at elevated temperatures (Mussatto et al. 2008).

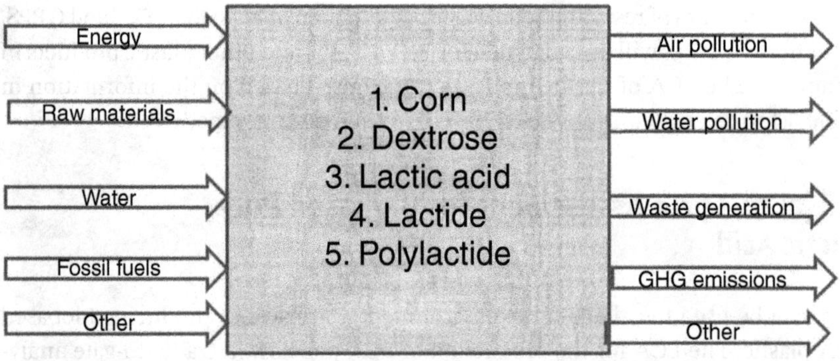

Figure 3.5 "Cradle-to-gate" process of producing Ingeo PLA plastic pellet.

Most of lactic acid is produced from biochemical processes and polymerized to PLA. PLA is usually produced from lactic acid through fermentation of the sugars in corn (Natureworks 2013). PLA can also be produced from sugars from other carbohydrate sources, including organic waste. l-Lactic acid was produced from spent grain by controlling the pH and other process parameters (Shindo and Tachibana 2004). Researchers in Japan successfully produced lactic acid from spent grains with immobilized lactic acid bacterium (*Lactobacillus*). They were able to produce 60 g of sugar (glucose, xylose, and arabinose) from 210 g of spent grains (79% water), which was converted to lactic acid after five days.

The air and water emissions, waste generations, and pollution can be calculated based on the software developed with Plastics Europe and a series of published EcoProfiles for traditional petroleum-based polymers. The same methodology, software, and core databases developed an EcoProfile for Ingeo. All calculations followed ISO 14040 and 14044 requirements (Vink et al. 2010).

The assumptions in the LCA analysis are as follows:

Assumptions

- One kilogram of Ingeo.
- Cradle-to-gate analysis was used.
- End-of-life options, conversion to plastic products, product use, and transportation to retail outlets are not considered.
- Boustead 5.0 methodology and software were used to calculate the LCI and LCA of PLA (Boustead Model 5.0 2013).
- All LCI and LCA calculations based on ISO 14040 and 14044.

The LCA considers environmental impacts of the energy, materials, processes, and end-of-life. For appropriate comparisons with other LCA studies, we will consider only the environmental effects of the process of each material. The environmental categories considered are:

- GHG emissions
- Energy usage
- Waste generation
- Water usage

Table 3.2 provides the environmental impacts of producing 1 kg of Ingeo PLA.

Additional LCAs will be calculated for PLA in Chapter 7. The LCA of plastic products can be used based on the information in Table 3.2 to calculate the cradle-to-grave LCA of plastic products made with Ingeo plastic.

3.7.1 LCA Sensitivity Analysis

The analysis has limitations for cradle to gate and not cradle to grave. One limitation is that the energy usage is for one year and not the averages of three to five years.

Table 3.2 Environmental impacts to produce 1 kg of Ingeo PLA.

Environmental impact	Ingeo
Mass (kg)	1
Density	1.22
Energy consumed (GJ)	0.06784
Carbon footprint (kg CO_2eq)	1.24
Solid waste generated (kg)	0.266968
Water consumed (l)	48.787674
Eutrophication, water (gPO_4eq)	0.0000316

3.8 Summary

LCA is a methodology to assess the environmental impacts of a product, process, or service. ISO 14040 and 14044 are international standards for developing LCA, which has four steps that include definition of goal or scope, inventory of relevant material and energy inputs and relevant environmental outputs, evaluation of environmental impacts per functional unit, and interpretation of results. "Cradle-to-gate" LCA calculates the environmental impacts of products, processes, or systems based on energy, material, transportation, and other inputs. "Cradle-to-gate" LCA is one where the materials and energy required to produce a product from raw materials to final product are used in the LCA calculations. Likewise, the "cradle-to-gate" LCA will provide environmental emissions, waste generation, and pollution generated during the production of a product from raw materials to final product. "Cradle-to-grave" LCA includes all of "cradle-to-gate" calculations, but also includes product use and end-of-life areas of the life cycle.

LCA for four plastics demonstrates that PP and HDPE produces lower GHG emissions per kilogram of resin and solid waste per kilogram of resin than PET and GPPS.

Chapter 3

LCI for PLA

- LCI results for 1 kg of plastic for cradle-to-gate analysis was used.
  - End-of-life options, conversion to plastic products, product use, and transportation to retail outlets are not considered.
  - Boustead 5.0 methodology and software was used to calculate the LCI and LCA of PLA (Boustead 5.0 2013).
  - All LCI and LCA calculations based on 1SO 14040 and 14044.

- Lowest energy consumed
  - PLA
- Lowest carbon footprint
  - PLA
- Lowest solid waste generation
  - PLA

Environmental impact	PET	GPPS	PLA	PHA
Mass (kg)	1	1	1	TBD
Specific gravity	1.37	1.06	1.22	
Energy consumed (GJ)	0.0704	0.0952	0.06784	
Carbon footprint (kg CO_2eq)	2.733	3.242	1.24	
Solid waste generated (kg)	0.141	0.110	0.266968	

LCI for PLA

- LCI results for 1 kg of plastic for cradle-to-gate analysis was used.
  - End-of-life options, conversion to plastic products, product use, and transportation to retail outlets are not considered.
  - Boustead 5.0 methodology and software was used to calculate the LCI and LCA of PLA (Boustead 5.0 2013).
  - All LCI and LCA calculations based on ISO 14040 and 14044.

- Lowest energy consumed
  - PLA
- Lowest carbon footprint
  - PLA
- Lowest solid waste generation
  - PLA

Environmental impact	PET	GPPS	PLA	PHA
Mass (kg)	1	1	1	TBD
Specific gravity	1.37	1.06	1.22	
Energy consumed (GJ)	0.0704	0.0952	0.06784	
Carbon footprint (kg CO_2eq)	2.733	3.242	1.24	
Solid waste generated (kg)	0.141	0.110	0.266968	

LCI for PLA

- LCI results for 1 kg of plastic for cradle-to-gate analysis was used.
  - End-of-life options, conversion to plastic products, product use, and transportation to retail outlets are not considered.

  - Boustead 5.0 methodology and software was used to calculate the LCI and LCA of PLA (Boustead 5.0 2013).
  - All LCI and LCA calculations based on ISO 14040 and 14044.
- Lowest energy consumed
  - PLA
- Lowest carbon footprint
  - PLA
- Lowest solid waste generation
  - PLA

Environmental impact	PET	GPPS	PLA	PHA
Mass (kg)	1	1	1	TBD
Specific gravity	1.37	1.06	1.22	
Energy consumed (GJ)	0.0704	0.0952	0.06784	
Carbon footprint (kg CO_2eq)	2.733	3.242	1.24	
Solid waste generated (kg)	0.141	0.110	0.266968	

References

ACC LCA (2011). Cradle-to-gate life cycle inventory of nine plastic resins and four polyurethane precursors. American Chemistry Council. http://plastics.americanchemistry.com/LifeCycle-Inventory-of-9-Plastics-Resins-and-4-Polyurethane-Precursors-APPS-Only (accessed July 2013).

Boustead Model 5.0 (2013). Tool the Boustead Model 5.0.12. http://lca.jrc.ec.europa.eu/lcainfohub/tool2.vm?tid=186 (accessed July 2013).

Cellura, M., Longo, S., and Mistretta, M. (2011). Sensitivity analysis to quantify uncertainty in Life Cycle Assessment: the case study of Italian tile. *Renew. Sustain. Energy Rev.* 15: 4697–4705.

Cradle to Gate Life Cycle (2011). Cradle to gate life cycle inventory for nine plastic resins and four polyurethane precursors. Final Report, American Chemistry Council, Franklin and Associates. http://plastics.americanchemistry.com/Education-Resources/Publications (accessed June 2013).

GaBi Software Product Sustainability (2013). http://www.gabisoftware.com/ (accessed June 2014).

ISO 14040:2006 (2006). ECS, Environmental management – life cycle assessment – principles and framework. European Standard ISO 14040. Brussels: European Committee for Standardisation.

ISO 14044:2006 (2006). ECS, Environmental management – life cycle assessment – requirements and guidelines. European Standard ISO 14044. Brussels: European Committee for Standardisation.

Life Cycle Initiative (2013). http://www.estis.net/sites/lcinit/default.asp?site=lcinit&page_id=15CFD910-956F-457D-BD0D-3EF35AB93D60 (accessed June 2013).

Madival, S., Auras, R., Singh, S., and Narayan, R. (2009). Assessment of the environmental profile of PLA, PET and PS clamshell containers using LCA methodology. *J. Clean. Prod.* 17: 1183–1194.

May, J. and Brennan, D. (2003). Application of data quality assessment methods to an LCA of electricity generation. *Int. J. Life Cycle Assess.* 8 (4): 215–225.

Mussatto, S., Fernandes, M., Manciha, I., and Roberto, I. (2008). Effects of medium supplementation and pH control on lactic acid production from brewer's spent grain. *Biochem. Eng. J.* 40: 437–444.

Natureworks (2013). How Ingeo is made. http://www.natureworksllc.com/The-Ingeo-Journey/Eco-Profile-and-LCA/How-Ingeo-is-Made (accessed July 2013).

Shindo, S. and Tachibana, T. (2004). Production of l-lactic acid from spent grain, a by-product of beer production. *J. Inst. Brew.* 110 (4): 347–351.

SimaPro (2013). LCA software. http://www.pre.nl/simapro/ (accessed June 2013).

Sustainable Minds (2013). LCA software. http://www.sustainableminds.com/software (accessed July 2013).

US Department of Energy (2009). chrome-extension://efaidnbmnnnibpcajpcglclefindmkaj/viewer.html?pdfurl=https%3A%2F%2Fwww.nrel.gov%2Fdocs%2Ffy09osti%2F45153.pdf&clen=1385506&chunk=true (accessed February 2022).

U.S. NREL (2014). U.S. life cycle inventory database. National Renewable Energy Laboratory. http://www.nrel.gov/lci/ (accessed June 2014).

Vink, E., Davies, S., and Kolstad, J. (2010). The eco-profile for current Ingeo® polylactide production. *Ind. Biotechnol.* 6 (4): 212–225.

4

Bio-Based and Biodegradable Plastics

The chapter will discuss the bio-based and biodegradable plastics and reasons for the classification. The plastics will include polylactic acid (PLA), polyhydroxyalkanoate (PHA), polybutylene succinate (PBS), Ecoflex or polybutylene adipate terephthalate (PBAT), thermoplastic starch (TPS), and other bio-based plastics. The discussion will include manufacturing processing to make the plastic pellet, mechanical and thermal properties, shrinkage, and typical automotive applications with these bio-based plastics.

4.1 Bio-Based Plastics Definition

Bio-based and biodegradable polymers have two different meanings. Bio-based products were defined in the 2002 Farm Bill as commercial or industrial products that are composed in whole, or in significant part, of biological products, renewable agricultural materials, or forestry materials. The definition was expanded with the 2008 Farm Bill that incorporated bio-based intermediate ingredients or feedstock. The U.S. Department of Agriculture (USDA) says bio-based products must contain at least 99% materials from organic sources (Bio-based 2018).

Bio-based and biodegradable polymers have two different meanings. Biodegradable polymers are converted to biomass, carbon dioxide (CO_2), and water through a thermochemical process in a specified time frame and in a specified disposal environment. Bio-based and biodegradable plastics are more fully explained in a *Sustainable Plastics* book (Greene 2014). Biodegradable polymers are those in which the carbon in the polymer:

- Is converted into gases
- Over a specified time frame

Sustainable Plastics: Environmental Assessments of Biobased, Biodegradable, and Recycled Plastics, Second Edition. Joseph P. Greene.

- In a specified environment:
  - Compost ("compostable" if meets American Society of Testing and Materials (ASTM) D6400 standards)
  - Marine (ASTM D7081 biodegradation standards)
  - Landfill
  - Soil
  - Anaerobic digester

Bio-based polymers are those made from:

- Natural or organic ingredients, such as starch from corn, potato, tapioca, rice, or wheat, and
- Oils, such as palm seed, linseed, soy bean, etc., or fermentation products, like PLA, PHA, and polyhydroxybutyrate (PHB).

BPI World provides a listing of compostable plastic resins, bags, cutlery, and packaging (BPI World 2018). BPI also certifies the testing for industrial compost per ASTM D6400 and provides a listing of products that pass the D6400 standard.

4.2 Bagasse

Bagasse is sugarcane-based polymer that can be made into paper-like products for packaging, disposable tableware, and containers. Sugarcane can generate 280 kg of bagasse fiber from 1 ton of sugarcane. The composition of bagasse is as follows:

- Cellulose: 42%
- Hemicellulose: 25%
- Lignin: 20%
- Ash: 1.64%
- Extractives (water + ethanol): 8.38%
- Other: 3.98%

The chemistry of bagasse is made from mostly cellulose. Cellulose has glucose groups in its linear crystalline structure. Hemicellulose has a branched amorphous structure and can contain many sugar groups including xylose, glucose, and galactose. The molecular structure of lignin has aromatic rings and consists of various substructures that appear to repeat in a random pattern. The chemical structure of cellulose is $(C_6H_{10}O_5)_n$, as shown in Figure 4.1.

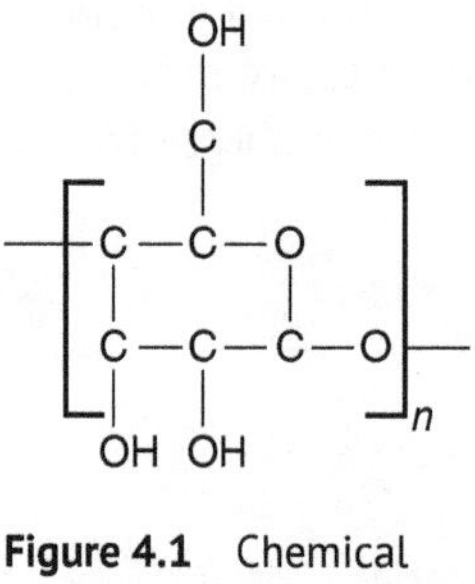

Figure 4.1 Chemical structure of cellulose.

Bagasse is converted into a paper product with a modified paper-pulping process, as shown in Figure 4.2. Typically, strong acids or bases are used to remove the lignin from the bagasse mixture. The properties of bagasse-based polymers are similar to cardboard.

Figure 4.2 Bagasse fiber.

The mechanical properties of bagasse are dependent in part on the aspect ratio of the bagasse fiber. The average diameter and length of bagasse fibers were measured to be 13.0 and 61.0 μm, respectively. The aspect ratio of these results would be 4.69, which is much less than a typical short glass fiber of approximately 100. Wood fibers have an average diameter between 0.016 and 0.030 mm and an average length between 1 and 3 mm resulting in an average aspect ratio between 60 and 100. Typically, an aspect ratio of 100 is needed to obtain structural properties of a fiber.

4.3 Polyhydroxyalkanoates (PHAs)

PHA can be made from over 100 monomers based on polyhydroxybutyrate (P3HB, P4HB, PHB), and polyhydroxy valerate (PHV). PHA is produced in the cells of bacteria from at least five different PHA biosynthetic pathways. PHA is harvested from the cells and made into plastic pellets. P3HBs are the most common PHA and are the basis of PHAs produced by Metabolix Company and Tianjin Company. The P3HB can be copolymerized with P4HB to produce a flexible polymer based on PHAs. P3HB–P4HB is the most common form of PHA. PHAs have hydroxy-acid repeat units that are produced by bacterial fermentation with sugars or lipids to create linear polyesters.

PHA plastics can be made into bottles, bags, containers, and other consumable plastic applications. PHA is not clear but opaque for bottles and bags. PHA is biodegradable under industrial composting conditions (compostable), is marine biodegradable, and soil biodegradable. The chemical structure of PHA is shown in Figure 4.3.

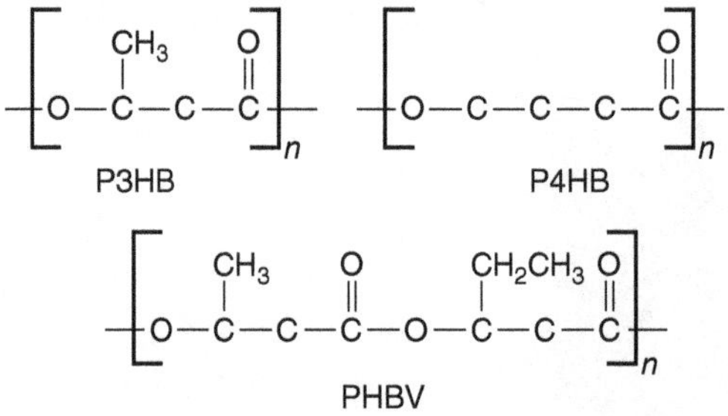

Figure 4.3 Chemical structure of common PHA molecules.

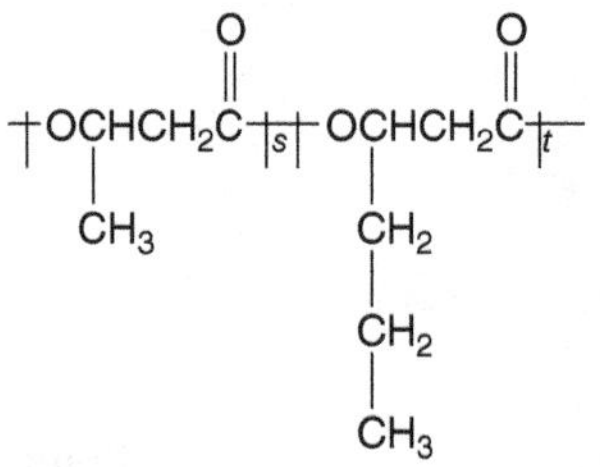

Figure 4.4 Chemical structure of PHBH.

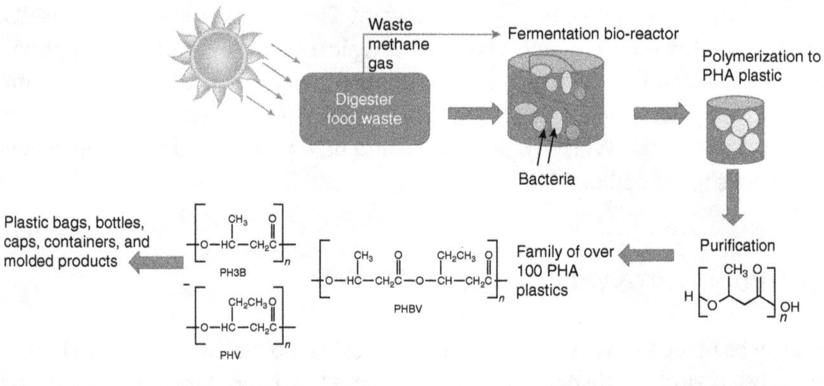

Figure 4.5 Production of PHA from waste methane. Source: PHA Production (2020).

PHA is family of several plastics that includes P3HB, P4HB, Poly hydroxy butyrate Valerate (PHBV), and others. P3HB and P4HB are the most common form of PHA and the chemical basis of commercially available PHAs. New PHA can be made into Poly hydroxy butyrate Hexanate (PHBH) with a hexanate, poly-3-hydroxybutyrate-co-3hydroxyhexanoate, as shown in Figure 4.4.

PHAs can be made from over 100 monomers based on P3HB, P4HB, PHB, and PHV. PHA is produced in the cells of bacteria from at least five different PHA biosynthetic pathways as shown in Figure 4.5. PHA is harvested from the cells and made into plastic pellets. P3HBs are the most common PHA and are the basis of several PHAs. The P3HB can be copolymerized with P4HB to produce a flexible polymer based on PHAs. PHAs have hydroxyacid repeat units that are produced by bacterial fermentation with sugars or lipids to create linear polyesters.

PHAs are polyesters produced in nature by numerous microorganisms, including through bacterial fermentation of sugars or lipids (PHA Wikipedia 2020).

When produced by bacteria they serve as both a source of energy and as a carbon store. More than 150 different monomers can be combined within this family to give materials with extremely different properties (Ibid). These plastics are biodegradable and are used in the production of bioplastics. They can be either thermoplastic or elastomeric materials, with melting points ranging from 40 to 180 °C. The mechanical properties and biocompatibility of PHA can also be changed by blending, modifying the surface, or combining PHA with other polymers, enzymes, and inorganic materials, making it possible for a wider range of applications (Ibid).

To produce PHA, a culture of a microorganism such as *Cupriavidus necator* is placed in a suitable medium and fed with appropriate nutrients so that it multiplies rapidly. Once the population has reached a substantial level, the nutrient composition is changed to force the microorganism to synthesize PHA. The yield of PHA obtained from the intracellular granule inclusions can be as high as 80% of the organism's dry weight (Ibid).

The biosynthesis of PHA is usually caused by certain deficiency conditions (e.g. lack of macro elements such as phosphorus, nitrogen, trace elements, or lack of oxygen) and the excess supply of carbon sources (Ibid).

Polyesters are deposited in the form of highly refractive granules in the cells. Depending upon the microorganism and the cultivation conditions, homo- or copolyesters with different hydroxyalkanoic acids are generated. PHA granules are then recovered by disrupting the cells. Recombinant *Bacillus subtilis* str. pBE2C1 and *B. subtilis* str. pBE2C1AB were used in production of PHAs and it was shown that they could use malt waste as carbon source for lower cost of PHA production (Ibid).

PHA synthases are the key enzymes of PHA biosynthesis. They use the coenzyme-A-thioester of (r)-hydroxy fatty acids as substrates (Ibid). The two classes of PHA synthases differ in the specific use of hydroxy fatty acids of short- or medium-chain length.

The resulting PHA is of the two types:

1) Poly hydroxyalkanoates Short-chain length (PHA SCL) from hydroxy fatty acids with short-chain lengths including three to five carbon atoms are synthesized by numerous bacteria, including *C. necator* and *Alcaligenes latus* (PHB) (Ibid).
2) Poly hydroxyalkanoates Medium-chain length (PHA MCL) from hydroxy fatty acids with medium chain lengths including 6–14 carbon atoms can be made, for example, by *Pseudomonas putida* (Ibid).

A few bacteria, including *Aeromonas hydrophila* and *Thiococcus pfennigii*, synthesize copolyester from the above two types of hydroxy fatty acids, or at least possess enzymes that are capable, or part, of this synthesis (Ibid).

Another even larger-scale synthesis can be done with the help of soil organisms. Even with lack of nitrogen, it can produce 1 kg of PHA/3 kg of sugar.

The simplest and most commonly occurring form of PHA is the fermentative production of poly-beta-hydroxybutyrate P3HB, which consists of 1000–30 000 hydroxy fatty acid monomers.

Researchers have tried for decades to reduce plastic pollution in the oceans and in the landfills by creating plastics made from plants. A new approach being taken is the development of an efficient production process for PHA with a process from waste methane gas.

Scientists at Stanford University and a Palo Alto, California, based start-up company called Mango Materials have come up with a new way to make PHA from waste methane gas. And, with funding from the National Science Foundation (NSF), Mango Materials is advancing the process toward commercialization (PHA from waste methane 2020).

Traditionally, microorganisms feed on plant-derived sugars, from usually corn-based products, and produce PHA. The PHA is then separated from the bacteria and made into pellets that can be molded into plastic products. This approach requires the use of agricultural land and other inputs to produce feedstock, and it competes with the food supply.

Mango Materials developed a new process that uses bacteria grown in fermenters to transform methane and oxygen, along with added nutrients (to supply excess carbon), into PHA (Mango PHA 2020). Eventually, the PHA-rich bacteria – now literally swollen with PHA granules – are removed from the fermenters, and the valuable polymer is separated via proprietary techniques from the rest of the cell mass. The PHA is then rinsed, cleaned, and dried as needed.

This unique approach addresses challenges that have derailed previous attempts at PHA commercialization. Other research projects produced PHA from waste methane (Wang et al. 2015; Fergala et al. 2018).

PHAs are natural biodegradable polymers (biopolymers) showing properties similar to those of commonly produced petroleum-based nondegradable polymers. The utilization of cheap substrates for the microbial production of PHAs is crucial to lower production costs. Feedstock not competing with human nutrition is highly favorable. Syngas, a mixture of carbon monoxide, carbon dioxide, and hydrogen, can be obtained by pyrolysis of organic waste and can be utilized for PHA synthesis by several kinds of bacteria. Up to now, the biosynthesis of PHAs from syngas has been limited to short-chain-length PHAs, which results in a stiff and brittle material. In this study, the syngas-utilizing bacterium *Rhodospirillum rubrum* was genetically modified to synthesize a polymer that consisted of medium-chain-length constituents, resulting in a rubber-like material. This study reports the establishment of a microbial synthesis of these so-called medium-chain-length PHAs from syngas and therefore potentially extends the applications of syngas-derived PHAs (Helnrich et al. 2016).

Table 4.1 Mechanical properties of Mirel PHA, PP, and HDPE.

Property	Mirel PHA 1004	PP	HDPE
Specific gravity	1.3	0.91	0.95–0.98
Yield strength (MPa)	24	33	21–35
Elongation at break (%)	7	14	3–2000
Flexural modulus (GPa)	1.3	1.34	0.28–1.8
Heat distortion temperature (°C) (66 psi)	123	101	47–93

Source: Mirel PHA (2020).

The mechanical properties of P(3HB–4HB) and PHBV are similar to polypropylene as shown in Table 4.1. Mirel PHA was from the company Metabolix that went out of business in 2016. The results in the table are from the work that was done at Chico State University from 2015 to 2016. PP refers to polypropylene and HDPE refers to high density polyethylene.

PHA plastics can be molded via injection molding, extrusion, thermoforming, and blow molding thermoplastic processes. Typical injection molding parameters for P(3HB–4HB) are shown in Table 4.2. Mirel P(3HB–4HB) can be extruded into sheets with single-screw extruders and compounded with additives in a twin-screw extruder.

Table 4.2 Injection molding conditions for Mirel PHA.

Zone	Value
Feed section	175 °C
Compression section	170 °C
Metering section	170 °C
Nozzle	165 °C
Injection speed	25 mm/s
Injection time	1.25 s
Injection pressure	7200 kPa
Hold time	10.0 s
Hold pressure	4000 kPa
Screw speed	55 rpm
Back pressure	500 kPa
Cooling time	10.0 s
Mold temperature	60 °C

Source: Greene PHA conditions (2016).

Table 4.3 Extrusion molding conditions for Mirel PHA in a twin-screw extruder.

Zone	Value
Rear	190 °C
Middle	180 °C
Front	170 °C
Nozzle	160 °C
Screw speed	60 rpm
Side stuffer speed	30 rpm
Water bath temperature	40 °C

Source: Greene PHA conditions (2016).

Table 4.4 Extrusion molding conditions for Mirel PHA in a single-screw extruder.

Zone	Value
Rear	180 °C
Middle	175 °C
Front	170 °C
Nozzle	160 °C
Screw speed	100 rpm
Length/Diameter (L/D) ratio	24 : 1
Screw type	Low shear

Source: Greene PHA conditions (2016).

Mirel P(3HB-4HB) can be extruded into sheets with single-screw extruders and compounded with additives in a twin-screw extruder as shown in Tables 4.3–4.5.

4.4 Polylactic Acid (PLA)

PLA, which is manufactured and supplied by NatureWorks LLC, is a very important bio-based polymer. NatureWorks LLC applies its proprietary technology to process natural plant sugars into a family of polylactide biopolymers, which are marketed under the Ingeo™ brand name (PLA 2020). Ingeo™ biopolymers are

Table 4.5 Extrusion molding conditions for Mirel PHA in an extrusion blow molder.

Zone	Value
Rear	180 °C
Middle	175 °C
Front	170 °C
Nozzle	160 °C
Screw speed	100 rpm
L/D ratio	24 : 1
Screw type	Low shear

Source: Greene PHA conditions (2016).

used for rigid packaging, food service ware, films and cards, fibers and nonwoven applications, and durables. NatureWorks LLC has a 150 000 metric ton manufacturing capacity in Blair, Nebraska, and is exploring building a second polymer plant in South East Asia. NatureWorks is the largest producer of biodegradable plastic in the world. The process to make PLA is shown in Figure 4.6.

NatureWorks provides a corn-based bio-based and biodegradable plastic, PLA, that has a lower carbon footprint than traditional petroleum-based polymers and is biodegradable and compostable in commercial compost. PLA is sustainable since it has lower carbon footprint and lower waste than commercial plastics.

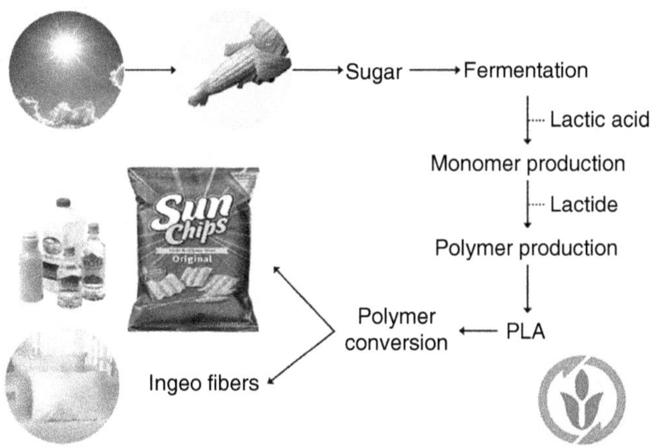

Figure 4.6 Production of PLA. Source: PLA (2020).

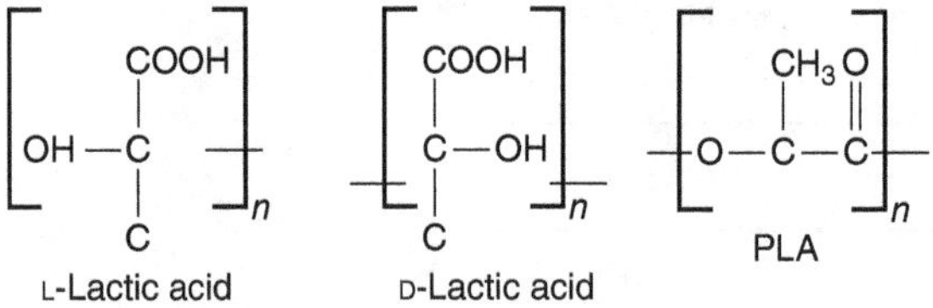

Figure 4.7 Chemistry of PLA.

The chemistry of polylactide $(C_3H_4O_2)_n$ is similar to organic-based polyesters and is shown in Figure 4.7.

Polylactide can occur in two forms, L and D. Poly-L-lactide is the most common form of PLA and is biodegradable while exposed to industrial composting conditions. PLA is created in the United States from corn. The corn starch is converted to glucose and then to lactic acid through bacterial fermentation. The lactic acid is fermented to lactide that is polymerized through a ring-opening process. The production of PLA is carried out in four steps:

1) Pretreatment of the carbohydrate to release sugars.
2) Fermentation of sugars to lactic acid.
3) Conversion of lactic acid lactide.
4) Polymerization of lactide to PLA.

Polylactides are polyesters and are similar to polycarbonate and PET, which also are polyesters. Polyesters have the C=O—CH linkages. The plastics with similar polyester linkages are PLA, PC, and PET. All three plastics have similar properties of high strength, clear plastic, low moisture absorption, and good impact strength.

The mechanical properties of PLA are similar to PET as shown in Table 4.6 (Ibid).

The injection molding processing conditions are shown in Table 4.7. Note the plastic must be dried for four hours at 75 °C.

Table 4.6 Mechanical properties of injection grade PLA.

Description	Indego 3801X	PET
Specific gravity	1.25	1.41
Tensile modulus (GPa)	2.98	2–2.7
Tensile yield strength (MPa)	25.9	14–33
Elongation at break (%)	8.1	10–775
Flexural modulus (GPa)	2.85	0.5–1.8
Heat distortion temperature (°C) (66 psi)	65	85–134

Source: Greene PHA conditions (2016).

Table 4.7 Injection molding processing conditions.

Zone	Value
Feed section	177 °C
Compression section	188 °C
Metering section	188 °C
Nozzle	188 °C
Injection pressure	7200 kPa
Hold pressure	4000 kPa
Screw speed	125 rpm
Back pressure	1700 kPa
Mold temperature	85–105 °C

Source: Greene PHA conditions (2016).

Table 4.8 Extrusion molding processing conditions.

Zone	Temperature
Rear (°C)	160–175
Center (°C)	168–185
Front (°C)	168–185
Die (°C)	170–185
Chill roll top (°C)	21
Chill roll middle (°C)	49
Chill roll bottom (°C)	57

Source: Greene PHA conditions (2016).

The extrusion processing conditions are shown in Table 4.8.

The blow molding processing conditions are shown in Table 4.9.

4.5 Thermoplastic Starch (TPS)

Starch-based polymers can be produced from potato, corn, wheat, cassava, tapioca, vegetable oils, and other renewable sources. In the United States and Europe, corn starch is the predominate source for starch-based polymers. Corn starch is mostly used for animal feed. Starch-based polymers can be processed on traditional thermoplastic forming operations of injection molding, extrusion, blow molding,

Table 4.9 Blow molding processing conditions.

Description	Indego 3801X
Specific gravity (g/cc)	1.24
Crystalline melt temperature (°C)	145–155
Glass transition temperature (°C)	52–58
Crystalline temperature (°C)	100–120
Transmission rates, Oxygen	550 cc-mil/m^2/24-h atm
Transmission rates, CO_2	3000 cc-mil/m^2/24-h atm
Water vapor	325 cc-mil/m^2/24-h atm
Clarity	Transparent

Source: Greene PHA conditions (2016).

compression molding, rotational molding, etc. The most common plasticizers for starch are water, glycerol, and Ecoflex biodegradable plastic. TPS is not considered bio-based since it does not meet the ASTM requirement of 99% made from renewable resources. TPS is compostable since it meets ASTM D6400 standards.

TPS can be added to PP, HDPE, and other plastics for some environmental benefits.

However when TPS is added to plastics, the problems that arise are:

- The compound is not recyclable with the Polyethylene terephthalate (PET) of HDPE recycling operations.
- The compound is not compostable since it will not biodegrade with industrial operations.
- The TPS compound should be disposed off in a landfill.

TPS can be added to bioplastics and biodegradable plastics. Starch, though, can be blended with aliphatic polyesters, like polycaprolactam (PCL), polylactide or poly lactic acid (PLA), PHA, or polyesteramide. The composition of TPS is starch, aliphatic polyester, glycerol, and water. Linear aliphatic polyesters that are compostable are added to starch to create compostable plastics for film, sheet, plastic bags, and liners. Starch, though, can be blended with aliphatic polyesters, like polycaprolactam (PCL), polylactide (PLA), PHA, or polyesteramide.

The chemistry of starch involves blends of two molecular structures of amylose and amylopectin, as shown in Figure 4.8. Amylose and amylopectin have very similar monomer repeating unit. Amylopectin is highly branched molecule that dissolves very quickly in enzymes due to the multiple end points. Amylose is a linear molecule with very little branching and very limited end points. High-amylose starch can be produced with grades of up to 90% amylose.

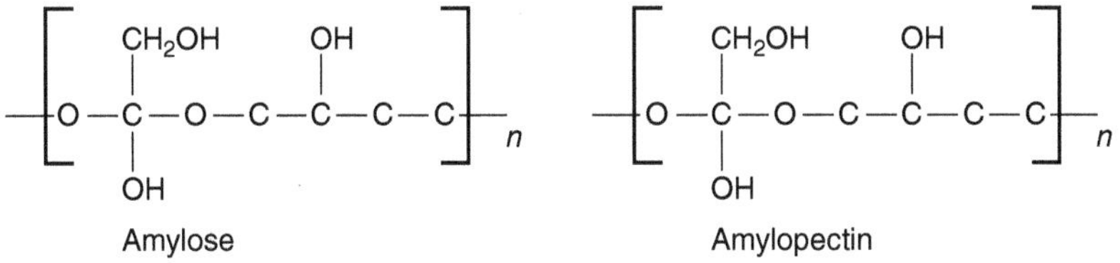

Figure 4.8 Chemical structure of amylose and amylopectin.

TPS is manufactured with corn starch. Corn starch can be produced with a six-step process (International Starch Organization 2013). The steps include:

- Steeping: Cleaned corn is placed into steeping tanks where the corn is soaked in hot water for 30–48 hours. The kernel swells to twice its original size with moisture content between 15 and 45%.
- Steep water evaporation: Steep water containing approximately 10% solids is drained from the kernels and condensed with a multistep condenser. The steep water is condensed to a solid product and then used commercially in the fermentation industry.
- SO_2 addition: SO_2 can be added to the process to produce SO_2 that improves fermentation and reduces destructive molds, yeast, and fungi.
- Germ separation: Softened kernels are broken up with attrition mills to loosen the hulls and break bonds between the germ and endosperm.
- Germ drying: Surface water is removed with the use of a tapered screw press and then dried in a rotary steam bundle dryer. The moisture level is reduced to approximately 4%.
- Storage in germ silo: The germ is transported to a germ silo.

TPS is produced in a twin-screw extruder and then cut into resin pellets. Novamont Mater-Bi® grades of TPS are compostable. Blends of starch and polyethylene films are not compostable, however, due to the polyethylene plastic.

TPS has the following when compared to Low Density Polyethylene (LDPE) plastic:

- Higher melt flow
- Higher specific gravity (density)
- Higher tensile modulus
- Higher tensile strength
- Lower elongation

The mechanical properties of TPS are given in Table 4.10. TPS starch can be blended with PLA for improved properties. The addition of PLA increased the ductile properties of the PLA but it caused a phase separation due to the lack of strong interactions between the PLA and TPS. The glass transition temperature, T_g, was also reduced about 10% (Ferri et al. 2016).

Table 4.10 Mechanical properties of Novamont TPS and LDPE plastic films.

Description	Mater-Bi AB05H TPS	LDPE film
Melt flow index (g) (10 min)	3 at 190 °C and 2.16 kg	0.1–22 at 190 °C and 2.16 kg
Specific gravity	1.28	0.92–0.95
Tensile modulus (MPa)	1103.2	100–200
Tensile strength (break) (MPa)	18.409	8–10
Tensile elongation (break) (%)	160	150–600

Source: TPS Properties (2020).

4.6 Petroleum-Based Compostable Polymers

Compostable polymers or plastics can be made from petroleum sources as well. These can include:

- Ecoflex-film
- Ecovio-film
- Polycaprolactone (PCL)-injection molded parts

All are biodegradable with industrial composting at:

- 58 °C
- 50% moisture
- Mature soil
- 90% of carbon in plastic

The plastic is converted to carbon dioxide in 180 days.

4.6.1 Ecoflex

Ecoflex is an aromatic–aliphatic copolyester that is produced from petroleum-based products, e.g. terephthalic acid, adipic acid, and 1.4-butanediol. The compostable plastic is an excellent biodegradable material for blown film and sheet products, e.g. grocery bags, lawn and leaf bags, agricultural mulch films, etc. It can be blended with other compostable plastics like PLA, PHA, and TPS. Ecoflex is produced by BASF since 1998 (Ecoflex; https://plastics-rubber.basf.com/global/en/performance_polymers/products/ecoflex.html).

The chemical structure of Ecoflex® is a polyester type and shown in Figure 4.9.

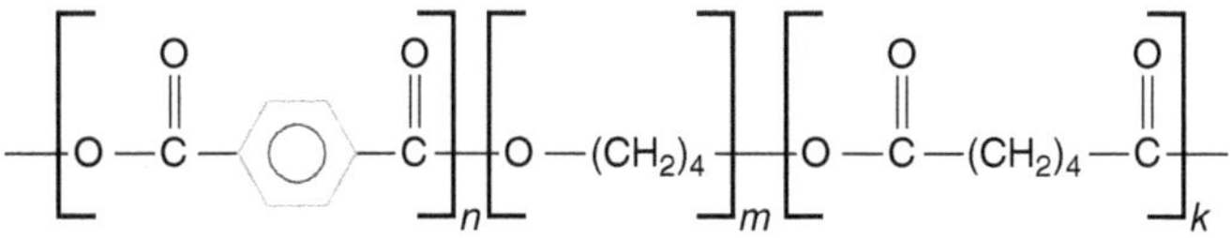

Figure 4.9 Chemical structure of Ecoflex.

Ecoflex processes in blown film extruders like LDPE.

Ecovio is a new biodegradable plastic film that is a combination of Ecoflex and PLA (Ecovio; https://plastics-rubber.basf.com/global/en/performance_polymers/products/ecovio.html). The main uses for Ecovio® are plastic films such as organic waste bags, fruit and vegetable bags, carrier bags with dual-use bags that can be used as a trash can liner, or agricultural films. Ecovio can be used for paper-coating, shrink, and cling films, as well as injection molding and thermoformed products. Ecovio adds bio-based component to Ecoflex. Ecovio can add film-type applications to PLA.

The mechanical properties of Ecoflex when compared to LDPE film include the following:

- Higher melt flow
- Higher specific gravity (density)
- Similar melting temperature
- Lower tensile modulus
- Higher tensile strength
- Higher impact strength
- Higher elongation

Ecoflex and Ecovio can be processed on conventional thermoplastic extrusion, injection molding, and blown film equipment.

4.6.2 Poly-ϵ-Caprolactone, (PCL)

Poly-ϵ-caprolactone (PCL) is an aliphatic polyester produced from petroleum products. PCL is compatible with human tissues and an excellent additive for starch polymers. PCL is used by Novamont with the Mater-Bi biodegradable plastic. PCL can be used for adhesives, compatibilizers, plasticizers, and films for the packaging and for the biomedical industries.

Polycaprolactone (PCL) is a biodegradable polyester with a low melting point of around 60 °C and a glass transition temperature of about −60 °C. Polycaprolactones impart good resistance to water, oil, solvent, and chlorine. It can be blended with

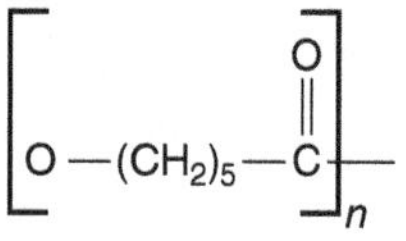

Figure 4.10 Chemical structure of PCL.

polyurethane and be used with 3-D printer polymers. It can be used in biomedical applications, especially for long-term implants and controlled drug release applications (https://en.wikipedia.org/wiki/Polycaprolactone).

The chemistry of polycaprolactone is produced from ring-opening polymerization of ϵ-caprolactone, as shown in Figure 4.10.

The mechanical properties of PCL when compared to LDPE film include the following:

- Higher density
- Similar elongation
- Higher tensile strength
- Higher tensile modulus

The mechanical properties of PCL are provided in Table 4.11.

4.6.3 Poly(Butylene Succinate) (PBS)

PBS is an aliphatic polyester produced from petroleum products. PBS can be used to produce biodegradable sheets, film, bottles, and molded products. PBS can be blended with PLA, PHA, and TPS.

Table 4.11 Mechanical properties of PCL.

Specimens subjected in vacuum	Elastic modulus (MPa)	Yield stress (MPa)	Tensile strength (MPa)	Elongation at break (%)
PCL	238.8±12.5	13.7±0.3	19.5±1.5	>413.35
PCL 15 days	185.3±16.3	10.7±0.6	22.5±1.1	464.0±91.0
PCL 30 days	228.6±9.2	13.4±0.4	17.8±0.7	364.3±22.2
PCL 45 days	211.0±4.8	12.5±0.4	22.7±1.5	507.8±86.2
PCL/MMT	227.5±16.1	13.4±0.6	22.8±0.9	>563.19
PCL/MMT 15 days	218.1±3.3	13.7±1.0	19.5±1.1	321.9±53.4
PCL/MMT 30 days	243.4±21.5	14.8±0.9	19.8±1.2	328.4±97.4
PCL/MMT 45 days	233.6±14.8	14.3±0.6	19.7±0.7	256.9±56.9
PCL/OMMT	217.4±23.9	13.7±1.2	22.7±1.1	>359.5
PCL/OMMT 15 days	252.9±19.8	14.5±0.5	19.9±0.9	234.8±106.7
PCL/OMMT 30 days	297.2±19.0	17.0±0.8	21.3±1.1	191.4±76.2
PCL/OMMT 45 days	268.0±17.5	15.5±0.7	20.0±0.7	255.9±84.7

Source: Poly Caprolactam/montmorillonite (PCL/MMT); Poly Caprolactam/oriented montmorillonite (PCL/OMMT) Franca et al. (2016).

Table 4.12 Mechanical properties of PBS compared to PP and LDPE.

Description	Bionelle PBS	PP	LDPE
Specific gravity	1.260	0.90	0.92
Glass transition temperature (°C)	−32	−5	−120
Melting temperature (°C)	114	163	110
Tensile strength (MPa)	34	33	10
Tensile elongation (break) (%)	560	415	300
Izod impact strength (J/m)	300	20	>400
Crystallinity (%)	35–45	56	49

Source: Liu (2009).

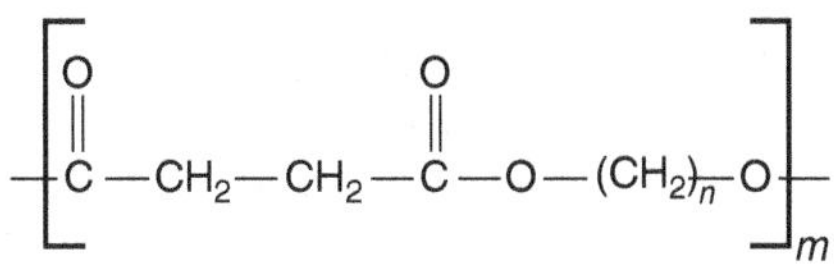

Figure 4.11 Chemical structure of PBS.

The chemistry of PBS can be produced from a poly condensation reaction of succinic acid and butanediol in the presence of catalysts. The mechanical properties, as listed in Table 4.12, when compared to LDPE film, include the following:

- Similar melting temperature
- Higher density
- Higher elongation
- Higher tensile strength
- Lower impact strength

Bionelle PBS had higher density than PP or LDPE. It, though, has similar tensile strength and elongation. It also has higher impact strength, but lower crystallinity.

The chemical structure of PBS is shown in Figure 4.11.

References

Fergala, A., AlSayed, A., Khattab, S. et al. (2018). Development of methane-utilizing mixed cultures for the production of polyhydroxyalkanoates (PHAs) from anaerobic digester sludge. *Environ. Sci. Technol.* 52 (21): 12376–12387.

Ferri, J.M., García García, D., Sánchez Nacher, L. et al. (2016). The effect of maleinized linseed oil (MLO) on mechanical performance of poly(lactic acid)-thermoplastic starch (PLA-TPS) blends. *Carbohydr. Polym.* 147: 60–68.

Franca, D.G., Bezerra, E.B., De Souza Morais, D.D. et al. (2016). Hydrolytic and thermal degradation of PCL and PCL/bentonite compounds. *Mater. Res.* 19: 618–627.

Greene, J.P. (2009). Biodegradable and Oxodegradable Plastics Degradation in Compost and Marine Environments. *Proceedings of the 8th World Congress of Chemical Engineering*, Montreal, Canada (accessed August 2009).

Greene, J.P. (2014). *Sustainable Plastics: Environmental Assessments of Biobased, Biodegradable, and Recycled Plastics*. Hoboken, NJ, ISBN: 978-1-118-10481-1: Wiley.

Greene, J.P. (2017). A review of biodegradation of biodegradable plastics under industrial compost, marine, soil, and anaerobic digestion environments. *Proceedings of the BioPolymers and BioPlastics Conference* (accessed October 2017).

Greene PHA conditions (2016). California State University Plastics Lab Notes. Dr. Joseph Greene, Professor (June 2016).

Helnrich, D., Heinrich, D., Raberg, M. et al. (2016). Synthesis gas (syngas)-derived medium-chain-length polyhydroxyalkanoate synthesis in engineered *Rhodospirillum rubrum*. *Appl. Environ. Microbiol.* 82 (20): 6132–6140.

Liu, L. (2009). Mechanical properties of poly(butylene succinate) (PBS) biocomposites reinforced with surface modified jute fiber. *Compos. Part A Appl. Sci. Manuf.* 40 (5): 669–674.

Nayaran, R. (2022). Carbon footprint of bioPlastics using biocarbon content analysis and life-cycle assessment. chrome-extension://efaidnbmnnnibpcajpcglclefindmkaj/viewer.html?pdfurl=http%3A%2F%2Fwww.agropolocampinasbrasil.org%2Farquivos%2F8_workshop_bioeconomia%2FBrazil_Campinas_2017_workshop.pdf&clen=1376685&chunk=true (accessed February 2022).

Nayaran, R. and Pettigrew, C. (December 1999). ASTM standards help define and grow a new biodegradable plastic industry. *ASTM Standardization News*.

Wang, D., Yang, G., Fu, J., and Zhang, G. (2015). How does poly(hydroxyalkanoate) affect methane production from the anaerobic digestion of waste-activated sludge? *Environ. Sci. Technol.* 49 (20): 12253–12262.

Websites

ASTM (2020). www.astm.org (accessed June 2020).

Bio-based (2020). https://www.usda.gov/media/press-releases/2016/02/18/fact-sheet-overview-usdas-biopreferred-program (accessed June 2018).

BPI World (2018). http://www.bpiworld.org/ (accessed June 2018).

BPI's *Compostable Logo* effort is featured in the American Environmental Review Series, (23 April 2003). <http://chemical.press-world.com/v/9085/bpi-s-compostable-logo-effort-is-featured-in-the-american-environmental-review-series.html> (accessed December 2011).

Carbon reservoirs (2018). https://en.wikipedia.org/wiki/Radiocarbon_dating#/media/File:Carbon_exchange_reservoir_2.svg (accessed June 2018).

International starch organization. https://internationalstarch.org/ (accessed August 2022).

Liquid Scintillation Counting LSC (2020). http://diposit.ub.edu/dspace/bitstream/2445/32102/1/BT12%20-%20Liquid%20and%20solid%20scintillation_ed2.pdf (accessed June 2020).

Mango PHA (2020). https://www.mangomaterials.com/ (accessed September 2020).

Mirel PHA (2020). http://www.matweb.com/search/datasheettext.aspx?matguid=910e09388b934b5aa5f8938c43b727d1 (accessed June 2020).

PHA from waste methane (2020). https://www.nsf.gov/discoveries/disc_summ.jsp?cntn_id=131553#:~:text=PHA%20is%20a%20biodegradable%20polyester,carbon%20and%20limited%20nutrient%20availability.&text=Mango%20Materials'%20process%20uses%20bacteria,excess%20carbon)%2C%20into%20PHA (accessed June 2020).

PHA Production (2020). http://en.wikipedia.org/pha (accessed June 2020).

PHA Wikipedia (2020). http://en.wikipedia.org/pha (accessed June 2020).

Phytotoxicity Tests for Soil Amendments (2011). Woods End Research Laboratory, Inc. http://www.woodsend.org/aaa/phytox.html (accessed December 2011).

PLA (2020). http://www.natureworks.com/ (accessed June 2020).

Title 40: Protection of the Environment (2011). Electronic Code of Federal Regulations. http://ecfr.gpoaccess.gov/cgi/t/text/text-idx?type=simple;c=ecfr;cc=ecfr;idno=40;region=DIV1;q1=503;rgn=div5;sid=0a83cc85ba6faf8267c49539d2c34027;view=text;node=40%3A30.0.1.2.41 (accessed 2 November 2011).

TPS (2020). Properties. https://www.novamont.com/eng/mater-bi (accessed June 2020).

5

Bio-Based and Recycled Petroleum-Based Plastics

As defined in Chapter 1, sustainable plastics are those plastics made with lower energy, lower carbon footprint, lower waste, and lower pollution than conventional plastics. Plastics that are made from plants and bio-based sources or from recycled plastics can be considered sustainable plastics.

5.1 Bio-Based Conventional Plastics

Several conventional plastics can be produced from organic sources. Bio-based polyethylene and bio-based polyethylene terephthalate (PET) resins are currently available. Other plastics can be produced with bio-based sources. Currently, bioplastics represent approximately 1% of the about 360 million tons of plastic produced worldwide annually. But, this should increase due to increased demand for more sophisticated biopolymers from a growing market (European Bioplastics 2021). According to the European Bioplastic Council, the total bioplastics production capacity is set to increase from around 2.11 million tons in 2019 to approximately 2.43 million tons in 2024. Figure 5.1 shows the expected growth for bioplastics in the next three years. Both bio-based non-biodegradable plastics and bio-based biodegradable plastics should grow in the next several years.

Figure 5.1 lists all of the bio-based plastic that can include the following:

- Bioplastics that are based on renewable resources that are biodegradable, like starch, polylactide (PLA), polyhydroxyalkanoate (PHA), polyhydroxybutyrate (PHB), cellulose polymers, proteins, lignin, etc.
- Bioplastics that are based on petroleum resources and are biodegradable, like polycaprolactone (PCL), Ecoflex, polybutylene adipate terephthalate (PBAT),

Sustainable Plastics: Environmental Assessments of Biobased, Biodegradable, and Recycled Plastics, Second Edition. Joseph P. Greene.

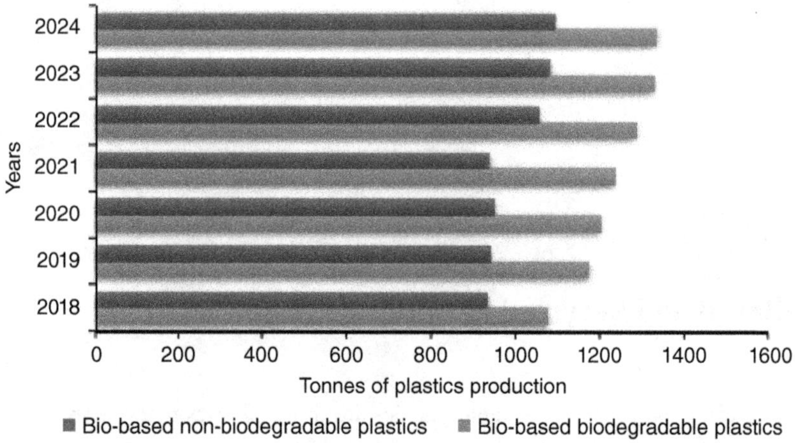

Figure 5.1 Bio-based non-biodegradable plastics and Bio-based biodegradable plastics production from 2018 to 2024 in tonnes of plastics produced. Source: European Bioplastics (2021).

polybutylene succinate (PBS), polybutylene adipate (PBA), polyvinyl alcohol (PVOH), etc.

- Bioplastics that are based on renewable resources and are not biodegradable, like polyethylene, polypropylene (PP), PET, polybutylene terephthalate (PBT), nylon, polyurethane, polyvinyl chloride (PVC), polyester, epoxy, etc.

The list should be expanding in the future. This chapter will consider the bioplastics in the third area.

Table 5.1 lists the bio-based plastics that are commercially available.

In particular, the global production capacities of bio-based non-biodegradable plastics account for about 45% of the total, while bio-based biodegradable plastics account for about 55% of the total, as reported in Figure 5.2, by material type.

Table 5.1 Bio-based conventional plastics.

Material	Type	Resin supplier	Products
Bio-based polyethylene	Polyethylene	Braskem, DOW Chemical	Blown film, sheet, bottles, packaging products, fibers, etc.
Bio-based PET	Polyethylene terephthalate	Many	Bottles, containers, packaging products, fibers, etc.
Bio-based polypropylene	Polypropylene	Braskem	Blown film, sheet, bottles, packaging products, fibers, etc.

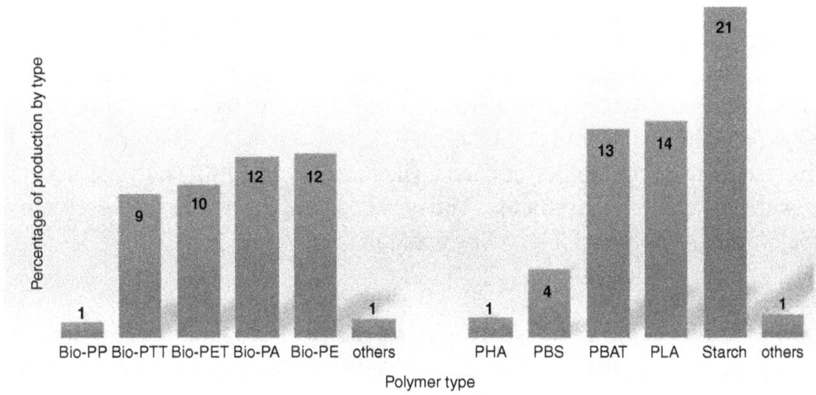

Figure 5.2 Percentage of global bio-based plastics production by type. Source: European Bioplastics (2021).

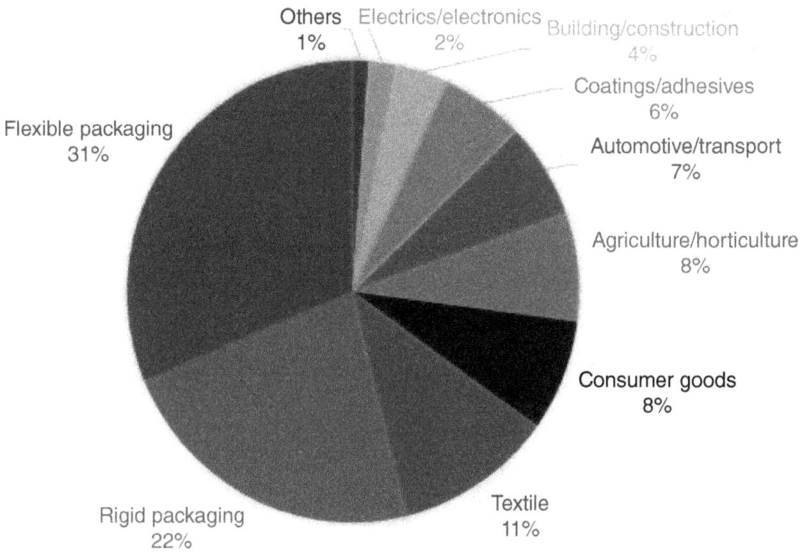

Figure 5.3 Market segments for bio-based plastics.

Packaging is the largest area of use for bioplastics as shown in Figure 5.3. This is for both flexible and rigid packaging. This accounts for over 50% of 1.14 million tons of bio-based plastic production (in 2019), followed by textile, consumer goods, agriculture and horticulture, automotive and transport goods, coatings and adhesives, building and construction, electrics and electronics, and others like toys (European Bioplastics 2021). The percentage distribution is reported in Figure 5.3.

5.1.1 Bio-Based Polyethylene

Bio-based polyethylene is currently available from Braskem and DOW Chemical Company. Bio-based polyethylene can be made from sugarcane or other agricultural materials. DOW Chemical partnered with UPM Biofuels to commercialize the bio-polyethylene in 2019 (DOW and UPM 2020). DOW hopes to produce plastic packaging from the bio-polyethylene. The process will significantly reduce CO_2 emissions by carbon sequestration. The plastic packaging can be fully recyclable as well through commercial recycling operations.

5.1.1.1 Composition

The composition of bio-based polyethylene is the same as conventional polyethylene. Typically, polyethylene is produced from natural gas or petroleum products, for example, naphtha. Bio-based polyethylene is produced from bio-based ethanol as depicted in Figure 5.1. Ethanol can be produced from natural products like sugarcane, corn, oats, etc.

5.1.1.2 Chemistry

Bio-based ethylene can be produced from bio-based ethanol. Ethanol can be fermented from sugars found in organic sources like corn, sugarcane, potatoes, etc. The two common sources of bioethanol are corn in the United States and sugarcane in Brazil. Bioethanol is converted to ethene with an aluminum oxide catalyst. The ethene is polymerized to polyethylene. Figure 5.4 lists the molecular formula of ethanol, ethane, and polyethylene.

Figure 5.5 shows a detailed process for producing bio-polyethylene from sugarcane, sugar beets, starch, and lignocellulose. The process has six steps that include the following:

1) Cleaning, slicing, and shredding organic crops of sugarcane, sugar beet, starch, or lignocellulose.
2) Fermenting the sugars derived from the first step.
3) Distillation of the sugars to bioethanol (CH_3CH_2OH and CO_2).
4) Dehydration at high temperature of the azeotropic mixture of hydrous ethanol and vinasse.
5) Creation of bio-polyethylene.
6) Polymerization of bio-polyethylene.

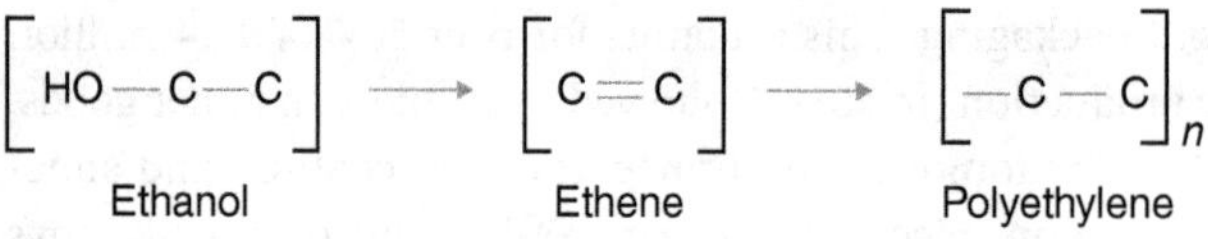

Figure 5.4 Molecular structures of ethanol, ethane, and polyethylene.

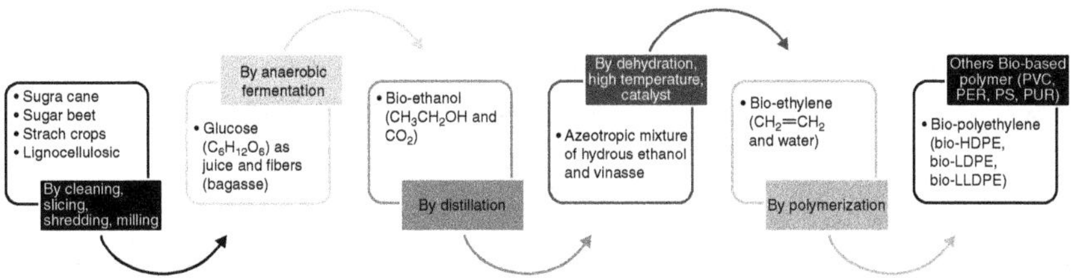

Figure 5.5 Process for producing bio-polyethylene from sugarcane, sugar beets, starch, and lignocellulose. Source: Siracusa and Blanco (2020).

5.1.1.3 Mechanical Properties

The mechanical properties of polyethylene made from sugarcane or corn are the same as conventional polyethylene. Polyethylene is a very versatile plastic and can be made into many different types of products that are injection molded, extruded into profile or sheet, blown into film or bottles, rotation molded into hollow parts, or compression molded into molded products. Braskem Company produces seven grades of plant-based high-density polyethylene (HDPE) and eight grades of plant-based low-density polyethylene (LDPE) for use in blow molding, injection molding, extrusion, and thermoforming applications. Table 5.2 lists some mechanical properties of one grade of sugarcane-based HDPE (MatWeb 2014). Also, Braskem has a site describing its green initiatives, highlighting bio-based polyethylene (Braskem I'm green 2022).

The mechanical properties are similar to DOW polyethylene plastic. The Braskem bio-based polyethylene has similar density, but higher tensile modulus, tensile strength, and tensile elongation than the DOW polyethylene.

Table 5.2 Mechanical properties of bio-based HDPE injection molding grade plastics.

Description	Bio-based Braskem SHC7255 HDPE	Petroleum-based DOW 17450N grade HDPE
Specific gravity	0.954	0.950
Melt flow index (g) (10 min)	4.5	17
Tensile modulus (MPa)	1270	593
Tensile strength (yield) (MPa)	27	18.6
Tensile elongation (break) (%)	880	660

Source: MatWeb (2021).

Table 5.3 LCA of bio-based and petroleum-based polyethylene per 1000 kg.

Polymer type	Energy (MJ)	GHG ($kgCO_2/kg_{HDPE}$ CO_2eq)
Bio-based polyethylene	10 (−85%)	2 (−74%)
Petroleum-based polyethylene	65	7.8

5.1.1.4 Life Cycle Assessment for Bio-Based Polyethylene

Life cycle assessment (LCA) can be used to determine the environmental impacts of producing the bio-based polyethylene. The LCA will consider the energy and greenhouse gas (GHG) emission for producing bio-based polyethylene from the raw materials to the plastic pellet. The "cradle-to-factory gate" approach can be useful for plastic packaging, bags, and other products. The "cradle-to-factory gate" LCA of bio-based polyethylene and petroleum-based polyethylene is listed in Table 5.3 (Hunter et al. 2008). The results show that bio-based polyethylene has 85% less energy and 74% less carbon footprint than petroleum-based polyethylene.

The LCA results for bio-based polyethylene demonstrate the ability of sugarcane to absorb CO_2 during the growing season resulting in a net reduction in GHGs. Additional LCA calculated including end-of-life options are presented in Chapter 7.

A comprehensive study on the environmental assessments of producing bio-based LCA was completed to include LCAs, land-use, and water consumption of bio-based polyethylene (Environmental Assessment 2013).

The report found that the bio-based PE polymer resin leads to a net removal of CO_2 from the atmosphere, averaging −2.15 kg CO_2eq/kg bio-based HDPE (slurry). The emissions from bio-based PE production (sugarcane production, ethanol production, Green Ethylene production, and bio-based PE production) are more than outweighed by the CO_2 removed from the atmosphere and embodied into the bio-based PE resin. The petroleum-based polyethylene produced approximately 1.7 kg CO_2eq/kg. The acidification potential (kg SO_2 eq/kg PE) was approximately 0.036 vs 0.011 kg CO_2eq/kg for petroleum-based PE. The eutrophication potential (kg PO_4 eq/kg PE) was 0.017 vs 0.002 kg PO_4eq/kg PE. The photochemical ozone creation was 0.0033 kg C_2H_4eq/kg PE. The fossil fuel energy demand (MJ/kg PE) was significantly less for the bio-based PE (18 MJ/kg PE) versus petroleum-based PE (85 MJ/kg PE). These pollution values are consistent with farming practices in generating SO_2, PO_4, and ozone. The report found that sufficient land is available to meet Braskem's needs for production of bio-based PE. Bio-based PE was found to have good performance for water usage and abiotic depletion potential.

5.1.2 Bio-Based Polypropylene

PP can also be produced from sugarcane and other organic plant-based materials. Braskem Company planned to produce PP from sugarcane in 2013 (Braskem Green Products 2013). Also, Borelis Company started to produce PP based on Neste-produced renewable feedstock in its production facilities in Kallo and Beringen, Belgium, in 2019 (Borelis bio-PP 2021). Borelis is projected to produce three million tons from renewable propane.

5.1.2.1 Composition

The composition of bio-based PP is the same as conventional PP. Typically, PP is produced from natural gas or petroleum products. Bio-based PP is produced from ethanol.

5.1.2.2 Chemistry

Bio-based PP can be produced from bio-based ethanol as depicted in Figure 5.5. Ethanol can be fermented from sugars found in organic sources like corn, sugarcane, potatoes, etc. PP can be produced by several different techniques. One method includes converting sugarcane to PP by fermenting the sugars in sugarcane to ethanol, then to butylene as an intermediate, and finally to PP through a metathesis reaction of butylene and ethylene. The molecular structure is shown in Figure 5.6.

Figure 5.7 shows a detailed process for producing bio-polyethylene from sugarcane, sugar beets, starch, and lignocellulose. The process has six steps that include the following:

1) Cleaning, slicing, and shredding organic crops of sugarcane, sugar beet, starch, or lignocellulose.
2) Fermenting the sugars derived from the first step.
3) Distillation of the sugars to isobutanol.
4) Dehydration at high temperature to bio-butylene.
5) Creation of bio-propylene.
6) Polymerization of bio-poly propylene.

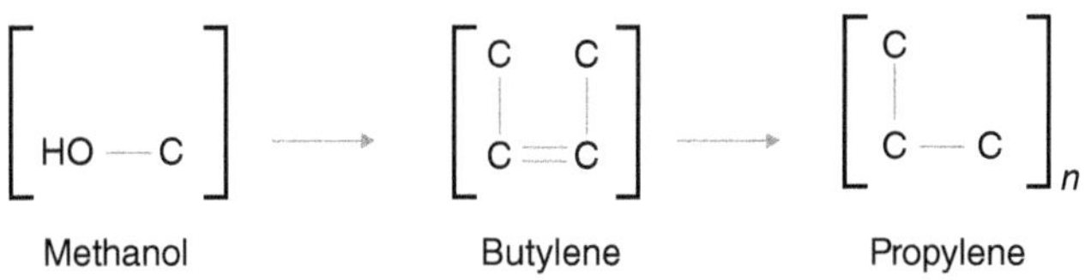

Figure 5.6 Molecular structures of methanol, butylene, and polypropylene.

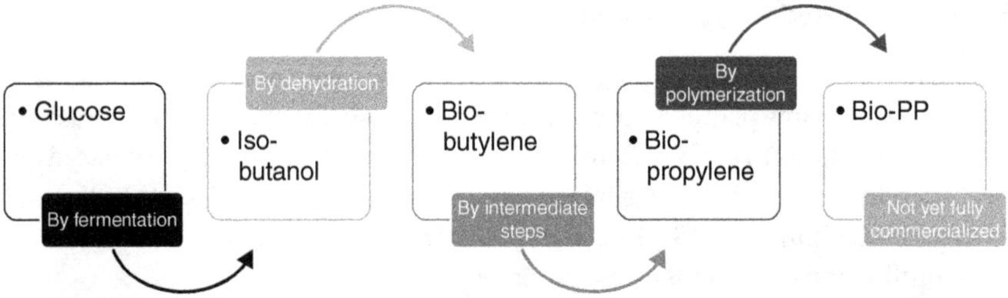

Figure 5.7 Chemical production of bio-based polypropylene. Source: Siracusa and Blanco (2020).

$$\underset{\text{Butene}}{C_4H_8} \longrightarrow \underset{\text{Polypropylene}}{C_3H_6} + \underset{\text{Hydrogen}}{H_2}$$

Figure 5.8 Molecular structures of butene and polypropylene.

Another method includes converting propane to PP by creating propene, which is polymerized to polyethylene. Figure 5.8 shows the process. Propane can be obtained through fracking.

Another method involves gasifying sugarcane to syngas and methanol that is polymerized to PP via "methanol-to-olefins" technology (Liu and Liang 1999).

5.1.2.3 Mechanical Properties

The properties of PP made from sugarcane or corn are the same as conventional polyethylene. PP is a very versatile plastic and can be made into many different types of products that are injection molded, extruded into profile or sheet, blown into film or bottles, rotation molded into hollow parts, or compression molded into molded products. The properties of petroleum-based PP are listed in Table 5.4

Table 5.4 Mechanical properties of petroleum-based polypropylene injection molding grade plastics.

Description	PP
Specific gravity	0.954
Melt flow index (g) per (10 minutes)[a]	4.5
Tensile modulus (MPa)	1270
Tensile strength (yield) (MPa)	27
Tensile elongation (break) (%)	880

[a] Tested at 190 °C with 2.16 kg plunger mass.

(MatWeb 2014). The properties of bio-based PP will be very similar to these values. Researchers today will have to confirm this, though.

5.1.3 Bio-Based Ethylene Vinyl Acetate

Braskem produced ethylene vinyl acetate (EVA) from sugarcane in 2018. The new bio-based EVA can be used to produce foot-ware. It also can be used with adhesives, films, toys, wires, cables, tatami, and general foams (Braskem bio-EVA 2021). The chemical structure of EVA is shown in Figure 5.9.

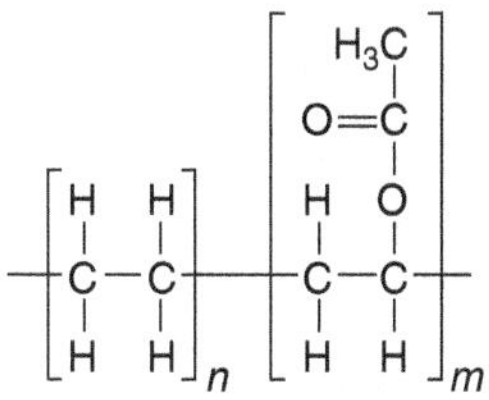

Figure 5.9 Chemical structure of ethylene vinyl acetate. Source: EVA chemical structure (2021).

5.1.4 Bio-Based Polyethylene Terephthalate

PET plastic can also be made from plant sources. PET is typically produced from terephthalic acid and monoethylene glycol (MEG) that are made from petroleum products. MEG can also be fermented from sugarcane, corn, soy, or other organic ingredient.

Bio-based PET, then, can be produced with approximately 30% from organic sources and 70% from petroleum sources. Coca-Cola, Ford Motor, Heinz, Nike, and Proctor & Gamble combined forces in a consortium to develop a 100% plant-based PET product. The plant-based PET consortium will collaborate on PET research projects with universities, suppliers, and research and development companies.

The plant-based PET can be used to produce plastic bottles, plastic packaging, clothing, shoes, automotive carpets, fibers, and other durable goods made currently from traditional PET (Nova 2013).

Figure 5.10 shows a detailed process for producing bio-polyethylene from sugarcane, sugar beets, starch, and lignocellulose. The process has six steps that include the following:

1) Cleaning, slicing, and shredding organic crops of sugarcane, sugar beet, starch, or lignocellulose.
2) Fermenting the sugars derived from the first step.
3) Distillation of the sugars to bioethanol.
4) Dehydration at high temperature to bio-ethylene and bio-ethylene glycol sugar.
5) Creation of bio-ethylene glycol and para-xylene.
6) Polymerization of bio-poly purified terephthalic acid (PTA) and bio-PET.

5.1.4.1 Composition

The composition of bio-based PET is the same as conventional PET. Typically, PET is produced from natural gas or petroleum products. Bio-based PET is produced from MEG and PTA. Recently, terephthalic acid can be produced from

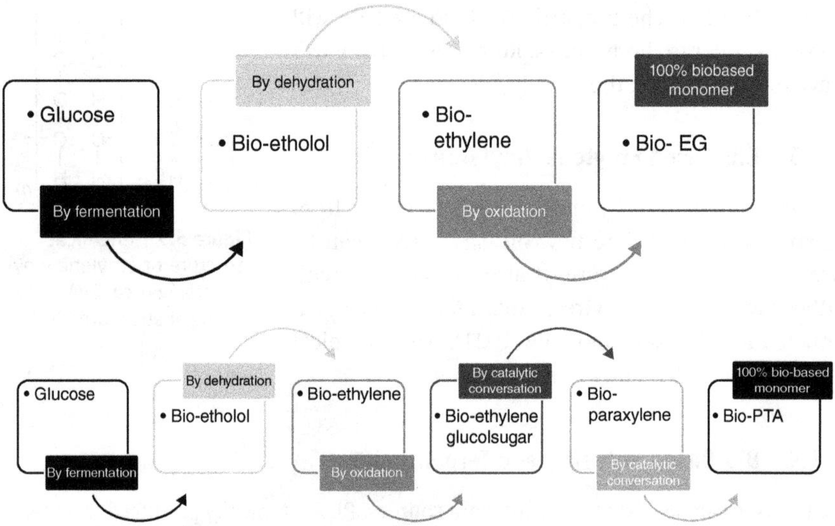

Figure 5.10 Production process to create bio-PET.

organic sources through selective oxidation with a special Co/Mn catalyst (Chavan et al. 2001).

Toray Company in Japan and Virent Company in the United States have produced terephthalic acid from para-xylene and MEG, which were made from organic sources (Toray 2011; Virent 2011). Gevo Company in the United States can produce terephthalic acid and MEG from sugars, starches, and cellulosic materials via para-xylene and isobutanol intermediates (Gevo 2011).

5.1.4.2 Chemistry

Bio-based PET can be produced from bio-based MEG and terephthalic acid. Figure 5.11 lists the molecular structures of MEG, terephthalic acid, and PET.

5.1.4.3 Mechanical Properties

The properties of PET made from sugarcane or corn should be the same as conventional PET. PET is a very versatile plastic and can be made into many different types of products that are injection molded, extruded into profile or sheet, blown into film or bottles, or compression molded into molded products. The properties of PET are listed in Table 5.5. Bio-based PET should have the same properties as petroleum-based PET.

Researchers found that terephthalic acid can be replaced with a sugar-derived monomer named furandicarboxylic acid (FDCA) to produce a 100% bio-PET. This new polymer is labeled as polyethylene furanoate (PEF). The properties of PEF

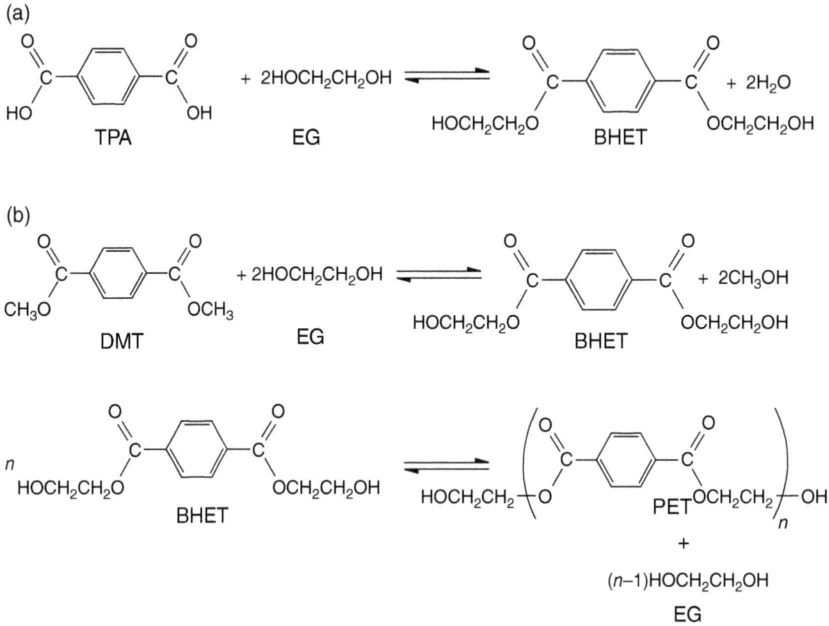

Figure 5.11 Molecular structures of (a) terphthalic acid (TPA), (b) monoethylene glycol (EG), and PET. Source: Bartolome et al. (2021).

Table 5.5 Mechanical properties of petroleum-based PET: blow molding grade plastics.

Description	PET
Melt flow index (g) (10 min)[a]	4.5
Specific gravity	0.954
Tensile modulus (MPa)	1270
Tensile strength (yield) (MPa)	27
Tensile elongation (break) (%)	880

Source: MatWeb (2014).
[a] Tested at 190 °C with 2.16 kg plunger mass.

versus PET are shown in Figure 5.12. The figure shows that PEF has 6–10 times better gas permeability of oxygen, 2–5 times better CO_2 gas permeability, twice as good moisture resistance, and 50% better carbon footprint. It also has a higher T_g, Young's modulus, and field strength. The density and melting points are similar.

Performance	PEF	PET
Gas permeability O_2	6–10 × better than PET	
Gas permeability CO_2	2–5 × better than PET	
Moisture resistance H_2O	2 × better than PET	
CO_2-footprint	50% better than PET	
Glass transition temperature T_g	84–90 °C	67–81 °C
Melting point T_m	195–265 °C (stable up to 325 °C	250–270 °C
Density	1.43 g/cm^3	1.36 g/cm^3
Young's modulus E	3.0–3.5 GPa	2.1–3.1 GPa
Field strength	90–100 MPa	50–60 MPa

Figure 5.12 Properties of PEF and PET. Source: PEF and PETs (2022).

5.1.4.4 LCA of Bio-Based PET

LCA can be used to determine the environmental impacts of producing the bio-based PET. The LCA will consider the energy and GHG generation for producing bio-based PET from the raw materials to the plastic pellet. The "cradle-to-factory gate" approach can be useful for plastic packaging, bags, and other products. Currently, 30% of the PET can be made from bio-based sources. Future research can increase the development of a 100% bio-based terephthalic acid and 100% PET. Very few LCA studies have been conducted on bio-based PET but it is an active area of research.

5.2 Recycled Petroleum-Based Plastics

Petroleum-based plastics can also be sustainable if they are made with plastics that produce less GHG emissions, less waste, and less pollution than conventional plastics. Recycled plastics can be produced with lower energy, lower GHG emissions, and lower pollution than conventional plastics since they are converting the recycled plastics into pellets and not polymerizing the plastic from petroleum or natural gas sources (Franklin and Associates 2010).

Braskem Company decided to produce a recycled PP rather than a bio-based PP. Figure 5.13 describes the recycling process for polypropylene. Braskem uses PP twine typically used for hay bales in the agricultural sector, which would otherwise be directed to landfills after use. In redirecting these PP twine waste streams from landfills, the twine is reused as a raw material offering a circular source of feedstock with consistent material characteristics. The twine is processed, dyed black for product color uniformity, and then tested for purity

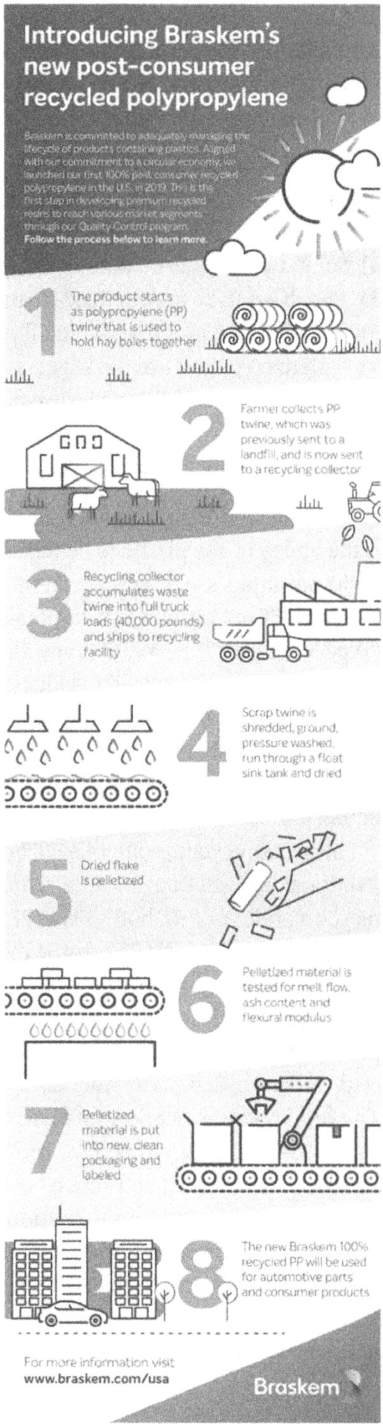

Figure 5.13 Braskem's Recycled polypropylene (Braskem's Recycled polypropylene 2022).

and quality. The resulting sustainably focused homopolymer PP pellets are well suited for use in compounding and packaging applications including uses in automotive, housewares, and consumer goods. Recycled PP has properties similar or as good as virgin PP. The process involves the following eight steps:

1) PP twine holds the straw bales together.
2) PP is cutoff of the bale and collected by the farmer and sent to PP recycling center.
3) Recycling collector ships the twine to a materials recycling facility (MRF).
4) Scrap twine is cleaned, crushed, pressure washed, and dried.
5) The dried flake is pelletized in an extruder.
6) PP pellets are tested with melt flow, density, ash content, and tensile properties.
7) Pelletized pellets are packaged and labels are added.
8) The new recycled PP can be used for automotive, packaging, and other products.

5.2.1 Mechanical Recycling

Plastics are seemingly ubiquitous in the world today. A key feature of thermoplastics is the ability of the plastic to be heated and formed multiple times. Recycling is one of the advantages of managing thermoplastics at the end-of-life. Thermoplastic types include PET or Poly ethylene terpthalate ester (PETE) (Type 1), HDPE (Type 2), PVC (Type 3), LDPE (Type 4), PP (Type 5), polystyrene (PS; Type 6), and other (Type 7).

The majority of recycled plastic is PET (or PETE) or HDPE. PET is used for bottles of soda pop and other beverages. Recycled PET can be remolded for strapping materials and fibers for clothing or carpeting. HDPE is most commonly used for plastic milk jug containers. HDPE can be used as a plastic material in plastic pipe, bottle, and lumber applications.

The gross recycling rate of PET bottles was 29.3% in the United States for 2011 (National Association for PET Container Resources [NAPCOR] 2011). The US recycling rate for PET bottles was 28.9% in 2018 and 27.9% in 2019, notes the 2019 PET Recycling Report (PET bottle recycling 2019). In California, the recycling rate of PET bottles is greater than 60% (CalRecycle PET 2013).

PET and HDPE bottles continue to comprise over 96% of the United States plastic bottle market. In 2011, the total annual postconsumer plastic bottles recycled in the United States increased to an all-time high of 2624 million pounds. The total bottle collection rate was 28.9% for 2011. Plastic bottles with recycling codes #3–7 make up only 3.5% of the plastic bottle market. Of these bottles, PP bottles were recycled at a rate of 2.2% and PVC recycled bottles were recycled at a rate of 0.8% in 2011 (Association of Postconsumer Plastic Recyclers and the American Chemistry Council 2011). In 2018, total collection rate was 28.9% with 52 million more pounds of postconsumer bottles collected in 2018 than in 2017. PET and HDPE bottles comprise 97.1% of the United States plastic bottle market, with PP at 1.8% of plastic bottles produced, LDPE at 0.7%, and PVC at 0.3%.

Together, PET and HDPE comprise 98.9% of the bottles recycled, with PP bottles constituting 1.1%. The total recycled PET increased due to increased usage in food/beverage and non-food bottle markets, with other major end markets – fiber, sheet, and strapping – down slightly over 2017. In 2018, HDPE postconsumer recycled bottles came from non-food application bottles, such as for detergent, motor oil, household cleaners, etc., and for film. Other HDPE postconsumer recycled products come from pipe, lawn products, and non-food application bottles. In 2018, plastic lumber continued to use plastic materials including recycled HDPE, LDPE, mixed rigid containers, and wide-specification virgin (PET bottles USA 2022).

Plastics can be recycled with two methods: mechanical recycling or chemical recycling. Chemical recycling will be discussed in Chapter 6. The mechanical recycling is the most common recycling method for plastics and is discussed in the following section.

5.2.1.1 Plastics Mechanical Recycling Process

The manufacturing process of converting recycled plastic into a new plastic product is daunting. Converting recycled plastics to plastic pellets involve the following: sorting, washing, drying, and pelletizing. The most common processing steps include granulation, air classification, washing, separation, rinsing, and drying. The plastics are sorted by either manual or automated identification methods.

Mechanical recycling makes great sense for petroleum-based plastic and bio-based plastics. Mechanical recycling has the appropriate recycling infrastructure established in the United States, Europe, Australia, and Asia. The recycling system requires a collection system for plastic products labeled as #1–7.

The recycling process has two major steps. The first step is with the MRF that collects the recycled materials and separates them into bales of plastic products by resin type. The second step is with the recycling company that converts the bales of plastic products into plastic pellets.

The first step can be broken down into the following recycling sequences:

- Consumer places plastic item in a recycling container.
- Waste management company collects recycled materials from the recycling container.
- Materials recovery facility collects the recycling container and segregates the plastics by type into collapsed pallets.

The segregation process to segregate the plastic from the debris can be broken down into the following steps:

- Remove large cardboard items from the recycled stream.
- Remove other paper products from the waste stream.

- Separate the glass and ferrous metals from the plastic bottles, bags, and other plastic products.
- Separate aluminum from the plastic stream with an automated "eddy current" process.
- Segregate the plastic stream into plastic type.
- Separate bottles from other plastics and segregate into green bottle (PET), opaque bottle (HDPE), and clear bottle (PET) streams with an automatic method.
- Compress plastic bottles of each type into a baled form with a metal strap.
- Compress other plastics that are not bottles into a baled form with a metal strap.

The recycled plastic bales are sent to recyclers for the next step in the process. The recyclers convert the bales of plastic bottles into plastic pellets of HDPE, PET, and PP. The pallets with the mixed plastics of #3–7 are sent to recyclers for additional sorting and then conversion to pellets of plastics from #3 to #7.

The manual sorting method is labor intensive and requires operators to monitor an assembly line and sort out clear plastic bottles (PET) from the milk containers (HDPE) and colored plastic containers (LDPE, PP, PVC). The automated method can employ one of several analytical techniques, including X-ray fluorescence, mass spectroscopy, Fourier Transform Near Infrared (FT-NIR) spectroscopy, Fourier Transform Medium Infrared (FTIR) spectroscopy, or tribo-electric analysis, on the recycled plastic materials. The state of Rhode Island has a single stream process to sort out the recycled materials with optical sorting technology. The recycled materials include cardboard, paper, glass, metal, and plastics (Rhode Island 2013). Recently, the state's recycling rates have improved modestly in 2019. The average annual municipal recycling rate of 21.5% is 16% higher than the year before the Rhode Island Resource Recovery Corporation (RIRRC) made the $16.9 million upgrade of its MRF. The recycling rate has improved with the use of optical scanners, magnets, and electric pulses. The high-tech sorting machine succeeds at separating recycling materials from the waste stream and bailing it into multi-ton cubes that are ready for shipping to processors worldwide (Rhode Island 2021).

The automated sorting method efficiently and quickly sorts the plastic. Researchers reported the speed at which spectroscopic techniques can identify plastics with the use of a computer and tabulated spectra. Hundreds of identifications per second can help sort plastics with more than 99% accuracy (Wienke 1995). Throughput rates can be significantly increased with automated sorting techniques (Dvorak et al. 2000). The sorting efficiency was improved with the development of an automated sensor cleaning system.

The second step in the recycling process transforms the recycled plastic products into plastic pellets. The bales of plastic bottles are converted into pellets with the following process:

- Plastic bottles are washed to remove the ink, labels, and glue from the outside of the bottle.
- The washed bottles are chopped into small pieces and then floated in a soak tank to wash the inside of the bottle plastic and to separate the PP caps from the PET bottles.
- The PET is then dried and then sent to an extruder for pelletizing into PET pellets.
- The PP cap small pieces are dried and then sent to an extruder and transformed into pellets.
- The HDPE small pieces are also dried and then sent to an extruder and converted into pellets.
- The pellets of HDPE, PP, and PET are evaluated for purity, density, and melt index for quality control.

The plastic pellets then can be sold as recycled pellets for plastic manufacturers of bottles, containers, caps, and plastic packaging. The mechanically recycled plastic experiences several thermal cycles to the melting temperature that can reduce the molecular weight of the polymer and reduce its mechanical properties. Mechanical recycled plastics can have two forms of recycled sources: postindustrial or postconsumer plastics. Chemical recycling is possible for PET but not currently used in high volumes.

Postindustrial recycled plastics occur in most plastic products, wherein the plastics from the sprues runners, extrudate, etc. are added back into the virgin plastic at the plastic manufacturing plant. Postconsumer resin (PCR) refers to plastics that were made into products, for example, bottles, bags, film, packaging, etc.; used by a consumer; collected by a waste disposal company; and processed into recycled pellets at an MRF, where the recycled plastic parts are converted into recycled plastic pellets. The plastic pellets are then sold to plastic companies.

5.2.2 California Plastics Recycling

According to the 2008 Waste Characterization Study, the amount of plastics disposed in California's waste stream was 3 807 952 tons, or 9.6% by weight (CalRecycle 2009). The Rigid Plastic Packaging Container Program of California, administered by the CalRecycle, mandates that companies whose products are sold in California must meet several recycling or product specifications, which includes: be made of at least 25% PCR, and produce the container with reduced amount of material. If using 25% PCR is technologically infeasible, such containers could be eligible for waivers (CalRecycle Container Recycling 2013).

California's Trash Bag Recycled Content Act, passed in the early 1990s, requires plastic trash bag manufacturers selling trash bags in California to meet one of the following requirements: either the plastic trash bags contain actual postconsumer material equal to at least 10% by weight of the regulated bags, or the trash bags must contain actual postconsumer material of at least 30% of the weight of material used in all of its plastic products (CalRecycle 2013).

Light weighting of plastic bags and containers can also achieve compliance with the laws in California, if the plastic products are made with a thinner gauge.

5.2.3 Society of Plastics Industry Recycling Codes

The plastic materials can be collected and sorted based on the Society of Plastics Industry (SPI) recycling code. The recycling code was created in the late 1980s. The recycling code can help keep the recycled plastics segregated by plastic type. Then, waste management companies can sort the plastics into bins of each recycled plastic.

Typically, recycling collection companies sort the plastics into clear bottles and containers (number 1), opaque bottles (number 2), and colored bottles and plastics (numbers 3–7). The recycled plastic bottles and containers are crushed and then placed in pallets and shipped to MRF.

Plastic recyclers sort the plastics by plastic type and then covert the recycled product into plastic pellets. The recyclers can also provide melt index and density measurements of the recycled plastic pellets to assist the plastic converter in using the recycled plastics to produce plastic bottles, bags, packaging, or other plastic products.

American Society of Testing and Materials (ASTM) D7611 committee provides a standard practice for coding plastic manufactured articles for resin identification (ASTM D7611 2013). The ASTM committee utilizes a resin identification code system to identify plastic products into resin family categories. ASTM D7611 provides codes for the six most commonly used resin types. The seventh code is reserved for plastic resins that are not in the six other plastic codes. The numeric code identifying the plastic type shall be placed in the middle of an equilateral triangle. The recycling code for plastics is listed in Table 5.6.

5.2.4 LCAs of Recycled Plastics

Recycling of plastics can make them sustainable with lower carbon footprint, lower waste generation, and lower pollution than virgin plastics.

Table 5.6 SPI recycling codes for plastics.

Code	Plastic	Typical products
1	PET: polyethylene terephthalate	Plastic beverage containers, food containers
2	HDPE: high-density polyethylene	Plastic grocery bags, beverage containers
3	PVC: polyvinyl chloride	Plastic containers
4	LDPE: low-density polyethylene	Plastic bags and stretch film
5	PP: polypropylene	Plastic containers
6	PS: polystyrene	Plastic containers, beverage containers
7	Other: can be a mixture of plastics above or polycarbonate	Plastic containers, beverage containers

5.2.4.1 Life Cycle Inventory

The environmental impacts of recycling can be found by accounting or the accumulation of waste, CO_2 emissions pollution, during each step in the recycling process.

Steps in mechanical recycling process include the following:

1) Collection by single stream, dual stream, or curbside collection.
2) Sorting and separation manually, or by conveyers and separators, separator technology, or computer-aided technology.
3) Reclaiming and pelletizing operations.

The energy usage, GHG emissions, waste generation, and pollution productions are calculated for each of the above steps and combined for an overall life cycle inventory. The recycling operations can involve four types of collection, including cutoff weight-based, cutoff volume based, open-loop weight-based, or open-loop volume-based collections. The open-loop collection methods include virgin plastic collection and will be omitted from this analysis.

Thus, for cutoff collection systems, the LCA is listed in Table 5.7 (Franklin and Associates 2010).

As shown in Table 5.7, recycled plastics require less energy and produce less GHGs than virgin plastics. Recycled plastics, though, require more water and produce more solid waste than virgin plastics. The table shows that the 100% recycled PET uses significantly less energy, produces less GHG, but uses more water and produces more waste than virgin PET. Similarly, 100% recycled HDPE uses significantly less energy, produces less GHG, but uses more water and produces more waste than virgin HDPE. These results will be used in LCA studies in Chapter 7.

Table 5.7 Life cycle assessment of recycled PET and HDPE per 1000 kg.

LCA category	Virgin PET	100% recycled PET	Virgin HDPE	100% recycled HDPE
Energy (GJ)	15.3	3.5	17.1	1.8
GHG (kg CO_2eq)	2746	1136	1822	609
Water (l)	27.5	81.2	30.7	91.5
Waste (kg)	142	385	74.6	212

5.2.4.2 Sustainable Recycled Plastic Products

Thus, plastic products can be made with recycled plastics that can require lower energy and produce lower GHG and lower waste than traditional plastic products that are made with virgin plastics. The plastic products made with recycled product, though, will have higher water usage due to the large amounts of water used in the washing process of recycled plastics. The recycled PET can be blended with virgin PET plastic to produce plastic containers and products. The recycled HDPE pellets can be blended with virgin HDPE to produce plastic products.

5.3 Oxodegradable Additives for Plastics

Oxodegradable additives, typically transition metals, can be added to plastics to cause fragmentation in the plastic. Temperature and relative humidity are the key factors in polymer fracture and reducing the molecular weight in polyethylene plastic (Chiellini et al. 2006). Oxidative degradation of polyethylene samples with proprietary pro-oxidant additives resulted in increased mass, oxygen uptake, and wettability. Two bacterial strains showed increased activity with polyethylene samples that contained antioxidants and pro-oxidants. The microorganisms gained energy from the plastic substrates on the biofilm surface with slow rate over a three-year testing period (Koutny et al. 2006). Oxo-additives can cause the plastic to fragment into smaller pieces over a period of time depending on temperature and humidity in the disposal environment. Oxodegradable additives did not cause polyethylene samples to biodegrade under industrial compost environment according to EN 13432 or ASTM testing standards (Greene 2009; Thomas et al. 2010). The length of time for degradation of oxodegradable plastics cannot be predicted accurately due to environmental conditions. Oxodegradable plastics are not suitable for mechanical recycling with PET, HDPE, or other petroleum-based plastics.

A position paper was released by the SPI concerning the biodegradability claims made by producers of degradable additives (SPI 2013). SPI asserts that the degradable additives were not found to result in biodegradation of the plastic, fragmentation is not equivalent with biodegradation, oxo-additives do not reduce litter, and accumulation of plastic fragments can cause risks to the environment. The plastic additive, however, can cause some concern with contamination with the plastic bottle recycling process.

5.4 Summary

Sustainable plastics are those plastics made with lower energy, lower carbon footprint, lower waste, and lower pollution than conventional plastics. Plastics from plants or bio-based sources and recycled plastics can be made with lower energy, lower carbon footprint, lower waste, and lower pollution than conventional plastics.

Bio-based polyethylene, propylene, and PET can be made from sugarcane or other agricultural materials. Bio-based plastics can be made with nearly identical mechanical properties as conventional petroleum-based plastics and can be manufactured on identical plastics processing equipment.

Recycled plastics can be produced with lower energy, lower GHG emissions, and lower pollution than conventional plastics since they are converting the recycled plastics into pellets and not polymerizing the plastic from petroleum or natural gas sources. Mechanical recycling is the most common recycling method for plastics. Converting recycled plastics to plastic pellets involves sorting, washing, drying, and pelletizing.

The recycling process has two major steps. The first step is with the MRF that collects the recycled materials and separates them into bales of plastic products by resin type. The second step is with the recycling company that converts the bales of plastic products into plastic pellets.

Through LCA results, recycled plastics require less energy and produce less GHGs than virgin plastics. Recycled plastics, though, require more water and produce more solid waste than virgin plastics.

References

Association of Postconsumer Plastic Recyclers and the American Chemistry Council (2011). 2012 national postconsumer recycling reports. http://www.moorerecycling.com/Publications (accessed June 2014).

ASTM D7611 (2013). http://www.astm.org/DATABASE.CART/HISTORICAL/D7611D7611M-10.htm (accessed July 2013).

Bartolome, L., M. Imran, B.G. Cho et al. (2021). Recent development in the chemical recycling of PET, p. 66. https://cdn.intechopen.com/pdfs/32561/InTech-Recent_developments_in_the_chemical_recycling_of_pet.pdf (accessed April 2021).

Borelis bio-PP (2021). Emballator launches renewable bio-PP product with Borealis. https://www.emballator.com/en/2021/10/19/emballator-launches-renewable-bio-pp-product-with-borealis/ (accessed February 2022).

Braskem bio-EVA (2021). https://www.braskem.com.br/imgreen/novidades-detalhe/braskem-supplies-im-greentm-bio-based-eva-for-new-dansko-footwear-line (accessed August 2022).

Braskem Green Products (2013). http://www.braskem.com.br/site.aspx/green-products-USA (accessed June 2014).

Braskem's Recycled polypropylene (2022). https://www.braskem.com.br/usa/news-detail/braskem-america-launches-im-green-recycled-polypropylene (accessed August 2022).

Braskem I'm green (2022). https://www.braskem.com.br/imgreen/home-en (accessed February 2022).

CalRecycle (2009). California 2008 statewide waste characterization study. http://www.calrecycle.ca.gov/Publications/Documents/General%5C2009023.pdf (accessed June 2013).

CalRecycle (2013). Container compliance options. http://www.calrecycle.ca.gov/plastics/rppc/Enforcement/Compliance.htm (accessed August 2013).

CalRecycle Container Recycling (2013). Recycled-content trash bag program. http://www.calrecycle.ca.gov/buyrecycled/trashbags/ (accessed June 2013).

CalRecycle PET (2013). Biannual report of beverage container sales, returns, redemption, and recycling rates. http://www.calrecycle.ca.gov/bevcontainer/Rates/BiannualRpt/12MonPeriod.htm (accessed August 2013).

Chavan, S., Srinivas, D., and Ratnasamy, P. (2001). Selective oxidation of para-xylene to terephthalic acid by μ3-oxo-bridged Co/Mn cluster complexes encapsulated in zeolite–Y. *J. Catal.* 204: 409–419.

Chiellini, E., Corti, A., D'Antone, S., and Baciu, R. (2006). Oxo-biodegradation carbon backbone polymers – oxidative degradation of polyethylene under accelerated test conditions. *Polym. Degrad. Stab.* 91 (11): 2739–2747.

DOW and UPM (2020). https://corporate.dow.com/en-us/news/press-releases/dow-and-upm-partner-to-produce-plastics-made-with-renewable-feedstock.html (accessed April 2020).

Dvorak, R., Kosior, E., Lovenitti, P., and Masood, S., (2000). Factors influencing the sorting efficiency of commingled post-consumer bottles using an automated sorting system. *Proceedings of ANTEC 2000 Conference*, Society of Plastics Engineers, Bethel, CT, pp. 2941–2944.

Environmental Assessment (2013). Environmental assessment of Braskem's bio-based PE resin. chrome-extension://efaidnbmnnnibpcajpcglclefindmkaj/viewer.html?pdfurl=https%3A%2F%2Fwww.braskem.com%2FPortal%2FPrincipal%2FArquivos%2FModuloHTML%2FDocumentos%2F1204%2F20131206-enviro-assessment-summary-report-final.pdf&clen=943433&chunk=true (accessed February 2022).

European Bioplastics (2021). Report bioplastics. https://docs.europeanbioplastics.org/publications/market_data/Report_Bioplastics_Market_Data_2019.pdf (accessed 5 April 2021).

Franklin and Associates (2010). Life cycle inventory of 100% postconsumer HDPE and PET recycled resin from postconsumer containers and packaging. Franklin and Associates, Prairie Village, KS April 7 2010. http://www.napcor.com/pdf/FinalReport_LCI_Postconsumer_PETandHDPE.pdf (accessed June 2013).

Gevo (2011). Bio-based isobutanol to enable Coca-Cola to develop second generation PlantBottle™ packaging. http://gevo.com/?casestudy=bio-basedReferencesisobutanol-to-enable-coca-cola-to-develop-second-generation-plantbottlepackaging (accessed July 2013).

Greene, J. (2009). Biodegradable and oxodegradable plastics degradation in compost and marine environments. *Proceedings of the Eighth World Congress of Chemical Engineering*, Montreal, Canada, August 2009.

Hunter, S., Pereira, B., and Helling, R. (2008). Life cycle assessment of sugarcane polyethylene. The DOW Chemical Company. http://www.lcacenter.org/LCA8/presentations/274.pdf (accessed June 2013).

Koutny, M., Sancelme, M., Dabin, C. et al. (2006). Acquired biodegradability of polyethylene containing pro-oxidant additives. *Polym. Degrad. Stab.* 91: 1495–1503.

Liu, Z. and Liang, J. (1999). Methanol to olefin conversion catalyst. *Curr. Opin. Solid State Mater. Sci.* 4: 80–84.

MatWeb (2014). http://www.matweb.com (accessed June 2014).

MatWeb (2021). Braskem Bio-polyethylene. http://www.matweb.com/search/DataSheet.aspx?MatGUID=09290df8155747bfb20605b0d6489b79&ckck=1 (accessed April 2021).

National Association for PET Container Resources (NAPCOR) (2011). 2011 report on postconsumer PET container recycling activity. http://www.napcor.com/pdf/NAPCOR_2011RateReport.pdf (accessed June 2013).

Nova (2013). Bio-based polymers-production capacity will triple from 3.5 million tonnes in 2011 to nearly 12 million tonnes in 2020. Nova-Institute GmbH. Hürth, 6 March 2013. http://www.nova-institute.eu (accessed July 2013).

PEF and PET (2022). PEF will not oust PET for beverage bottles anytime soon. https://www.packagingdigest.com/technologies/pef-will-not-oust-pet-beverage-bottles-anytime-soon (accessed February 2022).

PET bottle recycling (2019). https://napcor.com/reports-resources/ (accessed April 2019).

PET bottles USA (2022). https://napcor.com/news/ (accessed February 2022).

Rhode Island (2013). Rhode island to open single stream recycling plant. *Plastics News*. http://www.plasticsnews.com/article/20120605/news/306059970/rhodeisland-to-open-single-stream-recycling-plant# (accessed June 2013).

Rhode Island (2021). https://www.ecori.org/composting/2019/12/11/recycling-up-again-time-in-rhode-island (accessed February 2022).

Siracusa, V. and Blanco, I. (2020). Bio-polyethylene (BIO-PI), bio-polypropylene (BIO-PP) and bio-poly(ethylene terephthalate) (Bio-PET): recent developmens in bio-based polymers analogous to petroleum-derived ones for packaging and engineering applications. *Polymers* 12: 1641.

SPI (2013). Position paper on degradable additives. The Society of Plastics Industry (SPI), January 2013. http://www.plasticsindustry.org/files/about/BPC/Position%20Paper%20on%20Degradable%20Additives%20-%20012113%20-%20Final.pdf (accessed June 2013).

Thomas, N., Clarke, J., McLauchlin, A., and Patrick, S. (2010). EV0422:Assessing the Environmental Impacts of Oxo-degradable Plastics Across their Life Cycle. Report to the Department for Environment, Food, and Rural Affairs. Loughborough University. http://wales.gov.uk/topics/environmentcountryside/epq/waste_recycling/substance/carrierbags/oxydegradableplastic/?lang=en (accessed July 2013).

Toray (2011). Toray succeeds in production of the world's first fully renewable, bio-based polyethylene terephthalate (PET) fiber. https://asia.nikkei.com/Business/Materials/Exclusive-Toray-develops-first-100-plant-based-polyester (accessed February 2022).

Virent (2011). Virent's plant-based paraxylene paves the way for a 100% bio-PET bottle. https://www.virent.com/news/virent-bioformpx-paraxylene-used-for-worlds-first-pet-plastic-bottle-made-entirely-from-plant-based-material/ (accessed February 2022).

Wienke, D. (1995). Comparison of an adaptive resonance theory based neural network (ART-2a) against other classifiers for rapid sorting of post consumer plastics by remote near-infrared spectroscopic sensing using an InGaAs diode array. *Anal. Chem. Acta* 317: 1–16.

6

End-of-Life Options for Plastics

End-of-life options for plastics are important factors that influence the life cycle assessment (LCA) of plastics. Disposal options for plastics are to mechanically recycle the plastic, chemically recycle the plastic, compost the bio-based plastic, burn the plastic into energy, or bury the plastic in a landfill.

6.1 US EPA WARM Program

The Environmental Protection Agency (EPA) provides a Waste Reduction Model (WARM) to calculate the environmental impacts of end-of-life options for products, including plastics. The US EPA WARM calculates green-house-gas (GHG) emissions for source reduction, recycling, waste-to-energy, composting, and landfill end-of-life options. The US EPA WARM provides information about recycling, sources reduction, composting, waste-to-energy, and landfill processes. WARM calculations are available in web-based calculator and as a Microsoft Excel® spreadsheet. WARM has databases for over 45 material types and GHG emissions are provided in metric tons of CO_2eq or metric tons of carbon equivalent (EPA Waste Reduction Model 2013).

6.2 Mechanical Recycling of Plastics

Mechanical recycled plastics can have two forms, recycled sources, postindustrial or postconsumer plastics. Postindustrial recycling of plastics can occur with most plastic products, wherein the plastic from the sprues runners and extrudate is added back into the virgin plastic at the plastic manufacturing plant. Postconsumer recycling process occurs when plastics are made into products, for example,

Sustainable Plastics: Environmental Assessments of Biobased, Biodegradable, and Recycled Plastics, Second Edition. Joseph P. Greene.

bottles, bags, film, packaging, used by a consumer, collected by a waste disposal company, processed into recycled pellets at a materials recovery facility (MRF), and then sold to plastic companies.

6.2.1 US Plastics Recycling

The state of North Carolina developed a framework for the assessment of plastics recycling potential (Barlaz et al. 1993). It includes estimations of the quantity of waste that could be available for recycling, assessment of the state's recycling technology including the manufacturing capacity, and evaluation of potential markets for products manufactured from recycled plastics. The state's recycling manufacturing capacity was found to be adequate for the current recycling technology that does not include commingled plastics. Similarly, plastic bottle reclamation in the United States is a mature process for reclaiming plastic bottles and has a sufficient number of manufacturing companies to recycle plastic bottles.

6.2.2 Plastics Recycling Process

Mechanical recycling has the appropriate recycling infrastructure established in the United States, Europe, Australia, and Asia.

As explained previously, the recycling process is as follows:

- Consumer places plastic item in a recycling container.
- Recycling container is collected by a waste management company.
- Materials recovery facility collects the recycling container and segregates the plastics by type into collapsed pallets.
- Recycling company collects the segregated plastic from the MRF and converts the recycled plastic parts into plastic pellets.

The manufacturing process of converting recycled plastic into a new plastic product is daunting. Converting recycled plastics to plastic pellets involves sorting, washing, drying, and pelletizing. The most common processing steps include granulation, air classification, washing, separation, rinsing, and drying. The plastics are sorted by either manual or automated identification methods. The recycling process has two major steps.

1) The first step is with the MRF that collects the recycled materials and separates them into bales of plastic products by resin type.
2) The second step is with the recycling company who converts the bales of plastic products into plastic pellets.

The first step can be broken down into the following recycling sequences:

- Consumer places plastic item in a recycling container.
- Waste management company collects recycled materials from the recycling container.
- MRF collects the recycling container and segregates the plastics by type into collapsed pallets.

The segregation process to segregate the plastic from the debris can be broken down into the following steps:

- Remove large cardboard items from the recycled stream.
- Remove other paper products from the waste stream.
- Separate the glass and ferrous metals from the plastic bottles, bags, and other plastic products.
- Separate aluminum from the plastic stream with an automated "Eddy Current" process.

Segregate the plastic stream into plastic type:

- Separate bottles from other plastics and separated into green bottle (polyethylene terephthalate [PET]), opaque bottle (high density polyethylene [HDPE]), and clear bottle (PET) streams with an automatic method.
- Compress plastic bottles of each type into a baled form with a metal strap.
- Compress other plastics that are not bottles into a baled form with a metal strap.

Plastic recycling steps as shown in Figure 6.1.

- Collection
- Sorting
- Baling
- Crushing
- Washing
- Regeneration
- Extrusion
- Molding
- End products

Interesting facts about PET plastic recycling are as follows: (PET recycling rates 2020).

- PET was invented in 1973.
- There are about 7700 curbside collection and 10 000 drop-off locations in the United States.
- One cubic yard of landfill space is conserved by recycling.

Figure 6.1 Mechanical recycling process for plastics. Source: Mechanical recycling process (2020).

- 4800 of 16-ounce PET beverage bottles
- 3240 of 1-l bottles
- On average, a US household uses 45 lb of PET plastic bottles and jars in a year (PET bottles use 2020).
- If all of them were recycled, it would yield enough recycled PET fiber to make 12 dozen men's T-shirts or enough carpet for a 12-by-15 ft room.

The recycling rates of PET bottles are shown in Table 6.1 (PET recycling rates 2020).

The recycling process for PET includes the following and shown in Figure 6.2.

- Manual sort
- Paper or film, metals
- Metal detection
- Steel and aluminum
- Wash the bottles to remove labels and dirt
- NIR sort ($1 million)
- Color sort
- Inspection and manual sort
- Granular to flake

The reuses of recycled PET include plastic fiber, strapping, plastic sheet, plastic film, and plastic bottles. Table 6.2 shows the uses of recycled plastic PET bottles.

The recycled plastic bales are sent to recyclers for the next step in the process. The recyclers convert the bales of plastic bottles into plastic pellets of HDPE and PET. The pallets with the mixed plastics of 3 to 7 are sent to recyclers for additional

Table 6.1 Gross recycling rates of PET bottles from 2004 to 2014.

Year	Total US bottles collected (MMlbs)	Bottles on US shelves (MMlbs)	Gross recycling rate (%)
2004	1003	4637	21.6
2005	1170	5075	23.1
2006	1272	5424	23.5
2007	1396	5683	24.6
2008	1451	5366	27.0
2009	1444	5149	28.0
2010	1557	5350	29.1
2011	1604	5478	29.3
2012	1718	5586	30.8
2013	1798	5764	31.2
2014	1812	5849	31.0

Source: PET recycling rates (2020).

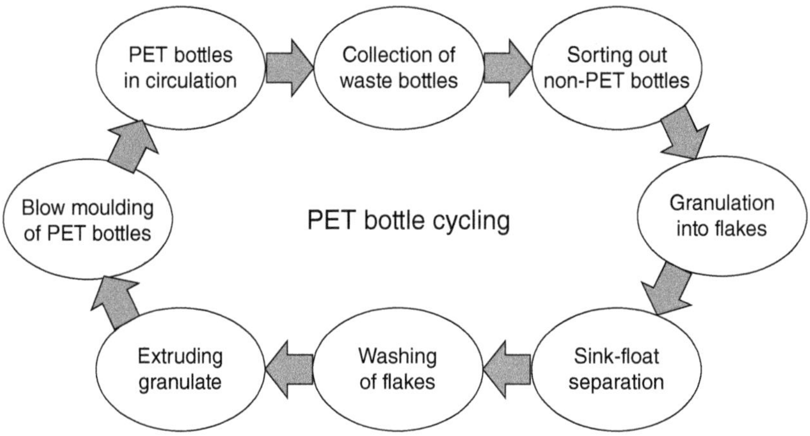

Figure 6.2 The recycling process for PET. Source: PET recycling (2022).

sorting and then conversion to pellets of plastics from 3 to 7. The manual sorting method is labor intensive and requires operators to monitor an assembly line and sort out clear plastic bottles (PET) from the milk containers (HDPE) and colored plastic containers [linear low density polyethylene (LDPE), polypropylene (PP), polyvinyl chloride (PVC)]. The automated method can employ one of several analytical techniques, X-ray fluorescence, mass spectroscopy, Fourier Transform Near Infrared (FT-NIR) spectroscopy, Fourier Transform medium

Table 6.2 The uses of recycled plastic PET bottles in MM of pounds.

Product category	2004	2005	2006	2007	2008	2009	2010	2011	2012	2013	2014
fiber	479	463	422	383	391	344	381	398	512	558	638
sheet and film	58	71	74	128	153	159	195	202	307	315	365
strapping	116	131	132	144	137	114	127	120	136	140	126
Engineered resin	12	8	9	11	7	10	9	See other	See other	See other	See other
Food and beverage bottles	126	115	139	136	141	203	216	242	276	425	351
Non-food bottles	63	63	49	60	55	65	58	57	50	50	57
Other	24	13	30	38	31	42	16	21	31	25	27
Total Converter Consumption	**878**	**864**	**855**	**900**	**915**	**937**	**1002**	**1040**	**1312**	**1513**	**1564**

Source: Uses of recycled PET bottles (2020).

Infrared (FTIR) spectroscopy, or tribo-electric analysis, on the recycled plastic materials.

The state of Rhode Island has a single stream process to sort out the recycled materials with optical sorting technology. The automated sorting method efficiently and quickly sorts the plastic. Researchers reported the speed at which spectroscopic techniques can identify plastics with the use of a computer and tabulated spectra. Hundreds of identifications per second can help sort plastics with more than 99% accuracy. Throughput rates can achieve of 2000 kg/h. The sorting efficiency was improved with the development of an automated sensor cleaning system. The second step in the recycling process transforms the recycled plastic products into plastic pellets. The bales of plastic bottles are converted into pellets with the following process:

- Plastic bottles are washed to remove the ink, labels, and glue from the outside of the bottle.
- The washed bottles are chopped into small pieces and then floated in a soak tank to wash the inside of the bottle plastic and to separate the PP caps from the PET bottles.
- The PET is then dried and then sent to an extruder for pelletizing into PET pellets.
- The PP cap small pieces are dried and then sent to an extruder and transformed into pellets.
- The HDPE small pieces are also dried and then sent to an extruder and converted into pellets.
- The pellets of HDPE, PP, and PET are evaluated for purity, density and melt index for quality control.

The plastic pellets then can be sold as recycled pellets for plastic manufacturers of bottles, containers, caps, and plastic packaging. The issues are mechanically recycling where the mechanically recycled plastic experiences several thermal cycles to the melting temperature that can reduce the molecular weight of the polymer and reduce its mechanical properties. Mechanical recycled plastics can have two forms recycled sources, postindustrial or postconsumer plastics. Postindustrial recycled plastics occurs in most plastic products, wherein the plastic from the sprue runners, extrudate, etc. are added back into the virgin plastic at the plastic manufacturing plant. Postconsumer resin (PCR) refers to plastics that were made into products, e.g., bottles or bags, or other, used by the consumer, collected by waste disposal company, sent to the MRF, and processed back into pellets. The amount of plastics disposed in the US waste stream in 2018 was 3 630 000 tons, or 12.2% by weight. (EPA plastics waste 2021). The MSW waste contents are shown in Figure 6.3. Note that paper, paperboard, food, and yard trimmings account for approximately 57% of the waste. These items should be composted. The total MSW waste in 2018 was 292.4 million tons.

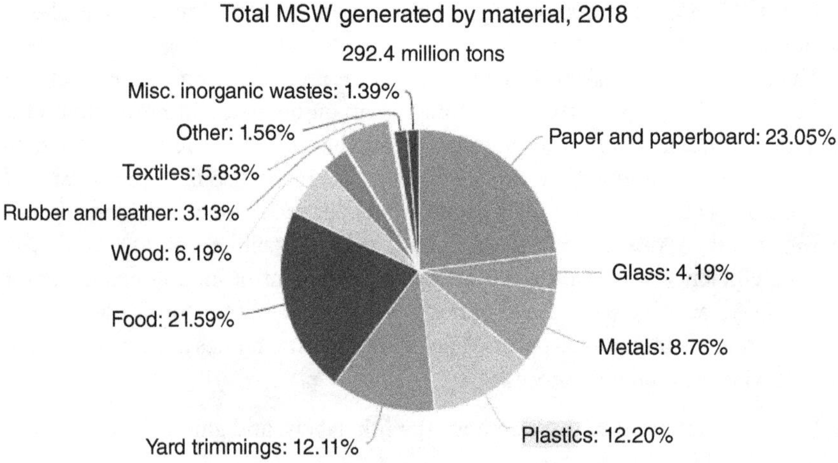

Figure 6.3 MSW waste in 2018. Source: EPA plastics waste (2021).

Recycling and composting almost 87 million tons of MSW saved more than 1.1 quadrillion BTU of energy; that's the same amount of energy consumed by almost 10 million US households in a year. (US Energy 2020).

6.3 Chemical Recycling

Chemical recycling is a process by which the polymer chemicals are retrieved from the plastic and then used for primary chemicals to produce new polymers. The chemical conversion process is usually done at high heats with the absence of oxygen. Chemical recycling is possible for PET but not currently used in high volumes. Some typical processes are as follows:

- Pyrolysis
- Glycolysis of PET to bis(2-hydroxyethyl) terephthalate (BHET) monomer
- Methanolysis of PET to dimethyl terephthalate (DMT) monomer
- BASF thermal cracking
- Degradative extrusion
- Steam gasification
- Polymer cracking in fluidized bed
- Battelle high temperature gasification
- Destructive distillation
- Catalytic cracking
- Cracking by hydrogenation
- Texaco gasification

Four chemical recycling methods from the above list are further explained in the following sections. Other resources can be used to describe the other chemical recycling processes.

Pyrolysis is a very common thermochemical decomposition of carbon-based materials at elevated temperatures of 700 °C in the absence of oxygen. Pyrolysis of plastic waste can yield gases and oils (Williams and Williams 1997; Pinto et al. 1999). The oils can be polymerized back to polymers. A new type of pyrolysis reactor utilizes a circulating fluidized bed boiler and is shown in Figure 6.4. Hot solids from the pilot boiler are led into an entrained flow reactor, in which solids (sand and char) are separated from the gas stream and returned to the boiler circulation loop. The formed pyrolysis gases are condensed in a two-stage scrubber and condenser system. The remaining non-condensable gases are recycled to the reactor as fluidization medium, while surplus NCGs are led to the boiler through a lance. Product oil from the condensing stage is led to an intermediate storage tank.

The reactor produced 80 tons of bio-oil that were tested for district heating burner applications. Bio-oils from plant residues are an alternative to fossil fuels. The use of bio-oil (bio crude, fast pyrolysis oil, flash pyrolysis oil, pyrolysis liquid) as a fuel oil in industrial kilns, boilers, diesel engines, and gas turbines were tested. Once produced, bio-oils can be shipped, stored, and utilized much like conventional liquid fuels. The project was unique in that it utilized hot sand in a fluidized bed boiler. This offers a considerable opportunity to integrate the pyrolysis and combustion processes. Considerable savings can be achieved in

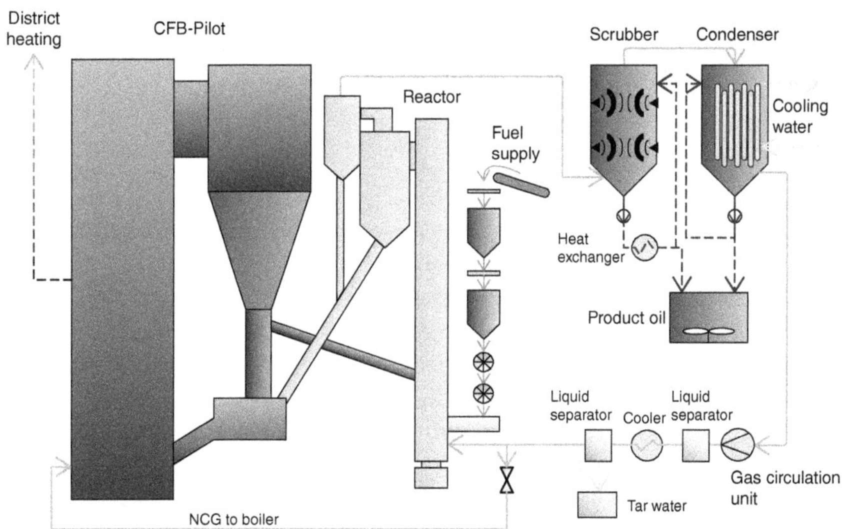

Figure 6.4 Pyrolysis reactor. Source: Autio et al. (2011).

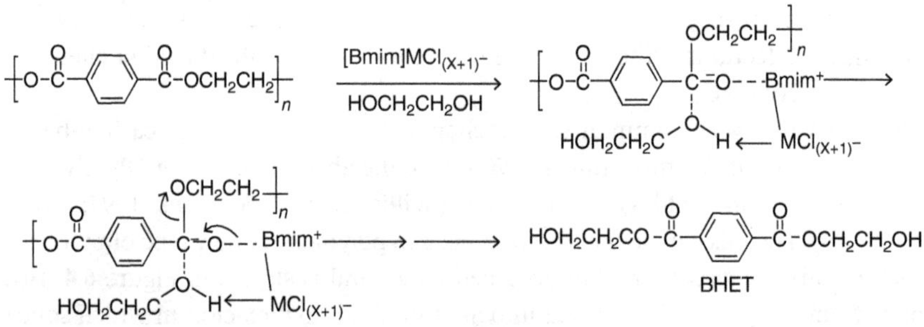

Figure 6.5 Chemical reaction of glycolysis of PET to BHET. Source: Yue et al. (2014).

operating costs and the price of the investment in both new boiler projects and retrofit solutions. The project successfully integrated pyrolysis and combustion processes. Pyrolysis process can be started, operated, and shut down without compromising boiler process. The process can be used to chemically recycle PET (Autio et al. 2011).

PET can be chemically recycled with glycolysis of PET to BHET monomer. The supercritical glycolysis reaction occurs at 450 °C and 15.3 MPa. The chemical reaction recovers BHET (Imran et al. 2010) and is listed in Figure 6.5.

PET can be chemically recycled with methanolysis to recover dimethyl terephthalate and ethylene glycol. The chemical reaction occurs at 200 °C and achieved yields of 64% (Kurokawa et al. 2003).

PLA can be chemically recycled in a process called LOOPLA hydrolysis (NatureWorks™ 2013). During hydrolysis, the Ingeo PLA plastic is broken down into lactic acid. The lactic acid can then be polymerized back into Ingeo PLA.

6.4 Composting

Industrial composting process is a thermophilic chemical process to convert organic materials to carbon dioxide, water, and biomass. Industrial compost occurs at thermophilic temperatures between 50 and 60 °C. The compost environment is hot and moist with a moisture content between 45 and 55%. The industrial compost uses a windrow method to biodegrade organic materials, for example, leaves, grass, sticks, yard waste food waste, and compostable plastics. The US Composting Council provides a wealth of information about composting (US Composting Council 2013).

Industrial compost is a viable end-of-life option for compostable plastics but not traditional plastics. Compostable plastics biodegrade under industrial composting conditions (Greene 2007b; Greene 2008; Greene 2009).

The composting process involves the following steps:

- Collect yard or food waste at an anaerobic compost facility.
- Place the compost with a proper mixture of green and brown organic materials in a row approximately 6 ft high and 6 ft wide at the bottom and 100 ft long.
- For in-vessel composting, an 8-mil plastic film is typically placed over the compost row.
  - Alternatively, in-vessel composting can include a concrete enclosure with control over temperature, airflow, moisture accumulate, and others (Greene 2007a).
- The composting row is flipped over or turned several times per week to introduce air and oxygen to the compost via mixing process.
- After 60–90 days, the compost row is sent to a trammel sorter to remove contaminants like plastic, rocks, large sticks, and other debris, and to produce a fine compost dirt.

PLA, PHA, starch, and other compostable plastics biodegraded in an industrial compost and an in-vessel industrial compost (Greene 2007a).

6.4.1 LCA of Composting Process

The use of green yard waste compost on farmland can lead to a positive environmental impact with lower water usage, lower fertilizer usage, lower herbicide usage, and sequestration. Life cycle impact assessments of environmental concerns from production and application of composted products provide a net positive environmental impact. The use of composting process and products provide a reduction in GHG, human toxicity potential, ecotoxicity potential, and eutrophication potential due to lower use of fertilizers, herbicides, water, and electricity (LCA for Windrow Compost 2006).

Industrial composting in Ireland had positive environmental impacts. Use of industrial compost rather than fertilizer can significantly reduce the global warming potential, eutrophication, and acidification. Irish composting operations had 50% less CO_2eq than equivalent fertilized systems. Sequestration occurs from the soil absorbing (Irish Composting Study 2013).

6.5 Waste to Energy

Solid waste can be burned to create energy in waste-to-energy facilities. The carbon sources in plastic can be a fuel source for several types of waste-to-energy processes. Four types of waste-to-energy facilities can use the plastic waste as a fuel in the combustion chamber. The types of waste-to-energy municipal solid waste (MSW) combustion include the following:

- Municipal solid waste combustion
- Blast furnace
- Cement kiln

6.5.1 Municipal Solid Waste Combustion

The US EPA provides a website for guidelines for generating electricity from MSW (EPA 2010). The MSW is unloaded at the waste-to–energy facility. Metals, glass, and other recyclables are separated out. The remaining burnable waste is fed into a combustion chamber and burned. The released heat produces steam that turns a steam turbine and generates electricity. Figure 6.6 shows an example of an MSW waste to energy reactor.

MSW is usually burned at special waste-to-energy plants that use the heat from the fire to make steam for generating electricity or to heat buildings. In 2019, 67 US power plants generated about 13 billion kW-h of electricity from burning nearly 25 million tons of combustible MSW. Biomass materials accounted for about 63% of the weight of the combustible MSW and for about 47% of the electricity generated. The remainder of the combustible MSW was non-biomass combustible material, mainly plastics. Many large landfills also generate electricity by using the methane gas that is produced from decomposing biomass in landfills (EPA

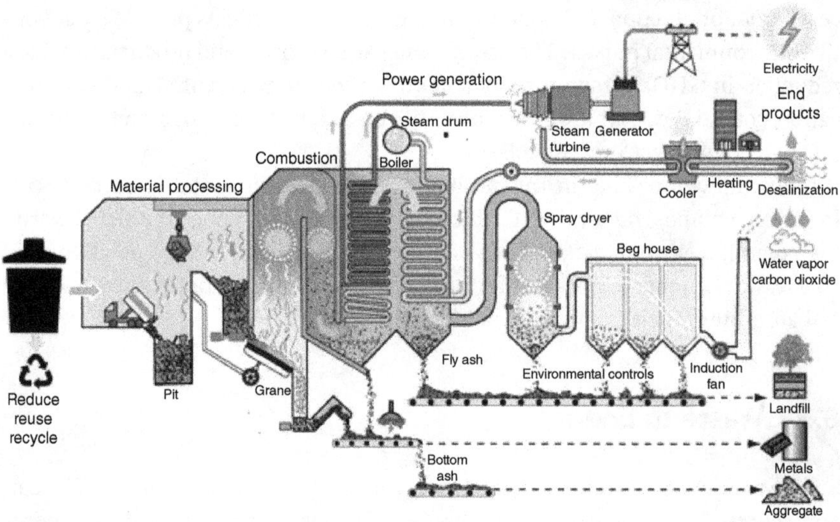

Figure 6.6 MSW waste to energy reactor. Source: MSW Biomas (2022).

Biomass 2021). The United States has approximately 67 MSW-fired power generation plants, generating approximately 2500 MW, or about 0.3% of total national power generation. The burning of MSW can create air emissions of 1671 kg of CO_2/MWh, 0.54 kg of SO_2/MWh, and 3.0 kg of nitrogen oxides/MWh. Emission and effluents can be minimized with efficient design (Fodor and Klemes 2012).

The energy potential of various sources of fuel in an MSW is provided in Table 6.3.

From Table 6.4 HDPE and polypropylene (PP) plastics have the highest heat content of 38 million BTU/ton of plastic. Mixed paper had the lowest heat content of 6.7 million BTU/ton of mixed paper. If the values were converted to Joules and then divided by the value for mixed paper we get the results shown in Table 6.4.

Table 6.3 Typical heat content of materials in MSW.

Materials	Million BTU/ton
Plastics	
Polyethylene terephthalate[c,e] (PET)	20.5
High density polyethylene[e] (HDPE)	38.0
Polyvinyl chloride[c] (PVC)	16.5
Low density polyethylene/linear low density polyethylene[e] (LDPE/LDPE)	24.1
Polypropylene[c] (PP)	38.0
Polystyrene[c] (PS)	35.6
Other[e]	20.5
Rubber[b]	26.9
Leather[d]	14.4
Textiles[c]	13.8
Wood[b]	10.0
Food[a,c]	5.2
Yard trimmings[b]	6.0
Newspaper[c]	16.0
Corrugated cardboard[c,d]	16.5
Mixed paper[e]	6.7

Source: DOE (2007).
[a] Includes recovery of other MSW organics for composting.
[b] Energy Information Administration (2005).
[c] Garth and Kowal (1993).
[d] Bahillo (2006).
[e] Utah State University Recycling Center Frequently Asked Questions. http://www.usu.edu/recycle/faq.htm.

Table 6.4 Ratio of MSW energy per yard trimmings.

Material	MM BTU	BB J	Ratio/paper
HDPE	38.0	40.1	5.7
PP	38.0	40.1	5.7
PS	35.6	37.6	5.3
Rubber	26.9	28.4	4.0
LDPE	24.1	25.4	3.6
PET	20.5	21.6	3.1
Other	20.5	21.6	3.1
PVC	16.5	17.4	2.5
Corrugated cardboard	16.5	17.4	2.5
Newspaper	16.0	16.9	2.4
Leather	14.4	15.2	2.1
Textiles	13.8	14.6	2.1
Wood	10.0	10.6	1.5
Mixed paper	6.7	7.1	1.0
Yard trimmings	6.0	6.3	0.9
Flour	5.2	5.5	0.8

The results show that HDPE and PP have the highest heat energy content for MSW. HDPE and PP have 5.7 times the energy of burning mixed paper. Flour and yard trimmings have less energy when burning than mixed paper.

Plastics have a stored energy for combustion due to their inherent material source is petroleum. One kilogram of plastics can generate twice as much energy as Wyoming coal and almost as much energy as fuel oil (SPI 2013). When combusted in a modern waste-to-energy facility, plastics can assist other solid waste burn more completely due to their intrinsic heat value and result in less residual ash in the furnace.

6.5.2 Blast Furnace

The blast furnace is used in the metal-processing industry. Plastics can be a fuel source and a reducing agent to replace coke as a carbon source. The blast furnace is used in the process of producing iron ore and has temperatures exceeding 2000 °C. The higher temperatures can generate less emissions and less particulate matter. The use of waste plastics in a blast furnace can reduce the energy and

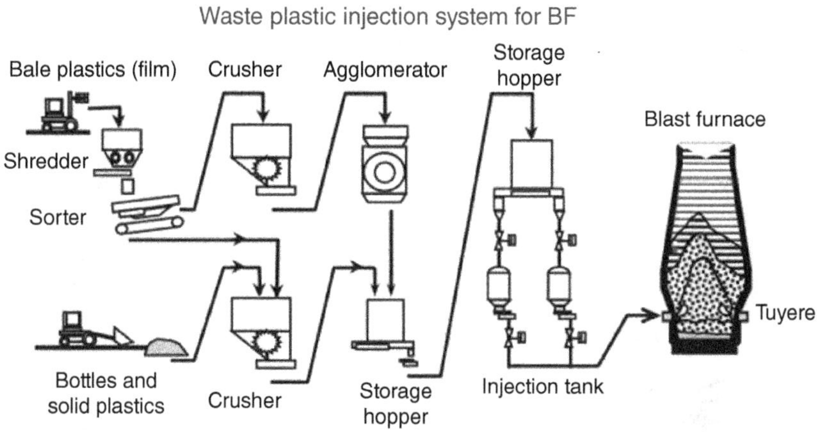

Figure 6.7 Blast furnace use with plastic waste. Source: Blast furnace (2021).

material costs and provide for improved combustion efficiencies (Ziebik and Stanek 2001; Kim et al. 2002). Polystyrene and PET were found to have negative effects on coke reactivity with CO_2 and with coke strength (Melendi et al. 2011). Polyolefins were also found to reduce coking pressure.

A typical blast furnace is shown in Figure 6.7. The process includes shredding and crushing the plastic and then sending it to an agglomerator, a storage hopper, and then the blast furnace. Plastic waste can be injected into the blast furnace to reduce the demand for coke. Plastics provide additional benefits that are similar to natural gas injection. Every ton of plastic used in a blast furnace can replace 750 kg of coke (Blast furnace 2021). Also plastic use can reduce the carbon emissions when compared to coke use.

6.5.3 Cement Kiln

Cement kilns are used to create cement under temperatures of approximately 1350 °C. Recently, cement kilns can also replace petroleum coke with refuse-derived fuel (RDF). Plastic waste can be a component of RDF. The quality of the cement was not affected by the use of blends of RDF and petroleum coke. The emissions from the blends of RDF and petroleum coke were within the industrial limits. Plastic waste that is used as a fuel source can offer environmental and economic advantages for the cement industry (Kara 2012). A cement kiln that uses plastic waste is shown in Figure 6.8. The plastic waste is an additive for cement production.

A US company has commercialized a process that uses low-value scrap plastic as fuel in cement production. The process is centered on a rotary kiln, which is

Figure 6.8 Cement kiln using plastic waste. Source: Cement kiln (2021).

frequently used in producing materials used in cement. Traditionally, this type of kiln would be fired using natural gas or liquified natural gas. Instead of using those fuels, Cement Lock takes in low-value, typically non-recyclable plastics for use in the rotary kiln. The plastic is shredded and combined with minerals such as silica, alumina, and others. The combination of minerals and composition of the plastic feedstock gives Cement Lock a formula for how long to run the furnace and the temperature range to run it at.

The product that comes out of the kiln is an aggregate, which is shipped to a pug mill and pulverized into a powder. The resulting material is used in creating a high-value pozzolan admixture for making cement. Pozzolans are silicate-based materials that are used to improve concrete by causing various chemical reactions during the manufacturing process.

6.5.4 Pollution Issues with Waste-to-Energy Process of Plastics

Incineration operations can produce toxic emissions including dioxins, furans, NO_x, SO_x, CO, CO_2, polychlorinated dibenzo-*p*-dioxins and dibenzofurans, and heavy metals. Incineration of plastic waste can produce HCl and other toxic emissions if PVC is burned. The World Health Organization provides best practices for the operators of incineration facilities, including small scale and large scale. The organization recommends removing PVC and vinyl products from the plastic

Table 6.5 United States and European Regulatory limits for pollutant

Pollutant	Units	US EPA limits	European limits
Cadmium	mg/m^3 (dry normal)	0.04	0.05
Carbon monoxide	ppm	40	100
Dioxins/furan	mg/m^3 (dry normal)	25	0.1
Hydrogen chloride	ppm	15	10
Lead	mg/m^3 (dry normal)	0.07	N/A
Mercury	mg/m^3 (dry normal)	0.55	0.05
Nitrogen oxides	ppm	250	200
Organics	mg/m^3 (dry normal)	N/A	10
Particulate matter	mg/m^3 (dry normal)	34	10
Sulfur dioxide	ppm	55	50

Note: (1) EPA defines a large incinerator as one that burns more than 227 kg/h. (2) European Council defines large incinerator as one that produces 50 MW or more of thermal output. (3) EU standards not shown for thallium, copper, manganese, nickel, arsenic, antimony, cobalt, vanadium, tin, oxygen.

waste stream to avoid generating HCl emissions. Continuous emissions monitoring is required on large incinerators. The emission standards in the United States are regulated by the EPA. The emissions limits for the combustion of waste materials in an incinerator are listed in Table 6.5 (World Health Organization 2013).

6.6 Landfill Operations

Plastics can be discarded and sent to local landfills for disposal. In 2010, the United States collected over 250 million tons MSW. In 2010, 34.1% or 85 million tons of solid waste were recycled or composted. Residential waste comprises approximately 64% of the MSW and commercial waste comprises 35% of the waste. Table 6.6 lists the materials collected at a typical landfill in the United States in 2010. Organics are the largest component of the waste stream sent to landfills. Paper and cardboard account for 29%, food and yard trimmings account for 27%, plastics comprise 12%, metals are at 9%, rubber and textiles are at 8%, wood at 6%, glass at 5% miscellaneous at 3% (EPA Municipal 2010). According to the EPA during 2010, 250 million tons of MSW were collected in the United States. Of that total, 34.1% (85.1 million tons) of the MSW were composted or recycled. Thus, 65.9% (164.9 million tons) of the MSW were sent to landfills or incinerated. The MSW before recycling or composting is comprised of the following materials

Table 6.6 Weight percentage of MSW before recycling or composting

Materials	Weight percentage of total waste (%)
Paper and paperboard	28.5
Food scraps	13.9
Yard trimmings	13.4
Plastics	12.4
Metals	9
Rubber, leather, and textiles	8.4
Wood	6.4
Glass	4.6
Miscellaneous	3.4

by weight percentage (EPA Municipal 2010). In general, landfills in the United States and Europe utilize a cap and seal strategy, wherein the bed of the landfill is lined with clay and then the top sealed to prevent contamination with ground water and to capture gases in the landfill. Landfill gases can be burned or collected to reduce dangers associated with overpressure. Methane can be captured in a bioreactor landfill for energy utilization.

6.7 Life Cycle Assessment of End-of-Life Options

LCA can be used to determine the environmental effects of various end-of-life scenarios for plastic waste. A review of a number of LCAs found that when single polymer plastic waste fractions with little organic contamination are recycled, and replace virgin plastic at a 1 : 1 ratio, mechanical recycling has the lower environmental burden than incineration of MSW (Lazarevic et al. 2010).

Landfill, incineration, pyrolysis–gasification, and anaerobic digestion are evaluated for environmental effects atmospheric emissions per unit electricity generation from solid waste using SimaPro LCA software (Zaman 2009). Landfill and incineration generated the highest global warming emissions. Incineration and pyrolysis–gasification had the significant impact on respiratory inorganics and acidification categories.

Anaerobic digestion had the lowest impacts on respiratory inorganics and acidification. Anaerobic digestion and pyrolysis–gasification had least overall environmental impact. LCA for the end-of-life was used for seven plastic components that are commonly used in automotive applications. The parts included the bumper cover made from PP, windshield washer fluid container made from polyethylene (PE), air-intake manifold made from 30% glass-filled nylon, air duct made from 20% talk-filled PP, seat cushion made from polyurethane foam, head lamp lenses made from polycarbonate, and mirror housing made from acrylonitrile butadiene styrene (Jenseit et al. 2003).

The LCA analysis included an eco-efficiency portfolio for each plastic part compared with a cost analysis. Plastics can be recovered as recycled plastics, used for energy source in a furnace, or sent to the landfills. The end-of-life scenarios included landfill, municipal incineration, cement kiln, blast furnace, syngas production, and mechanical recycling.

The LCA study evaluated the eco-environment portfolio and costs of sending the plastic waste to one of the end-of-life scenarios. The environmental burdens included considerations for 10 different environmental indicators including the following:

1) Fuel resources depletion potential
2) Mineral resources depletion potential
3) Cumulative energy requirement
4) Global warming potential
5) Photochemical ozone creation potential
6) Acidification potential
7) Water pollution
8) Final waste
9) Risk and misuse potential
10) Human toxicity potential

The LCA compared the environmental burdens and costs associated with each plastic part. The plastics are removed from the car for the recycling scenario versus being shredded into plastic fluff and sent to the waste-to-energy or landfill scenarios. The costs for dismantling were accounted for in the mechanical recycling option.

Mechanical recycling of the plastics obtained the lowest cost and lowest environmental burden than energy recovery and landfill options. The LCA work determined that in the cases with minimum disassembly time of plastic parts and 1 : 1 substitution with virgin plastics (PP bumper, HDPE fluid container, and glass-filled nylon air-intake manifold) mechanical recycling of the plastics obtained the lowest cost and lowest environmental burden than energy recovery and landfill options.

Mechanical recycling showed the worst eco-efficiency for parts with high disassembly costs. The parts included air duct, seat cushion, and headlamp lens. Eco-efficiency was also low if part thickness is increased for recycled plastics. The increased thickness of recycled plastic parts can led to increased weight of the plastic parts and ultimately the vehicle. In all cases of plastic parts, use of landfills was the highest cost and highest environmental burdens. Energy recovery of the plastic through a blast furnace process had the lowest cost and lowest environmental burden of all energy recovery methods that included cement kiln, municipal incineration, and syngas production. This is due to the higher temperatures in a blast furnace compared with other waste-to-energy processes.

6.8 Summary

Disposal methods for plastics have resulted in environmental impacts that can be evaluated with LCAs. The disposal options for plastics are to mechanically recycle the plastic, chemically recycle the plastic, compost the bio-based plastic, burn the plastic into energy, or bury the plastic in a landfill. Use of landfill is the most common disposal method for plastics.

Postconsumer recycling process occurs when plastics are made into products, for example, bottles, bags, film, packaging, used by a consumer, collected by a waste disposal company, processed into recycled pellets at an MRF, and then sold to plastic companies. Mechanical recycling is the most common recycling method for plastics. Chemical recycling is a process by which the polymer starting chemicals are retrieved from the plastic with high temperatures and in the absence of oxygen. The chemical recycling process can produce primary chemicals to produce new polymers. Chemical recycling process can be used for petroleum-based and bio-based plastics.

Industrial composting process is a thermophilic chemical process to convert organic materials to carbon dioxide, water, and biomass. The composting process can be used for bio-based plastics if they are certified as compostable.

Plastics can be burned to create energy in waste-to-energy facilities. The carbon sources in plastic can be a fuel source for several types of waste-to-energy processes, including municipal waste to energy, blast furnace, and cement kiln.

Plastics can be discarded and sent to local landfills for disposal. The landfill process is the most common end-of-life option for plastics and most solid waste.

Mechanical recycling of plastic parts can provide the least environmental burdens than waste to energy and sending the plastics to landfill. Use of landfill for plastic parts can provide the highest environmental burdens.

References

Autio, J., J. Lehto, A. Oasmaa et al. (2011). A pyrolysis unit integrated to a circulating fluidized bed boiler – experiences rom a pilot project. *10th International Conference on Circulating Fluidized Beds and Fluidization Technology*, Sun River, OR, Spring, 2 May 2011.

Bahillo, A., Armesto, L., Cabanillas, A., and Otero, J. (2006). NO_x and N_2O emissions during fluidized bed combustion of leather wastes. *J. Energy Resour. Technol.* 128 (2): 99–103.

Barlaz, M., Haynie, F., and Overcash, M. (1993). Framework for assessment of recycled potential applied to plastics. *J. Environ. Eng.* 119 (5): 798–810.

Blast furnace (2020). http://www.iipinetwork.org/wp-content/Ietd/content/plastic-waste-injection.html (accessed March 2020).

Borelis Bio-PP. https://www.plasticstoday.com/sustainability/borealis-targets-asia-bio-pp-market-japans-itochu (accessed September 2022).

Braskem bio-EVA (2022). https://www.braskem.com.br/imgreen/bio-based-en (accessed February 2022).

Braskem-green polyethylene (2022). https://www.braskem.com.br/imgreen/bio-based-en (accessed February 2022).

Cement kiln (2022). https://link.springer.com/article/10.1007/s11356-017-0429-9/figures/21 (accessed May 2022).

DOE (2007). Department of energy methodology for allocating municipal solid waste to biogenic and non-biogenic energy. http://www.eia.gov/totalenergy/data/monthly/pdf/historical/msw.pdf (accessed June 2013).

Energy Information Administration (2006). Average heat content of selected biomass fuels. *Renewable Energy Annual 2004*, Washington, DC, 2006.

EPA (2010). Municipal solid waste. chrome-extension://efaidnbmnnnibpcajpcglclefindmkaj/https://archive.epa.gov/epawaste/nonhaz/municipal/web/pdf/msw_2010_factsheet.pdf (accessed May 2022).

EPA Municipal (2010). EPA municipal solid waste generation, recycling, and waste generation. http://www.epa.gov/osw/nonhaz/municipal/pubs/msw_2010_rev_factsheet.pdf (accessed June 2013).

EPA plastics waste (2021). https://www.epa.gov/facts-and-figures-about-materials-waste-and-recycling/national-overview-facts-and-figures-materials (accessed April 2021).

EPA Waste Reduction Model (2013). Environmental protection agency. https://www.epa.gov/warm (accessed April 2021).

EVA chemical structure (2021). https://en.wikipedia.org/wiki/Ethylene-vinyl_acetate (accessed April 2021).

Fodor, Z. and Klemes, J. (2012). Waste as alternative fuel – minimizing emissions and effluents by advanced design. *Process Safety Environ. Prot.* 90: 263–284.

Garth, J. and Kowal, P. (1993). Penn State Agricultural College Agricultural and Biological Engineering and Council for Solid Waste Solutions. *Resource Recovery, Turning Waste into Energy*, University Park, PA, 1993.

Greene, J. (2007a). Performance evaluation of environmentally degradable plastic packaging and disposable service ware. Cal Recycle Publications. https://www2.calrecycle.ca.gov/Publications/Details/1266 (accessed May 2022).

Greene, J. (2007b). Biodegradation of compostable plastics in green yard-waste compost environment. *J. Polym. Environ.* 15 (4): 269–273.

Greene, J. (2008). To degrade or not to degrade: compostable plastics go under the microscope. *Resour. Recycl.* 27: 28–36.

Greene, J. (2009). Biodegradable and oxo-degradable plastics degradation in compost and marine environments. *Proceedings of the Eighth World Congress of Chemical Engineering*, Montreal, Canada, August 2009.

Imran, M., Kim, B.-K., Han, M. et al. (2010). Sub- and supercritical glycolysis of polyethylene terephthalate (PET) into the monomer bis(2-hydroxyethyl) terephthalate (BHET). *Polym. Degrad. Stab.* 95: 1686–1693.

Irish Composting Study (2013). A life cycle assessment of a standard irish composting process and agricultural use of compost. http://www.rx3.ie/MDG Uploaded Files/file/rx3publications/Compost_Life_Cycle.pdf (accessed June 2013).

Jenseit, W., Stahl, H., Wollny, V., and Wittlinger, R. (2003). Recovery options for plastic parts from end-of-life vehicles: an eco-efficiency assessment. APME, Brussels, Belgium, May 2003. http://www.oeko.de/oekodoc/151/2003-039-en.pdf (accessed June 2013).

Kara, M. (2012). Environmental and economic advantages associated with the use of RDF in cement kilns. *Resour. Conserv. Recycl.* 68: 21–28.

Kim, D., Shin, S., Sohn, S. et al. (2002). Waste plastics as supplemental fuel in the blast furnace process: improving combustion efficiencies. *J. Hazard. Mater.* B94: 213–222.

Kurokawa, H., Ohshima, M., Sugiyama, K., and Miura, H. (2003). Methanolysis of polyethylene terephthalate (PET) in the presence of aluminium triisopropoxide catalyst to form dimethyl terephthalate and ethylene glycol. *Polym. Degrad. Stab.* 79: 529–533.

Lazarevic, D., Aoustin, E., Bucklet, N., and Brandt, N. (2010). Plastic waste management in the context of European recycling society: comparing results and uncertainties in a life cycle perspective. *Resour. Conserv. Recycl.* 55: 246–259.

LCA for Windrow Compost (2006). Life cycle inventory and life cycle assessment for windrow composting systems. New South Wales, Australia. http://www.epa.nsw.gov.au/resources/warr/2006400_org_lcassesswindrowsys.pdf (accessed June 2014).

Mechanical recycling process for plastics (2020). https://www.researchgate.net/figure/Overall-mechanical-recycling-process-of-waste-plastics_fig2_354740283 (accessed May 2022).

Melendi, S., Diez, M., Alvarez, R., and Barriocanal, C. (2011). Plastic wastes, lube oils and carbochemical products as secondary feedstocks for blast-furnace coke production. *Fuel Process. Technol.* 92: 471–478.

MSW Biomass (2022). http://www.b-t-einc.com/municipal-waste-to-energy.htm (accessed May 2022).

NatureWorks™ (2013). Feedstock recycling. http://www.natureworksllc.com/The-Ingeo-Journey/End-of-Life-Options/Feedstock-Recycling (accessed June 2013).

PET recycling rates (2020). http://www.napcor.com/PET/funfacts.html (accessed August 2020).

Pinto, F., Costa, P., Gulyurtlu, I., and Cabrita, I. (1999). Pyrolysis of plastic wastes. 1. Effect of plastic waste composition on product yield. *J. Anal. Appl. Pyrolysis* 51: 39–55.

Recycling process for PET (2022). https://www.researchgate.net/figure/Typical-flow-sheet-of-single-use-polyethyleneterephtalate-PET-bottle-recycling_fig1_325653869 (accessed May 2022).

SPI (2013). Incineration of plastics. Society of Plastics Industry. http://www.plasticsindustry.org/AboutPlastics/content.cfm?ItemNumber=793&navItemNumber=1124 (accessed June 2013).

US Composting Council (2013). Compost: nature's way to grow. http://compostingcouncil.org/mission/council.org/.

US Energy (2020). https://www.epa.gov/sites/production/files/2015-09/documents/2012_msw_fs.pdf (accessed August 2020).

Uses of recycled PET bottles (2020). http://www.napcor.com/PET/funfacts.html (accessed May 2020).

Change Utah State to: Utah State Recycling Facilities (2022). https://www.usu.edu/facilities/recycling/recyclables (accessed September 2022).

Williams, E. and Williams, P. (1997). Analysis of products derived from fast pyrolysis of plastic waste. *J. Anal. Appl. Pyrolysis* 40–41: 347–363.

World Health Organization (2013). Best practices for incineration. http://www.who.int/water_sanitation_health/medicalwaste/en/smincinerators3.pdf (accessed June 2013).

Yue, Q.F., Yang, H.G., Zhang, M.L., and Bai, X.F. (2014). Metal-containing ionic liquids: highly effective catalysts for degradation of poly(ethylene terephthalate). *Adv. Mater. Sci. Eng.* 2014 (4): 1–6.

Zaman, A.U. (2009). Life cycle environmental assessment of municipal solid waste to energy technologies. *Glob. J. Environ. Res.* 3 (3): 155–163.

Ziebik, A. and Stanek, W. (2001). Forecasting of the energy effects of injecting plastic wastes into the blast furnace in comparison with other auxiliary fuels. *Energy* 26: 1159–1173.

Ziem, S., C. Chudziak, R. Taylor, A. Bauen (2021) Environmental assessment of Braskem's biobased PE resin. November 2013. chrome-extension://efaidnbmnnnibpcajpcglclefindmkaj/https://www.braskem.com.br/Portal/Principal/Arquivos/ModuloHTML/Documentos/1204/20131206-enviro-assessment-summary-report-final.pdf (accessed May 2022).

7

Sustainable Plastic Products

7.1 Introduction

As defined previously, sustainable plastic products can be created with lower carbon footprint, lower waste, and lower pollution than conventional plastic products. Plastic products can be used for sustainable plastic packaging, bottles, and bags. Life cycle assessments (LCAs) can be used to compare the different options for plastic packaging, bottles, and bags. Those three plastic products account for approximately 40% of all of the plastic products sold, but approximately 10% of the plastic products in a typical landfill and approximately 30% of the plastic waste in the oceans.

Sustainable plastic products are made from recycled or bio-based plastics and are produced with minimum environmental and social impacts. Sustainable plastic products can be defined as being produced with:

- Lower carbon footprint than virgin petroleum-based plastics.
- Lower waste generation than virgin petroleum-based plastics.
- Lower pollution generation than virgin petroleum-based plastics.
- Minimum levels of regulated heavy metals.
- Clean manufacturing principles.
- End-of-life using composting or recycling processes rather than being sent to landfill.
- Fair worker wages.
- Safe worker environment.

Sustainable Plastics: Environmental Assessments of Biobased, Biodegradable, and Recycled Plastics, Second Edition. Joseph P. Greene.

7.2 Sustainable Plastic Packaging

Plastic packaging can be produced with sustainable plastics through the use of the definitions presented above. Plastic packaging products account for approximately 30% of the plastics sold in the United States, and approximately 27% of the plastic products sold in Europe (Beswick and Dunn 2002). Sustainable plastic packaging can be made from recycled plastics or bio-based plastics, like polyhydroxy alkoanates (PHA), Poly lactic acid (PLA), starch, and others. LCAs can be used to compare environmental impacts of using recycled or bio-based plastic materials for plastic packaging products.

7.2.1 LCAs of Sustainable Plastic Packaging

Sustainable plastic packaging can include plastics made from recycled plastics and bio-based plastics. Bio-based plastics for packaging can be made from PLA, PHA, and thermoplastic starch. Recycled plastics for packaging can be made from PET, PS, PP, high density polyethylene (HDPE), or PLA. Currently, recycled plastics are predominately Polyethylene Terephthalate (PET) and High Density Polyethylene (HDPE). The following sections include LCA studies of Polyethylene Terephthalate (PET), PLA, Polypropylene (PP), High Impact Polystyrene (HIPS), and High Density Polyethylene HDPE.

7.2.1.1 LCA Step 1. Creation of the LCA Goal for Plastic Packaging

The goal of the LCA is to determine the LCAs of plastic packaging with bio-based, petroleum-based, and recycled plastics. Sustainable plastic packaging can be compared with virgin PET, PP, and PS for packaging plastic materials.

7.2.1.2 LCA Step 2. Creation of the Life Cycle Inventories for Plastic Packaging

Life cycle inventory (LCI) methodology breaks down the plastics manufacturing of bio-based plastics, recycled plastics, and virgin plastics.

The manufacturing process for PLA includes the following:

- Harvesting corn.
- Isolating starch.
- Converting starch to dextrose.
- Fermenting dextrose to lactic acid via bacteria.
- Polymerizing lactic acid to poly lactide pellets.
- Extruding PLA into plastic containers.
- Creating packaging of plastic trays.
- Transporting packaging boxes to retail outlets.

- Consumers using the PLA packaging containers.
- Collecting of compostable plastics for compost, incineration, recycling, or landfill.
- Incinerating, composting, or sending the PLA packaging to a landfill.

The manufacturing process for virgin PET containers includes the following:

- Mining of naphtha.
- Producing benzene and ethylene oxide.
- Producing ethylene oxide.
- Producing ethylene glycol and terephthalic acid.
- Polymerizing PET pellets.
- Extruding PET into plastic containers.
- Creating packaging of plastic trays.
- Transporting packaging boxes to retail outlets.
- Consumers using the PET packaging containers.
- Collecting of compostable plastics for incineration or landfill.
- Incinerating or sending the PET packaging to a landfill.

7.2.1.3 LCA Step 3. Creation of the LCAs for Plastic Packaging

The third step in the LCA process is to create a functional group and then recalculate the LCI into functional groupings for environmental impacts. For example, the LCI will calculate the energy consumed during the creation of the plastic packaging products. The LCA step will group the energy information into 10 000 containers as an example. Thus, the LCA will provide the energy consumed during the creation of 10 000 plastic packaging containers. The LCA will also repeat the process for water consumed, CO_2eq emission, waste generation, pollution generated, and other environmental impacts.

7.2.1.4 LCA Step 4. Interpretation of the Three Previous Steps for Plastic Packaging

The fourth step in the LCA process is to interpret the results in the previous three steps. For instance, the first step will be reviewed to make sure the goal was reached in the LCA process. The goal was to compare the LCA results for plastic packaging made from several types of plastic materials. The goal was not changed throughout the process and was accomplished through the creation of LCI and LCA for plastic packaging products. The assumptions used in the LCA step support the goal and purpose of the LCA study. The second step was the creation of the LCI. The LCI was properly created and supported with the assumptions in the LCA step. The LCA assumptions should be reviewed for consistency. For example, the assumptions in the third step must support whether to create a cradle-to-gate or cradle-to-grave analysis. This constraint is part of the goal creation step.

The following sections provide LCA examples with the four-step LCA process.

7.2.2 Literature Review of LCAs for Plastic Packaging

LCA can be used to evaluate the environmental impacts of several plastic materials used for clamshells. The LCAs will follow the EPA four-step process. The LCAs will provide a goal, LCI, LCA, and interpretation. The LCAs will be compared for the LCA results and the process to develop the LCAs.

7.2.2.1 Case 1: LCA of Plastic Food Service Products

The first LCA from Franklin and Associates is dated but can provide a starting point in discussing LCAs of plastics for packaging. The Franklin LCA compares the environmental impacts of PS, paper, and PLA packaging materials. Franklin and Associates use SIM Pro LCA software with a database of LCI data. Food service items were produced from polystyrene foam, paper, and PLA. The food service items included cups, plates, and clamshells. Only the clamshell data are analyzed in the following analysis (LCI Food Service Products 2006).

The assumptions are listed in two groups: product category and scope category. The assumptions are listed as follows:

7.2.2.1.1 Assumptions

Product category

- Ten thousand clamshell containers with equal carrying capacity.
- Ten thousand containers per year.
- General purpose polystyrene (GPPS) data came from Franklin and Associates' database.
- Mass of the GPPS container was 4.8 g and PLA was 23.3 g.
- NatureWorks LLC PLA Ingeo™ LCI data were provided by a PLA report reference from 2006.
- PLA sample mass was measured from actual PLA containers.
- Sandwich-size clamshell containers made from GPPS and PLA.

Scope category

- Composting of PLA is not considered.
- Cradle-to-grave analysis was used. End-of-life options include landfilling and waste to energy.
- Eutrophication, acidification, and other pollution impacts are not considered.
- Recycling of Polylactic Acid (PLA) and Expanded Polystyrene (EPPS) are not considered.
- Transportation was included for movement of containers to retail stores and for collection of postconsumer products.
- Eighty percent of the PLA and GPPS clamshells are sent to landfill at end-of-life and 20% are sent to waste-to-energy incineration.

The LCA considers environmental impacts of the process, fuel, end-of life, energy material resource, and end-of-life waste-to-energy credit. For appropriate comparisons with other LCA studies, we will consider only the environmental effects of the process of each material. The environmental categories considered are as follows:

- Greenhouse gas (GHG) emissions
- Energy usage
- Waste generation
- Water usage

7.2.2.1.2 LCA Results for Case 1

Table 7.1 lists the environmental impacts of PLA and foam PS clamshells. For 10 000 clamshells, foam PS had lower mass, energy consumption, carbon footprint, waste generation, and water consumption. The pollution aspects were not considered and no data were provided for eutrophication, acidification, smog generation, or release of toxic chemicals.

7.2.2.1.3 Sensitivity Analysis

The LCA did not use an equal carrying capacity of foam PS and PLA. The analysis assumed a specific gravity of foam PS was 0.8, and the specific gravity of PLA was 1.2. Thus, with a mass of 4.8 g for the foam PS and 23.3 g for the PLA, the volume of the foam container was 6 cm^3 and the volume of the PLA container was 19.3 cm^3. The LCA also did not include waste generation, pollution with acidification, eutrophication, or toxic generation. Additionally, transportation was not considered.

Table 7.1 Environmental impacts of PLA, EPS for 10 000 clamshells.

Environmental impact	Foam PS	PLA
Units	10 000	10 000
Mass (g)	4.8	23.3
Specific gravity	0.8	1.25
Energy consumed (MJ)	3034.662	13 125.7
Carbon footprint (tons CO_2eq)	0.024853	0.118367
Solid waste generated (kg)	40.41723	187.161
Water consumed (l)	423.9648	8611.785

7.2.2.2 Case 2: LCA of Plastic Packaging Products

The second LCA of packaging materials was provided by IFEU from Germany and was commissioned by LCA NatureWorks LLC (2006). The LCA is dated but can show the improvement in reduced environmental impacts of PLA in the third LCA study (LCA Packaging NatureWorks LLC 2006). The assumptions are listed in two groups: product category and scope category. The assumptions are listed as follows:

7.2.2.2.1 Assumptions

Product category

- Thousand clamshell containers with equal carrying capacity of 500 ml.
- Thousand containers per year.
- Sandwich-size clamshell containers made from oriented polystyrene (OPS), Polypropylene (PP), Polyethylene terephthalate (PET), and Polylactic Acid (PLA).
- The masses of the clamshells were 15.2 g (OPS), 16.9 g (PP), 19.9 g (PET), and 12.2 g (PLA).
- The specific gravities of the plastic materials are 1.06 (OPS), 0.9 (PP), 1.25 (PET), and 1.22 (PLA).
- Solid waste values were calculated based on LCA per kilogram information from Franklin and Associates (2011) and Madival et al. (2009), and then extrapolated for the mass of 1000 containers.
- Water consumed values were calculated based on LCA per kilogram information from Franklin and Associates (2011) and Madival et al. (2009), and then extrapolated for the mass of 1000 containers.

Scope category

- Composting of PLA is not considered.
- Cradle-to-grave analysis was used.
- End-of-life options include mechanical recycling of 80% for PP, PET, OPS and 20% energy recovery.
- End-of-life for PLA is 80% composted and 20% chemical recycling.
- Eutrophication, acidification, and other pollution impacts are considered.
- Transportation of plastics is considered.
- Eighty percent of the PLA and GPPS clamshells are sent to landfill at end-of-life and 20% are sent to waste-to-energy incineration.

The LCA considers fossil fuel consumption, global warming potential, summer smog potential, acidification, eutrophication, carcinogenic risk, and human toxicity. The LCA did not consider waste generation or water consumption. To provide a more complete LCA, the waste generation and water consumption were calculated for each plastic based on reference sources.

Table 7.2 Environmental impacts of PLA, PET, PP, OPS for clamshells.

Environmental impact	PLA	PET	PP	OPS
Units	1000	1000	1000	1000
Mass (g)	12.2	19.9	16.9	15.2
Specific gravity	1.22	1.25	0.9	1.06
Energy consumed (MJ)	1100	1700	1500	1500
Carbon footprint (tons CO_2eq)	0.03	0.08	0.045	0.065
Solid waste generated[a] (kg)	0.83	2.72	1.44	1.67
Water consumed[a] (l)	4.10	2.33	3.25	11.84
Eutrophication, water (g PO_4eq)	4	0.8	5.5	0.5
Eutrophication, terrestrial (g PO_4eq)	23	22	9	13
Acidification (kg SO_2eq)	0.25	0.33	0.13	0.19

[a] These values were calculated based on creation of pellets from raw materials.

The waste generation was calculated based on the mass of each material. If most of the waste is generated from the creation of the plastic clamshell, then the waste from transportation, use, and end-of-life conditions can be ignored.

7.2.2.2.2 LCA Results for Case 2

Table 7.2 provides the environmental impacts of clamshells made from plastic materials. The PLA clamshells had the lower mass, energy consumed, carbon footprint, and waste generation than clamshells made from PET, PP, and OPS. PLA clamshells consumed more water than clamshells made with PET and PP but less water than clamshells made with OPS. PET clamshells produced the least amount of chemicals that create eutrophication in the water. PP clamshells produced the least amount of chemicals that create eutrophication in the land and the least amount of chemicals that produced acidification in the water.

7.2.2.2.3 Sensitivity Analysis

The analysis included chemical pollution and production of chemicals that can lead to eutrophication, acidification, summer smog, and human toxicity. The mass of the PLA was lighter than the other plastics even though it had a higher density than PP and OPS. The LCA claims that PLA can be produced with thinner walls due to the high modulus of elasticity for PLA.

7.2.2.3 Case 3: LCA of Plastic Clamshell Products

A third LCA that calculated the environmental profiles of clamshells was created for PLA, PET, and PS plastic materials. The LCA was calculated for use as containers for strawberries. The functional units were 1000 containers that have packaging capacity to hold 0.4536 kg of strawberries. The LCA included transportation effects and provided environmental impact results for global warming, energy consumption, aquatic acidification, ozone layer depletion, aquatic eutrophication, respiratory organics, respiratory inorganics, land occupations, and aquatic eco-toxicity. SimaPro™ LCA software was used to calculate the environmental impacts of the plastic materials. The mass of the PET was calculated based on the same volume as the PLA clamshell and the density ratio of PET and PLA (Madival et al. 2009).

7.2.2.3.1 Assumptions

The assumptions are listed in two groups: product category and scope category. The assumptions are listed as follows:

Product category

- Thousand clamshell containers with equal carrying capacity of 0.4536 kg of strawberries.
- Sandwich-size clamshell containers made from GPPS, PET, and PLA.
- The masses of the clamshells were 24.2 g (GPPS), 32.55 g (PET), and 29.6 g (PLA).
- The dimensions of the clamshell containers, $19 \times 16.5 \times 7\,\text{cm}^3$, were the same for all three plastics.
- The specific gravities of the plastic materials are 1.052 (GPPS), 1.37 (PET), and 1.25 (PLA).

Scope category

- Composting of PLA is not considered.
- Consumer usage and environmental impacts are not considered.
- Cradle-to-grave analysis was used.
- End-of-life options include mechanical recycling, landfill, incineration, combinations of the three options.
- End-of-life scenario in the LCA includes: 23.5% of the clamshells were sent to incineration and 76.5% of the clamshells were sent to landfill operations.
- Eutrophication, acidification, and other pollution impacts are considered.
- Pollution is considered if they exceed governmental regulations in Greene Sustainability Index (GSI).
- Solid waste values were calculated based on LCA per kilogram information from Franklin and Associates (2011) and Madival et al. (2009), and then extrapolated for the mass of 1000 containers.
- Transportation of plastics is considered during the life cycle of the plastic clamshells.

Table 7.3 LCA of 1000 clamshells with transportation including end-of-life scenarios.

Environmental impact	PLA	PET	GPPS
Units	1000	1000	1000
Mass (g)	29.6	32.55	24.2
Specific gravity	1.246	1.37	1.052
Energy consumed (MJ)	2993	4560	4090
Carbon footprint (tons CO_2eq)	0.171	0.198	0.165
Solid waste generated[a] (kg)	2.02	4.59	2.66
Water consumed[b] (l)	0.12	0.08	0.29
Eutrophication, water (g PO_4eq)	0.01603	0.0753	0.0094
Eutrophication, terrestrial (g PO_4eq)	23	22	13
Acidification (kg SO_2eq)	0.25	0.33	0.19
Summer smog (POCP)	12	68	8

[a] Solid waste values were calculated based on LCA per kilogram information from Franklin and Associates (2011) and Madival et al. (2009).
[b] Water consumed values were calculated based on LCA per kilogram information from Franklin and Associates (2011) and Madival et al. (2009).

- Water consumed values were calculated based on LCA per kilogram information from Franklin and Associates (2011) and Madival et al. (2009) and then extrapolated for the mass of 1000 containers.

7.2.2.3.2 LCA Results for Case 3

Table 7.3 provides the environmental impacts of clamshells made from plastic materials when incineration and landfill end-of-life options are considered. The PLA clamshells had the lower energy consumed and waste generation than clamshells made from PET and GPPS. PLA clamshells consumed more water than clamshells made with PET but less water than clamshells made with GPPS. Clamshells made from GPPS produced the clamshells with the lowest carbon footprint and with the least amount of chemicals that create eutrophication in the oceans and land.

7.2.2.3.3 Sensitivity Analysis

The LCA is more complete with inclusion of end-of-life, waste generation, pollution, transportation. The LCA assumes equal stiffness of the clamshell containers and equivalent dimensions for each material. GPPS and PLA clamshells may have increased thickness and hence increased mass for the containers.

7.2.3 LCA of Sustainable Plastic Containers Made from Bio-Based and Petroleum-Based Plastics

Environmental impacts of sustainable plastic containers can be determined with the LCA information and the definitions of sustainable plastics. The sustainable plastic containers can be made of recycled plastic or bio-based plastics. The end-of-life for the plastic is either recycled or composted. Thus, polystyrene is not considered because sufficient recycled plastic is not available. As a comparison, virgin PET is evaluated for LCA. Recycled PET and bio-based PLA can be compared for clear sustainable plastic containers. The LCA information from Madival et al. (2009) is used as a starting point. Then, WARM model from EPA is used to estimate the environmental effects of composting the PLA and recycling the PET at their end-of-life.

Sustainable packaging and containers include plastics made from bio-based or recycled materials for clear packaging plastics. Currently, PS was not considered as the recycling rate is low which may be problematic for collecting of sufficient quantities of recycled PS. The following assumptions are used with the sustainable container LCA:

Product category

- Thousand clamshell containers of equal carrying capacity are used per year.
- The masses of the clamshells were 32.55 g (PET), and 29.6 g (PLA).
- The specific gravity of the plastic materials is 1.37 (PET), and 1.25 (PLA).
- Cradle-to-grave analysis was used.
- End-of-life options include 40% mechanical recycling of PET and 5% composting of PLA.
- EPAWARM model was used for composting of 100 tons of PLA resulting in 20 tons of CO_2eq savings.

Scope category

- Eutrophication, acidification, ozone layer depletion, and aquatic toxicity are equivalent for recycled PET and virgin PET.
- Transportation of plastics is considered for raw materials and containers to retail outlets.
- Consumer usage and environmental impacts are not considered.
- Pollution is considered if they exceed governmental regulations in GSI.
- No heavy metals are present in the plastic or in the inks or color concentrates.
- Five percent of the PLA clamshells are composted as an end-of-life scenario.
- Forty percent of the PET is recycled as an end-of-life scenario.
- The manufacturing process utilizes Operation Clean Sweep Practices to minimize pellet loss at plastic manufacturing operations.

Table 7.4 Cradle-to-grave LCA of 1000 sustainable plastic containers.

Environmental impact	PLA with 5% composting	rPET with 40% recycled content	Virgin PET
Container units	1000	1000	1000
Mass (g)	29.60	32.5	32.5
Specific gravity	1.25	1.37	1.37
Energy consumed (MJ)	4000	3107	4500
Global warming potential (tons CO_2eq)	0.1047	0.1531	0.198
Solid waste generated (kg)	2.02	6.54	4.59
Eutrophication, water (g PO_4eq)	0.01603	0.0753	0.0753
Acidification (kg SO_2eq)	1.82	1.34	1.34
Ozone depletion (kg CFC-11)	0.0000145	0.0000179	0.0000179
Freshwater eco-toxicity (kg, TEG)	33 000	41 600	41 600

- Solid waste and water consumption values were calculated based on LCA per kilogram information from Franklin and Associates (2011) and Madival et al. (2009), and then extrapolated for the mass of 1000 containers.

The goal of the sustainable packaging LCA is to develop an LCA for recycled and bio-based plastics from published data. The first step in the LCA is to incorporate data from a recent LCA that considered end-of-life options (Madival et al. 2009). The research provides data for PLA and PET with 100% landfill as an end-of-life scenario. The second step is to modify the data in the LCA report to include 5% composting of PLA and 40% recycling of PET. The EPAWARM model can be used to provide carbon credits for composting of PLA of 20 tons of CO_2eq savings/100 tons of PLA. The recycling of PET can provide reduced carbon footprint and energy usage, but results in increased waste generation and water consumption (Franklin and Associates 2010). Table 7.4 lists the LCA for 1000 plastic containers. The data show that PLA plastic containers had lower global warming potential and lower waste generation than virgin and recycled PET containers. Recycled PET containers had lower energy consumption than PLA or virgin PET containers. PET containers produced fewer chemicals that can cause eutrophication, acidification, and ozone depletion.

7.2.4 Greene Sustainability Index (GSI) of Sustainable Plastic Containers

The environmental impacts of the plastic materials can be calculated based on the data in Table 7.4. Environmental comparisons can be made based on the ratio of the environmental impact of the plastic material and a reference material as listed in Table 7.4. For our purposes, we will use PLA as the reference material. Thus, we will be able to compare the ways in which different plastic materials have more or less environmental impacts of global warming potential, solid waste generation, and pollution than the reference material.

The GSI can provide an overall sustainability measurement of the three sustainable plastics based on an arithmetic summing of weighted individual environmental factors (Basurko and Mesbahi 2014). An integrated quantitative approach was found to provide a holistic assessment of sustainability technologies. The methodology measured environmental, economic, and social assessments separately and then provided a single measurement of sustainability by summing the weighted assessments.

Similarly, the GSI for each of the materials can be found from the ratios of environmental impacts of the plastic material with a reference material. PLA can provide the reference material. The different environmental impact factors can be weighted with the following: 50% for GWP, 25% for waste generation, and 25% for pollution factor. The pollution factor is an arithmetic average of eutrophication, acidification, ozone depletion, and freshwater eco-toxicity. Table 7.5 provides a summary of the ratios of environmental impacts of rPET and virgin PET containers versus the reference containers made from PLA. Table 7.5 shows that containers made from PLA had lower global warming potential and lowest waste generation than containers made from recycled PET and virgin PET. Containers made from recycled PET produced less GHGs but generated more solid waste

Table 7.5 Greene Sustainability Index for sustainable containers using cradle-to-grave life cycle assessment.

Environmental impact	PLA with 5% composting	rPET with 40% recycled content	Virgin PET
Global warming potential ratio	1.000	1.115	1.456
Solid waste generation ratio	1.000	3.231	2.267
Eutrophication, water ratio	1.000	4.697	4.697
Acidification ratio	1.000	0.736	0.736
Ozone depletion ratio	1.000	1.234	1.234
Freshwater eco-toxicity ratio	1.000	1.261	1.261
Greene Sustainability Index—50 : 25 : 25	1.000	1.860	1.790
Greene Sustainability Index—60 : 20 : 20	1.000	1.711	1.723

than virgin PET containers. The GSI results are provided for 50 : 25 : 25 index of global warming ratio, solid waste generation ratio, and pollution ratio and for 60 : 20 : 20 index.

The GSI results reveal that for the 50 : 25 : 25 split, recycled PET containers had 86% higher overall environmental impacts than PLA clamshells. Similarly, virgin PET containers had 79% higher overall environmental impacts than PLA clamshells. The GSI results reveal that for the 60 : 20 : 20 split recycled PET containers had 71% higher overall environmental impacts than PLA clamshells. Similarly, virgin PET containers had 72% higher overall environmental impacts than PLA clamshells.

Containers made from recycled PET produce less GHG emission, less waste, and less pollution than containers made from virgin PET. Containers made from recycled PET produce equivalent chemicals that result in eutrophication, acidification, ozone depletion, and freshwater eco-toxicity than containers made from virgin PET. Polystyrene was not considered because of the lack of recycled PS materials available for production of clamshells.

7.3 Sustainable Plastic Grocery Bags

Sustainable grocery bags can be made with bio-based or recycled plastics, produced without regulated heavy metals or toxins, made with clean manufacturing principles, and produced fair employment and safety practices. LCA can be used to provide a sustainable plastic grocery bag. The LCA process can include key elements of LCA that features consistent functional units, energy, GHGs, waste, and pollution that include eutrophication, acidification, toxic chemical release, and end-of-life.

7.3.1 Literature Review of LCA of Plastic Bags

Environmental aspects of various bags have been analyzed by several researchers. Three LCA studies are summarized in the following. The three studies conducted the LCA per ISO standards. The first LCA study, from Boustead Consulting and Associates, compares the LCA of single-use plastic bags with single-use paper. The "cradle-to-gate" analysis includes the environmental impacts of plastic bags from the creation of the plastic from raw materials to plastic pellets. The second LCA study, from the paper industry in Hong Kong, compares the environmental effects of single-use paper and plastic bags. The "cradle-to-grave" analysis includes the environmental impacts of plastic and paper bags over the life cycle of the product. The third LCA study, from Hyder Consulting Pty Ltd of Victoria,

Australia, is a "cradle-to-grave" analysis that includes EOL and transportation impacts. It compares the LCA of single-use plastic and paper bags with reusable plastic and cotton bags. The fourth LCA study, from Scottish Executive of Edinburgh, Scotland, compares the effects of a bag tax on consumers and the LCA of single-use plastic, paper, and compostable bags versus reusable rPET bags.

7.3.1.1 LCA of Plastic Bags from Boustead Consulting

The Boustead report is comprehensive in its evaluation of the LCAs of paper and plastic bags. The Boustead report was funded by the American Chemical Council Plastics Division. The Boustead report compares LCAs of 1500 plastic bags with 1000 paper bags and 1000 compostable plastic bags. The 1.5 : 1 ratio was determined from a Franklin report from 1990, which pointed out that consumer bagging behavior illustrates that plastic-to-paper use ranged from 1 : 1 all the way to 3 : 1, depending on the situation (Council for SolidWaste Solutions 1990). The information for the LCI was taken from Boustead database and plastic suppliers (Chaffe and Yaros 2007).

The assumptions in the Boustead LCA study are as follows:

Product category

- One-and-a-half plastic bags had equal carrying capacity as one paper bag.
- Ten bags were used per week for 52 weeks.
- Plastic bags were made from HDPE.
- The mass of the plastic bag is 6 g and the paper bag is 52 g.
- Paper bag has 30% recycled content.

Scope category

- Cradle-to-gate analysis was used, ignoring end-of-life impacts.
- Eutrophication, acidification, and other pollution impacts are not considered.

Plastic bags require less energy, fossil fuel, and water than an equivalent number of paper bags. Also, this LCA reports that plastic bags generate less solid waste, acid rain, and GHGs than paper bags. Paper bags weigh significantly more than the traditional thin plastic bag and use an energy and water-intensive manufacturing process to produce paper bags. The results are listed in Table 7.6.

7.3.1.2 Sensitivity Analysis

The Boustead report provides an acceptable LCA analysis but failed to consider recycled content for plastics and the effects of using reusable bags instead of single-use bags. The report uses 30% recycled paper content which might be low. Other paper products have higher recycled content. The Boustead report claims equal carry capacity of 1.5 plastic bags for each paper bag. The ratio might not reflect actual usage of plastic bags that might reflect two or three plastic bags for every paper bag.

Table 7.6 Life cycle assessments of 1500 plastic bags and 1000 paper bags.

	1500 plastic bag industry average	1000 paper bag (30% recycled)	Paper/plastic bag ratio
Total energy (MJ)	763	2622	3.44
Fossil fuel used (kg)	15	23	1.53
Municipal solid waste (kg)	7	34	4.86
Greenhouse emission (tons CO_2)	0.04	0.08	2.00
Freshwater usage (gal)	58	1004	17.31
Mass (g, per paper)	6	52	8.67

7.3.2 LCA of Plastic Bags from the Paper Industry in Hong Kong

A second LCA report on plastic bags is provided by the paper industry from Hong Kong that investigated the environmental effects of paper and plastic bags. The results are based on commercial LCA software called SIMAPRO 7.1. The report used data from Franklin and Associates. The results show that paper bags had higher energy usage, higher carbon footprint, and higher material consumption than plastic bags. The study also investigated pollution generation from paper and plastic shopping bags with eco-indicator 91. The software considers creation of carcinogens, respiratory organic and inorganic, climate change, radiation, ozone layer, eco-toxicity, acidification, eutrophication land use, minerals, and fossil fuels. The LCA found that paper bags had higher pollution, higher carbon footprint, and higher waste generation. Table 7.7 describes the results of the LCA (Muthu et al. 2013).

Table 7.7 Life cycle assessment of plastic and paper bags.

Environmental impact	HDPE plastic bag two units	Paper bag (40% recycled content) one unit	Ratio, paper/ HDPE bag
Energy consumed (MJ)	1.47	1.68	1.14
Carbon footprint (factor)	0.45	1	2.22
Solid waste generated (kg)	14	50	3.57
Air pollution (kg)	1.1	2.6	2.36
Eutrophication/acidification (factor)	0.85	1	1.18
Greene Sustainability Index	1	2.445	

The assumptions in the LCA study are as follows:

Product category

- Two plastic bags had equal carrying capacity as one paper bag.
- Ten bags per week for 52 weeks were used.
- Plastic bags were made from HDPE and low density polyethylene (LDPE).
- The mass of the plastic bag is 6 g and the paper bag is 42.6 g.
- Paper bag has 40% recycled content.

Scope category

- Cradle-to-gate analysis was used, ignoring end-of-life impacts.
- Eutrophication, acidification, and other pollution impacts are considered.

The LCA found that the lighter plastic bag had less environmental impacts than paper bags, with less energy consumed, less carbon footprint, less solid waste generated, less air pollution, and less eutrophication/ acidification.

7.3.2.1 Greene Sustainability Index of Plastic Bags

The GSI can be calculated based on the data provided by Muthu et al. (2013). The GSI can provide an overall sustainability measurement of the plastic bag. The overall index for each of the materials can be found with the weighting factors of 50% for GWP, 25% for waste generation, and 25% for pollution factor. The pollution factor is an arithmetic average of eutrophication, acidification, and air pollution. The GSI results reveal that paper bags have approximately 245% more negative environmental effects than HDPE plastic bags.

7.3.3 Reusable Plastic Bags

Plastic bags can be designed and manufactured for reusable use. The reuse of plastic bags can reduce the environmental effects by the number of reuses that occurs. The GHG emissions, waste generation, energy usage, water usage, pollution generation, and other environmental impacts can be divided by the number of uses for the bag. The water usage may increase as reusable bags should be washed if they are used to carry meats of dairy. Washing of reusable plastic bags can reduce the level of bacteria present in the bag.

7.3.3.1 Australian LCA of Reusable rPET Bags

An LCA report on plastic bags, from Consulting Pty Ltd of Victoria, Australia, compares the environmental impacts of shopping bag alternatives for carrying goods in Australia. The Australian report was funded by Sustainability Victoria, which was created from the Sustainability Victoria Act 2005. The LCA data in the report were updated from an earlier LCA report from an Australian University of Design for RMIT in 2002 with more accurate values of recycling rates, bag mass, and bag capacity. The HDPE plastic grocery bag was compared with bags made from paper, compostable plastics, cotton, and polypropylene. The cotton and polypropylene bags were reusable bags (Dili 2007).

The assumptions in the Australian LCA study are as follows:

Product category

- Equal carry capacity of the bags that can carry seven grocery items.
- The masses of the plastic bags were HDPE of 7 g, Kraft paper of 43 g, PP of 95 g, and cotton of 85 g.
- Each shopping trip would require 10 bags/week for 52 weeks.

Scope category

- End-of-life included 75% of single-use HDPE plastic bags sent to landfill, 19% reused as trash liners then sent to landfill, 5% recycled, and 0.5% discarded as litter.
- End-of-life included 99.5% of reusable PP plastic bags sent to landfill and 0.5% discarded as litter.
- Cradle-to-grave analysis was used.
- Eutrophication, acidification, and other pollution impacts were not considered.

The Australian study used an LCA software called SimaPro 5.1 to assess the environmental impact of the carrier bags. The LCA analysis included production of raw materials, manufacturing of the bags, transportation of the bags to retailers, and disposal of the bags at the end-of-life. Australian data are used for energy production, material production, transportation, recycling, and waste disposal. The Australian study found that the reusable polypropylene bags had the least amount of environmental impact. The cotton reusable bag had low environmental impact except for high water usage. The results of the study are listed in Table 7.8 with relative ratings of 1 (preferred) to 5 (unacceptable).

Table 7.8 Environmental impact of grocery bags in Australia.

Bag material	Energy consumption	GHG (CO_2eq)	Water use	Disposal options
PP reusable bag	1	1	1	Recycle at major super markets
Cotton calico reusable bag	1	1	5	No recycling, sent to landfill
HDPE single-use bag	4	2	1	Recycle at major super markets
Kraft paper single-use bag	5	5	2	Reused as trash liner and sent to landfill

7.3.3.1.1 Sensitivity Analysis

The Australian report was a good evaluation of the importance of reusable bags. The report did not provide enough information on the assumptions of the data for the LCA. The report though is limited by the different carrying capacities of the plastic and paper bags. The report does show the importance of using recycled plastics in the manufacture of single-use bags, but does not show the use of recycled plastics for reusable bags.

7.3.3.2 Scottish LCA of Reusable rPET Bags

An LCA report on reusable plastic bags from Scotland and the United Kingdom studied the environmental effects of taxes on several plastic bag scenarios. The Scottish report acknowledged the Scottish Waste Strategy Team, Carrier Bag Consortium, Convention of Scottish Local Authorities, Friends of the Earth Scotland, Scottish Retail Consortium, and The Scottish Environment Protection Agency for direction and support during the project. The report used LCA to evaluate the environmental effects of grocery bag consumer choices. The report found that assessing a tax would reduce the use and prevalence of plastic in the environment and that consumption of non-renewable energy, solid waste, GHG emissions, and eutrophication of lakes and rivers would be significantly less (Cadman et al. 2005).

The assumptions of the Scottish LCA report are listed in the following:

Product category

- Two plastic bags had equal carrying capacity as one paper bag.
- Ten bags per week for 52 weeks were used.
- Volume of paper bag is 20.5 l (paper/plastic bag ratio is 1.46 : 1).
- Single-use plastic bags were made from HDPE.
- Reusable plastic bags were made from LDPE.
- Recycling content of the paper bag was 40%.
- The mass of the single-use plastic bag is 6 g and the paper bag is 52 g.

Scope category

- Cradle-to-grave analysis was used.
- End-of-life impacts assume 45% of paper bags are recycled, 25% of paper bags are incinerated, and 26% of paper bags resent to landfill.
- Plastics bags were incinerated or sent to landfills.
- Eutrophication, acidification, and other pollution impacts are considered.

The Scottish report uses LCA data from a French study (Carrefour 2004). The Carrefour LCA study examined energy, fuel, water, and other resource requirements for production, manufacture, use, and disposal of several plastic bags. The

Table 7.9 Environmental indicators for plastic and paper bags in Scottish report.

Indicator of environmental impact	Single-use HDPE plastic bag	Reusable LDPE plastic bag (used 4×)	Reusable LDPE plastic bag (used 20×)	Single-use paper bag
Non-renewable energy	1.0	0.7	0.1	1.1
GHG emissions	1.0	0.6	0.1	3.3
Solid waste	1.0	0.7	0.1	2.7
Water use	1.0	0.6	0.1	4.0
Acid rain	1.0	0.7	0.1	1.9
Eutrophication	1.0	0.7	0.1	14.0
Ozone formation	1.0	0.3	0.1	1.3

study considered plastic grocery bags, reusable polyethylene bags, Kraft paper bags with recycled paper content, and compostable plastic bags. The Carrefour LCA study assessed the environmental impact of the energy use, fuel and other resource use, waste generation, GHG emissions, and pollutant emissions. The results are summarized in Table 7.9 for eight environmental indicators with relative ratings of 1 (preferred) to 5 (unacceptable).

The report found that reusing plastic bags created comparably low environmental impact. The reusable bags must be used more than four times to have equivalent environmental impacts as using single-use plastic bags four times. After reusable plastic bags are used more than 20 times, the environmental impacts of water use, GHG emissions, acid rain, ozone formation, eutrophication, and solid waste can be reduced more than 90% rather than using single-use plastic bags.

The report found that most negative environmental impacts come from the production of the plastic pellets and paper from the raw materials in the first stage of manufacturing. The second manufacturing stage of conversion of the pellets and paper into plastic and paper products that are sent to retailers has less environmental impact but not negligible. The end-of-life scenarios for grocery bags can have significant impact on the creation of solid waste in the environment.

Other environmental indicators include eutrophication and acid rain generation. The environmental effects on polyethylene and polypropylene reusable bags would be similar due to the similar plastic chemistry and process to manufacture the bags. The Scottish report found that reusable bags have significantly less eutrophication and acid rain generation than single-use plastic.

The results from the Scottish report demonstrate that synthetic reusable bags have lower environmental impacts than all other types of lightweight carrier bags, including paper, plastic, or degradable plastic. The report did not list environmental indicators of reusable polypropylene plastic bag. The report could go further by studying a reusable plastic bag made from recycled plastics.

7.3.3.3 New LCA Development for Reusable Plastic Bags: Step 1 – Development of the Goal

The environmental impacts of reusable plastic bags can be compared with the environmental impacts of single-use plastic grocery bags. The environmental impacts of reusable and single-use plastic bags with equal carrying capacity can be evaluated with LCA. The reusable plastic bags are single-use HDPE, reusable LDPE with 40% postconsumer recycled content, recycled PP, and recycled PET. In our case, 1500 single-use plastic bags will be compared with 1000 reusable plastic bags for equal carrying capacity. The cradle-to-grave process steps for plastic bags manufacturing and use are shown in Table 7.10 (Greene 2011).

The consumer has the ability to recycle polyethylene bags because the recycling infrastructure is in place in the United States. Polyethylene plastic bags can be made from recycled LDPE from stretch wrap film recycling sources (Roplast Industries 2013). Reusable PP non-woven plastic bags are not readily recycled, due in part to the design of the PP non-woven plastic bag. rPET bags can be made from recycled PET plastic bottles (ChicoBag Company 2013).

Table 7.10 Cradle-to-grave process steps for plastic bags.

	Steps	Grocery bag: HDPE	PE Reusable PCR	PP non-woven
1	Produce plastic pellets from oil and natural gas	X	X	X
2	Ship pellet to convener	X	X	X
3	Convert pellet to film	X	X	X
4	Convert film to non-woven			X
5	Ship product to retail stores	X	X	X
6	Consumer uses bag first time	X	X	X
7	Consumer uses bag multiple times per year		X	X
8	Consumer washes 20% of the reusable bag weekly		X	X
9	Consumer recycles plastic bag	X	X	
10	Consumer throws plastic in trash for landfill			X

7.3.3.4 New LCA Development for Reusable Plastic Bags: Step 2 – LCI Development

The second step of the LCA process, LCI, tabulates the energy, fuel, water, and material inputs needed to produce and use plastic and paper bags and also lists solid waste that are created when the products are made, used, and thrown away for plastic and paper bags. Polyethylene can have three types of resins that are used for plastic bags, namely, Linear Low Density Polyethylene (LLDPE), LDPE, and HDPE. HDPE is commonly used for single-use grocery bags. LLDPE and LDPE are commonly used for trash bags and for thicker department store bags.

Each of the three polyethylene plastics can be used for reusable plastic grocery bags. Each of the process steps from Table 7.10 has environmental aspects that affect energy usage, water usage, GHG emissions, pollution, and solid waste generation.

Table 7.11 lists the cradle-to-gate aggregate US-averaged values of energy required, solid waste, and GHGs produced during the production of polyethylene and polypropylene. Polyethylene and polypropylene are made from natural gas and petroleum. The amount of energy and water that are needed to make polyethylene and polypropylene plastic pellets well as the amount of solid waste, pollution, and GHG generated during production is provided in Table 7.11. The polyethylene pellets are extruded and then blown into plastic bags with a blown film extrusion line. Similarly, polypropylene pellets are extruded in a sheet extruder and pressed into non-woven film that is sewn into a bag.

Table 7.11 shows that PP requires less energy to produce pellets, and also then produces less GHG due to the lower energy use. PP produces more solid waste than polyethylene during the manufacturing of plastic pellets. The solid waste and GHG information can be used to compare the environmental benefits of using recycled plastic as a source for plastic bags rather than virgin plastic. If recycled plastics are used for plastic bags, then the amount of energy needed to produce the virgin plastic can be saved when using recycled plastics because the plastic pellet is already available and does not need to be created from raw materials (Franklin and Associates 2011).

Table 7.11 LCA of plastic pellet manufacturing of four plastic materials per 1000 kg.

Plastic	Energy (GJ)	Solid waste (kg)	GHG (tons CO_2)
HDPE	69	78	1480
LDPE	74	79	1480
PET	70.4	141	2733
PP	63	83	1340

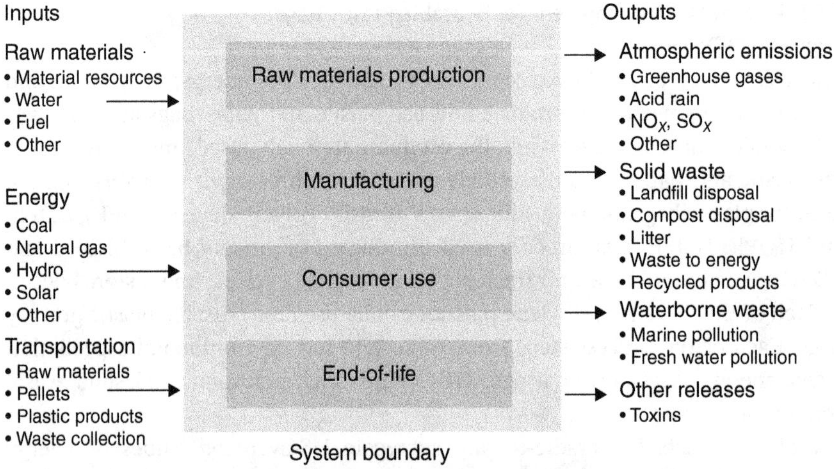

Figure 7.1 Process flow of inputs and outputs for plastic bag manufacturing, use, and end-of-life.

Figure 7.1 describes the energy and resource inputs during the production and use, and disposal of plastic bags as well as the waste, GHG, and pollution generation. The cradle-to-grave analysis calculates the environmental impacts of creating plastic pellets from raw materials, transporting them to the plastic bag converter, producing the plastic bags, and transporting the plastic bags to the retailers. The LCA is influenced by choices that consumers make on single-use versus reusable bags, and choices that consumers make on recycling, waste disposal, or waste-to-energy end-of-life options.

LCI of the plastic bag manufacturing process can be determined based on the data from the Boustead report, the Australian report, and the Scottish report. The methodology used in this report combines the data from the Boustead report with the reusable bag data from the Australian and Scottish report.

The methodology for the new LCA normalizes the Boustead data energy use, GHG emissions, water usage, and waste generation for polyethylene plastic bags to the mass of the bag. The functional unit for the LCA analysis is a plastic bag of equal carrying capacity that would be used in one-year timespan by consumers. The analysis assumes one trip per week that includes 10 bags.

The normalized Boustead data used in the LCA analysis include values of energy use, GHG emissions, water usage, and waste generation per kilogram of polyethylene. The reusable polyethylene and polypropylene bags will have the same dimensions without including handles. The reusable polyethylene thickness is 0.003 in., whereas the reusable polypropylene bag is 80 grams per square meter (GSM) of bag. The LCA of polypropylene is calculated based on combining the

Boustead data with the PP pellet manufacturing data from Table 7.8. The LCA of the PP per kilogram is calculated to include GHG emissions, energy usage, water usage, and waste generation per kilogram or PP.

The LCA of reusable polyethylene and polypropylene bags is calculated by multiplying the per kilogram LCA by the mass of the reusable bags. Thus, we can determine the energy use, GHG emissions, water usage, and waste generation of three bags, that is, HDPE grocery bag (Boustead data), reusable polyethylene bag (modified Boustead data), and reusable non-woven polypropylene bag (modified Boustead data).

Finally, the environmental credits for using recycled polyethylene in the reusable polyethylene bag are determined by subtracting the amount of energy use, GHG emissions, water usage, and waste generation from the virgin resin that was replaced by the recycled plastic, and adding the amount of energy and GHG produced by converting the recycled polyethylene to plastic pellets.

The assumptions of the new LCA are listed in the following:

Product category

a) One-and-a-half plastic bags had equal carrying capacity as one paper or reusable bag.
b) Ten bags per week for 52 weeks were used.
c) Single-use plastic bags were made from HDPE with a mass of 6 g each.
d) Reusable plastic bags were made from LDPE with 40% recycle PCR, rPET with 90% recycled PCR, and PP.
e) The masses of the reusable plastic bags were 44 g for LDPE, 42 for PP, and 45 g for rPET.
f) Paper bags were made with 40% recycled content with a mass of 52 g.
g) The Boustead data for single-use HDPE bag can be used to represent the manufacturing process of the thicker reusable polyethylene bag because it is made with the blown film extrusion process.
h) The Boustead data for single-use HDPE bag can be modified to represent the manufacturing process of the thicker reusable non-woven polypropylene bag because the non-woven PP bag is made with sheet extrusion process that requires similar energy use as blown film extrusion.
i) The production of PP non-woven bags has the same values for GHG, waste generation, energy usage, and water usage as HDPE blown film bags.
j) The non-woven PP bag is 80 GSM. The two options for polypropylene non-woven bags are 80 and 100 GSM based on industry standards.
k) Reusable polyethylene is manufactured in California and distributed throughout the United States.
l) Reusable rPET plastic bag is manufactured in California and distributed throughout the United States.

m) The rPET reusable plastic bag is made from 99% recycled PET.
n) The dimensions of the non-woven reusable bag are the same as the dimensions of the polyethylene reusable bag. The difference is the thickness of the bags.
o) For 40% recycled LLDPE and 99% recycled PET, the energy, GHG, waste, and water that are required from the pellet production are subtracted from the bag manufacturing minus the conversion costs of the recycled plastics to pellets.
p) Recycled PET bottles are converted into fiber and then into sheet for manufacturing into plastic bags.
q) 13 GJ of energy/ton is required to convert recycled PET bottles to fiber (Shen et al. 2011).
r) PET fiber to PET fabric assumes 131.7 MJ/kg PET fiber (Shen et al. 2011).
s) Energy usage conversion to GHG emissions was scaled from European data to USA LCA calculations with LCA information on plastic bags (Chaffe and Yaros 2007).
t) Eutrophication, acid rain, and ozone formulation values are created from the Scottish report (Cadman et al. 2005).
u) Eutrophication, acid rain, and ozone formulation values are equivalent for the reusable plastic bags.

Scope category

a) Cradle-to-grave analysis was used.
b) Plastics bags were incinerated or sent to landfills.
c) Eutrophication, acidification, and other pollution impacts are considered.
d) Transportation of non-woven polypropylene from China to Los Angeles has a distance of approximately 11 000 km. The GHG emissions from fuel consumptions are approximately 3% of the overall GHG emissions from the bag manufacturing based on data from the Australian report.
e) Recycled LDPE film into plastic bags does not include washing procedures in the recycling process.
f) Transportation of reusable non-woven polypropylene, rPET, and polyethylene bags throughout the United States accounts for 1% of the overall GHG emissions.
g) CO_2 emissions from the transportation of the rPET reusable bags from Vietnam/China to the United States account for 3% of the overall global warming potential of the bags.
h) The energy impacts of washing and drying reusable bags were not considered.

The LCI includes the manufacturing of plastic pellets and paper from raw materials, the conversion of plastic pellets into plastic bags, conversion of paper into paper bag, and transportation to the retail stores. The “cradle-to-grave” analysis can illustrate the environmental benefits of reusing the plastic bag and the benefits of using recycled plastic.

7.3.3.5 Bags Step 3: Life Cycle Assessment

The third step of the LCA process, life cycle impact assessment, takes the inventory of energy, fuel, water, materials, pollution, and waste and rearranges them in terms of the scope from the first step, to provide a comparison of environmental measures. In our case, the amount of energy, water, materials, and fuel needed to make 1500 plastic grocery bags can be compared with the amount of energy, water, materials, and fuel needed to make 1000 reusable plastic bags and 1000 paper bags.

Likewise, the pollution, GHGs, and solid waste produced to make 1500 plastic grocery bags will be compared with the waste produced from 1000 reusable plastic bags and 1000 paper bags.

Table 7.12 lists the cradle-to-gate LCI of single-use plastic bags, single-use paper bags, reusable non-woven polypropylene plastic bags, and reusable polyethylene (LLDPE) plastic bags. The table lists grocery bags with equal amount of carrying capacity for up to one year or 52 weeks.

Single-use paper bags are presented as a comparison. The LCA data for paper bags are from the Boustead report. Recycling content is included in the reusable polyethylene bag. The reusable bags are washed at a rate of 20% of the bags over the time period in the table. The single-use plastic bag is smaller than the reusable and paper bags. Thus, 1500 single-use plastic bags have similar carrying capacity as 1000 reusable plastic and single-use paper bags. This is consistent with the Boustead report.

The data in Table 7.12 represent the environmental impacts of using equal carrying capacity bags for one year. Table 7.12 illustrates that single-use reusable

Table 7.12 Cradle-to-gate LCA of single-use plastic and paper bags, and reusable plastic bags.

Environmental impact indicator	1500 HDPE single-use bag	1000 Reusable LLDPE with 40% PCR bag used 52 times	1000 Reusable PP non-woven used 52 times	1000 Reusable rPET used 52 times	1000 Paper bag single-use
Non-renewable energy (GJ)	763	57	72	125	2620
GHG emissions (CO_2eq)	0.04	0.003	0.005	0.007	0.08
Solid waste (kg)	7	0.106	0.106	0.333	34
Freshwater consumption (l)	99.6	18.9	18.9	11.16	1000
Mass (g)	6	44	42	45	52

Table 7.13 Normalized cradle-to-gate LCA of single-use plastic and paper bags, and reusable plastic bags.

Environmental impact indicator	1500 HDPE single-use bag	1000 Reusable LLDPE with 40% PCR bag used 52 times	1000 Reusable PP non-woven used 52 times	1000 Reusable rPET used 52 times	1000 Paper bag single-use
Non-renewable energy (GJ)	1	0.07	0.09	0.16	3.43
GHG emissions (CO_2eq)	1	0.08	0.13	0.18	2.00
Solid waste (kg)	1	0.02	0.02	0.05	4.86
Freshwater consumption (l)	1	0.19	0.19	0.11	10.04
Acid rain	1	0.1	0.1	0.1	1.9
Eutrophication	1	0.1	0.1	0.1	14
Ozone formation	1	0.1	0.1	0.1	1.3

bags made from polypropylene or polyethylene have significantly worse environmental impacts than the single-use polyethylene bags. The reusable bags have a better environmental impact if they are used more than eight times, which is an environmental crossover point for reuse. The reusable plastic bags have significantly better environmental impact if they are used 52 times (once a week for 12 months) or more.

Table 7.12 also illustrates that the reusable polyethylene bag has the lowest environmental impact than the reusable polypropylene bag due to the use of recycled polyethylene plastic or PCR. Paper bags have a negative environmental impact compared to single-use plastic bags and reusable plastic bags.

The data in Table 7.12 can be normalized to evaluate the environmental impacts of reusable and single-use plastic bags if the data for each bag type is divided by the data from single-use plastic bags. Table 7.13 lists the normalized values for reusable and single-use plastic bags.

7.3.3.6 Greene Sustainability Index (GSI) of Reusable Plastic Bags

The GSI can be calculated based on the data listed in Table 7.13. The data can be rearranged with the LLDPE reusable plastic bag as the reference. Table 7.14 lists the environmental impacts of reusable and single-use plastic bags with the reusable plastic bag as the reference material. The GSI can provide an overall sustainability measurement of the plastic bags. The overall index for each of the materials

Table 7.14 Normalized cradle-to-gate LCA with LLDPE as the reference material.

Environmental impact indicator	1000 Reusable LLDPE with 40% PCR bag used 52 times	1000 Reusable PP non-woven used 52 times	1000 Reusable rPET used 52 times	1500 HDPE single-use bag	1000 Paper single-use bag
Non-renewable energy (GJ)	1	1.26	2.19	13.39	45.96
GHG emissions (CO_2eq)	1	1.67	2.33	13.33	26.67
Solid waste (kg)	1	1.00	3.14	66.04	320.75
Freshwater consumption (l)	1	1.00	0.59	5.27	52.91
Acid rain	1	1.00	1.00	10	19
Eutrophication	1	1.00	1.00	10	140
Ozone formation	1	1.00	1.00	10	13
GSI	1	1.33	2.20	25.68	107.86

can be found with the weighting factors of 50% for GWP, 25% for waste generation, and 25% for pollution factor. The pollution factor is an arithmetic average of eutrophication, acidification, and air pollution. We can assume that the pollution impact would be equal for LDPE, rPET, and PP. The GSI can be calculated based on the data in the previous table. Table 7.13 lists the normalized environmental effects of single-use plastic and paper bags versus reusable plastic bags.

The GSI results reveal that after 52 uses, single-use plastic bags have approximately 25 times more negative environmental effects than reusable rLLDPE plastic bags with 40% PCR. Paper bags have approximately 108 times more negative environmental effects than a reusable rLDPE plastic bag made with 40% PCR. The PP reusable plastic bag has equivalent negative environmental effects as a reusable rLDPE plastic. The rPET reusable plastic bags have twice the negative environmental effects as a reusable LDPE plastic. Paper bags have approximately 12 times more negative environmental effects than a reusable rPET plastic bag.

7.4 Life Cycle Assessment of Sustainable Plastic Bottles

Sustainable plastic bottles can be made with bio-based or recycled plastics, produced without regulated heavy metals or toxins, and made with clean manufacturing principles, and fair employment and safety practices. LCA can be used to provide a sustainable plastic bottle. The LCA process can include key elements of LCA, which features consistent functional units, energy, GHGs, waste, and pollution that include eutrophication, acidification, toxic chemical release, and end-of-life.

LCA can be used to compare the sustainability of bottles made from PLA, 100% recycled PET, and virgin PET. The sustainability will be evaluated with criteria defined previously as reductions in global warming potential, reductions in waste, and reductions in pollution.

7.4.1 LCAs Literature Review of Plastic Bottles

The first LCA from Franklin and Associates compares the environmental impacts of PET, rPET, and PLA (LCI Summary for PLA 2013). Franklin and Associates use SIM Pro LCA software with a database of LCI data. The LCA used the same methodology as in the report, "Life Cycle Inventory of Five Products Produced from PLA and Petroleum-Based Resins" (LCI of Five products 2013).

The assumptions in the LCA study are listed as follows:

Product category

- Ten thousand water bottles with capacity of 12-ounces each.
- Caps and labels for each bottle were ignored in the analysis because they are equivalent for all bottles.
- Higher heating values for PLA and PET are 19 and 26 MJ/kg, respectively.
- Ingeo PLA resin data were referenced with 16% reduction in energy usage and 35% reduction in CO2eq per kilogram Ingeo PLA plastic (Vink et al. 2010).
- The mass of the PLA bottle was 21.0 g and the mass of PET bottle was 20.3 g.
- PET resin data were referenced from the US LCI database at http://www.nrel.gov/lci/.
- PLA resin data were referenced from "Life Cycle Inventory of Five Products Produced from PLA and Petroleum-Based Resins."

Scope category

- Composting of PLA is not considered.
- Cradle-to-grave analysis was used. End-of-life options include landfilling (80%), waste to energy (20%), and recycling of PET at a rate of 23.5%.
- Eutrophication, acidification, and other pollution impacts are considered.
- Transportation was included for the movement of plastic resins from resin producers to bottle converters, for example, PLA had 425 000 miles by combination truck, PET had 9000 miles by combination truck.
- Water eutrophication data and acidification data were modified from the IFEU-Heidelberg data for 1000 clamshells.

Table 7.15 Environmental impacts of PLA, PET, and rPET bottles.

Environmental impact	PLA with 2005 data	Virgin PET	rPET (25% PCR)	PLA data[a]
Energy consumed (MJ)	19 000	16 600	15 200	16 568
Global warming potential (tons CO_2eq)	0.744	0.757	0.71	0.536
Solid waste generated (kg)	168	163	144	168
Eutrophication[a], water (g PO_4eq)	99.26	38.87	34.79	80.79
Acidification[a] (kg SO_2eq)	5.30	3.47	3.11	4.59

[a] Data modified from LCA Nature Works LLC (2009).

The data were interpolated to equivalent mass of water bottles in the current study (LCA NatureWorks LLC 2009).

The LCA considers environmental impacts of the process, fuel, end-of-life, and energy material resource. The environmental categories considered are:

a) Greenhouse gas emission (CO_2 equivalent)
b) Energy usage (GJ)
c) Waste generation (kg)
d) Pollution (eutrophication and acidification)

Table 7.15 lists the environmental impacts of plastic bottles produced from PLA, PET, and rPET with 25% PCR recycled content.

7.4.2 Greene Sustainability Index of Sustainable Plastic Bottles

LCAs can compare the GHG generation, waste generation, and pollution generation for each scenario. The GSI can be used to compare the disposal options of plastic products. The weighting factors are 50% for generation, 25% for waste generation, and 25% for pollution generation. The weighting factors are factored with the most impact on cost for a Plastics manufacturing operation. Reductions in GHG, waste generation, and pollution can be done with lowering energy costs, lower disposal costs, and lower cleanup costs.

The GSI can be calculated based on the data in Table 7.16. The GSI will provide an overall sustainability measurement of the four sustainable plastics.

Table 7.16 lists the environmental impacts of plastic bottles. The GSI for each of the materials can be found with the weighting factors of 50% for GWP, 25% for waste generation, and 25% for pollution factor. The pollution factor is an arithmetic average of eutrophication and acidification.

Table 7.16 Normalized environmental impacts for sustainable bottles using cradle-to-grave life cycle assessment.

Environmental impact	PLA with 2010 data	PLA with 2005 data	Virgin PET	rPET (25% PCR)
Energy consumed (MJ)	1	1.15	1	0.92
Global warming potential (tons CO_2eq)	1	1.39	1.41	1.32
Solid waste generated (kg)	1	1	0.97	0.86
Eutrophication, water (g PO_4eq)	1	1.23	0.48	0.43
Acidification (kg SO_2eq)	1	1.16	0.76	0.68
GSI	1	1.24	1.1	1.02

The GSI results reveal that the new PLA resin bottles and recycled PET bottles have equivalent environmental impacts. The new PLA has 10% lower environmental impacts than virgin PET bottles and 24% lower environmental impacts as PLA from 2005 data.

7.4.3 Sensitivity Analysis

The LCA calculations on bottles provide an acceptable LCA analysis but failed to consider in-depth fuel data and the eutrophication data for PET was zero.

7.5 Summary

Sustainable plastic products can be created with lower carbon footprint, lower waste, and lower pollution than conventional plastic products. Plastic products can be used for sustainable plastic packaging, bottles, and bags. LCAs can be used to compare the different options for plastic packaging, bottles, and bags.

Sustainable plastic packaging can be made from recycled plastics or bio-based plastics, like PET, HDPE, PHA, PLA, starch, and others. Life cycle assessments show that containers made from PLA had lower global warming potential, lower pollution, and lower solid waste generation than containers made from recycled PET and virgin PET. Containers made from recycled PET produce less GHG emission, less waste, and less pollution than containers made from virgin PET. Containers made from recycled PET produce equivalent chemicals that result in eutrophication, acidification, ozone depletion, and freshwater ecotoxicity than containers made from virgin PET.

Sustainable plastic bags can be made from recycled plastics or made to be used multiple times. After 52 uses, reusable plastic bags can result in significantly lower GHGs, waste, and pollution. After 52 uses, single-use plastic bags have approximately 25 times more negative environmental effects than reusable LLDPE plastic bags with 40% PCR. Paper bags have approximately 108 times more negative environmental effects than a reusable rLDPE plastic bag made with 40% PCR. Sustainable plastic bottles can be made with bio-based PLA and recycle PET plastics. Plastics bottles made with PLA resin and recycled PET have equivalent environmental impacts. PLA and recycled PET bottles have 10% lower environmental impacts than virgin PET bottles.

References

Basurko, O. and Mesbahi, E. (2014). Methodology for the sustainability assessment of marine technologies. *J. Clean. Prod.* 68: 155–164.

Beswick, R. and Dunn, D.J. (2002). *Plastics and Packaging: Western Europe and North America*, 71. Shropshire: Rapra Technology Limited.

Cadman, J., Evans, S., Holland, M., and Boyd, R. (2005). Environmental Group Research Report. Proposed plastic tax levy extended impact assessment. http://www.scotland.gov.uk/Publications/2005/08/1993102/31039 (accessed July 2010).

Carrefour (2004). Evaluation des impacts environnementaux des sacs de caisse Carrefour. Analyse du cycle de vie de sacs de caisse en plastique, papier et mat´eriau biod´egradable. Report prepared for Carrefour, Ecobilan, February 2004. http://www.ademe.fr/htdocs/actualite/rapport_carrefour_post_revue_critique_v4.pdf (accessed July 2014).

Chaffe, C. and Yaros, B. (2007). Life cycle assessment for three types of grocery bags – recyclable plastic; compostable, biodegradable plastic; and recycled, recyclable paper, Bousted Consulting and Associates. http://heartland.org/sites/default/files/threetypeofgrocerybags.pdf (accessed June 2014).

ChicoBag Company (2013). http://www.chicobag.com/ (accessed July 2013).

City of Orilla Biodegradable Bag Study (2003). http://www.bpiworld.org/resources/Documents/City_of_Orillia_Compostable_Bag_Study_Dec_03.pdf (accessed September 2022).

Council for Solid Waste Solutions (1990). Resource and environmental profile analysis of polyethylene and unbleached paper grocery sacks. CSWS (800-243-5790), Washington, DC, June 1990. https://www.semanticscholar.org/paper/Resource-and-Environmental-Profile-Analysis-of-and-Sacks/ed041122b8a5dbef3d91d0c2953cc351908cbcb3 (accessed February 2022).

Dili, R. (2007). Comparison of existing life cycle analysis of shopping bags alternatives. Sustainability Victoria.http://www.zerowaste.sa.gov.au/upload/

resources/publications/plastic-bag-phase-out/F0008-AA002564-AAR-03%20 LCA%20of%20shopping%20bag%20alternatives.pdf (accessed June 2014).

Franklin and Associates (2010). Life cycle inventory of 100% postconsumer hdpe and pet recycled resin from postconsumer containers and packaging. Prairie Village, KS, April 7, 2010. http://www.napcor.com/pdf/FinalReport_LCI_Postconsumer_PETandHDPE.pdf (accessed June 2013).

Franklin and Associates (2011). Cradle-to-gate life cycle inventory of nine plastic resins and four polyurethane precursors. http://plastics.americanchemistry.com/LifeCycle-Inventory-of-9-Plastics-Resins-and-4-Polyurethane-Precursors-APPSOnly (accessed June 2013).

Greene, J. (2011). Life cycle assessment of reusable and single-use plastic bags in California. True Reusable Bags Company. http://www.truereusablebags.com/pdf/lca_plastic_bags.pdf (accessed June 2013).

LCA NatureWorks LLC (2006). Life cycle assessment of poly lactide. http://www.ifeu.de/oekobilanzen/pdf/LCA%20zu%20PLA%20erstellt%20fuer%20NatureWorks%20%28Okt%202006%29.pdf (accessed May 2022).

LCA NatureWorks LLC (2009). Life cycle assessment of food packaging made of Ingeo™ biopolymer and RPET. http://www.natureworksllc.com/~/media/The_Ingeo_Journey/Ingeo_vs_rPET/IFEU_LCA__Ingeo_Full_Report_012709_FINAL_pdf.pdf (accessed June 2013).

Cycle Inventory of Polystyrene Foam, Bleached Paperboard, and Corrugated Paperboard Foodservice Products. https://www.dartcontainer.com/media/1887/franklin2006_h.pdf (accessed September 2022).

LCA Packaging NatureWorks LLC (2006). Life cycle assessment of polylactide. http://www.ifeu.de/oekobilanzen/pdf/LCA%20zu%20PLA%20erstellt%20fuer%20NatureWorks%20%28Okt%202006%29.pdf (accessed June 2013).

LCI Summary for PLA (2013). LCI summary for PLA and PET 12-ounce water bottles. http://www.container-recycling.org/assets/pdfs/plastic/LCA-PETandPLA2007.pdf (accessed June 2014).

Life Cycle Inventory of Five Products (2013). Life cycle inventory of five products produced from PLA and petroleum-based resins. http://www.wincup.com/docs/franklin-plastic_products_lca_technical_rpt.pdf (accessed June 2014).

LCI Food Service Products (2006). Life cycle inventory of polystyrene foam, bleached paperboard, and corrugated paperboard food service products. Franklin and Associates, LTD, provided by the American Chemistry Council. http://plastics.americanchemistry.com/LifeCycle-Inventory-of-Polystyrene-Foam-Bleached-and-Corrugated-Paperboard-Foodservice-Products (accessed June 2014)

Madival, S., Auras, R., Singh, S.P., and Narayan, R. (2009). Assessment of the environmental profile of PLA, PET and PS clamshell containers using LCA methodology. *J. Clean. Prod.* 17: 1183–1194.

Muthu, S.S., Li, Y., Hu, J. et al. (2013). *An Exploratory Comparative Life Cycle Assessment Study of Grocery Bags – Plastic, Paper, Non-woven and Woven Shopping Bags*. Institute of Textiles and Clothing, The Hong Kong Polytechnic University, Kowloon, Hong Kong, China. *J. Fiber Bioeng. Inform.* 1 (4): 307–303.

Roplast Industries (2013). http://www.roplast.com/grocery-reusable-bags/ (accessed July 2013).

Shen, L., Nieuwlaar, E., Worrell, E., and Patel, M. (2011). Life cycle energy and GHG emissions of PET recycling: change-oriented effects. *Int. J. Life Cycle Assess.* 16: 522–536.

Vink, E., Davies, S., and Kolstad, J. (2010). The eco-profile for current Ingeo polylactide production. *Ind. Biotechnol.* 6 (4): 212–225.

8

Biobased and Biodegradation Standards for Polymeric Materials

8.1 Introduction

Biodegradable plastics are available throughout the world. Biodegradable plastics can be made with reduced carbon emissions, reduced waste, and reduced toxic pollution compared to traditional petroleum-based plastics. Typically, plastics account for 10% by weight or 20% of the volume of the landfill. Organic food waste can account for 20–30% of the landfill weight and 30–40% of the volume. Biodegradable plastics can reduce the amount of food waste and plastic waste if they are used to wrap or package the food items and then sent to industrial composting facilities rather than to landfills.

Biodegradation is an important feature of biodegradable plastics. Two essential components of the biodegradation process are that the material must be a food source for the bacteria in the disposal environment and that the biodegradation must take place within a short time period, typically six months. Therefore, biodegradation can occur in an industrial compost environment for biodegradable plastics if they are used as a food source for the bacteria in the compost and that they are consumed within a short time span. Likewise, biodegradation can occur in the marine environment if the bacteria in the seawater consume a major portion of the plastic within a short time span. Biodegradation standards are created to capture these essential components. These standards define the environment of biodegradation and the time of biodegradation. Thus, plastic materials can be defined as biodegradable in a compost environment if they biodegrade in one growing season or six months. Alternatively, materials can be properly defined as biodegradable in the marine environment if they partially biodegrade within six months. This chapter describes the worldwide biodegradation standards for biodegradable plastics, including starch-based plastics, in common disposal

Sustainable Plastics: Environmental Assessments of Biobased, Biodegradable, and Recycled Plastics, Second Edition. Joseph P. Greene.

environments, including compost, marine, anaerobic digestion, soil, and landfill. Compost environments include aerobic conditions within hot industrial compost environments. Marine environments include cold aerobic conditions. Landfill disposal environments include aerobic and anaerobic conditions. Anaerobic-digestion environments include hot anaerobic conditions.

8.1.1 Biodegradation Standards

Biodegradation standards for plastic materials are established based on two necessary categories for biodegradation: biodegradation testing method and biodegradation performance specifications. The first standard is a test method that accurately simulates the intended environment and specifies a method for measuring biodegradation. The second standard is a specification standard that assigns a minimum value to establish biodegradation. Both types of standards are necessary and sufficient to adequately establish the biodegradation performance of plastic materials. Solid waste disposal environments for plastic materials can include industrial compost, home compost, anaerobic digestion, landfill, litter, and ocean water. Only two disposal environments have both biodegradation standards for test methods and biodegradation performance standards, for example, industrial compost and marine biodegradation environments. Test method standards are available for anaerobic digestion, home compost, and landfill environments. The second necessary performance specification standard for biodegradation performance is not available for anaerobic digestion, home compost, or landfill environments. Therefore, plastic materials can claim to meet biodegradation performance standards for industrial compost and marine environments, but not for anaerobic-digestion, home compost, or landfill environments.

8.1.2 Worldwide Biodegradation

8.1.2.1 Standards Agencies

Several worldwide organizations as listed in Table 8.1 are involved in setting standards for biodegradable and compostable plastics, including American Society for Testing and Materials (ASTM), International Committee for Standardization (CEN), International Standards Organization (ISO), German Institute for Standardization (DIN), Japanese Institute for Standardization (JIS), and British Plastics Federation. The standards from these organizations have helped the industry create biodegradable and compostable products that meet the increasing worldwide demand for more environmentally friendly plastics (Narayan and Pettigrew 1999). International, American, and Japanese certification schemes are cooperating to enable international cross-certification of products so that a product certified in one of these countries would automatically be

Table 8.1 Biodegradation standards agencies.

Standards agency	Country	Web address	Certification agency
ASTM	USA	http://www.astm.org	http://www.bpiworld.org/BPI-certification-requirements
CEN	International Union	http://www.cen.eu	http://www.din.de/cmd
ISO	Worldwide	http://www.iso.org	http://www.din.de/cmd
JIS	Japan	http://www.jsa.or.jp	http://www.jbpaweb.net/english/english.htm

eligible for certification in other countries. Biodegradation standards are provided in the following sections. The standards are organized based on the disposal environment. Thus, the biodegradation standards from different worldwide standards organizations are presented in an industrial compost section, marine environment section, an anaerobic-digestion section, landfill section, and home compost section.

Standards for biobased materials are, also, presented in the following to establish the parameters of claims that plastic materials are biobased. Biodegradable plastics can be produced from organic materials or petroleum-based materials.

8.1.3 Certification

Certification is needed for biodegradable plastics to ensure that they meet the performance specification requirements in the biodegradation standards. Several certification companies, listed in Table 8.1, are available to certify the biodegradation performance of compostable plastics. In the United States, Biodegradable Products Institute (BPI) and the US Composting Council established the Compostable Plastics certification program in the United States to certify compostable plastics as meeting the ASTM D6400 or ASTM D6868 compostability standards. In Europe, the DIN certification organization was created to certify those biodegradable plastics (BPI Certified Compostable 2013) meet the International standard EN 13432 for compostability. Vincotte is also a certification company for certification of compostable plastics in Europe. Vincotte OK Biodegradable mark ensures that the plastic material will pass the performance biodegradable specification in a specific natural environment (soil, fresh water, seawater, etc.), and also ensures that the plastic material meets the performance specification of compostable plastics in EN 13432. In Japan, the BioPlastics Association (JBPA) was created to certify biodegradable plastics meet the

International Standard EN 13432 for compostability. JBPA in Japan established GreenPla certification and labeling system based on international biodegradation standards.

8.2 Biobased Standard Test Method

Many biodegradable plastics are made with biobased materials. These are derived from organic carbon sources such as cereal, corn, potato, rice, soy, sugar cane, wheat, vegetable oil, etc. Biobased plastics are

made with a renewable resource that can have lower environmental impacts than petroleum-based plastics. Replacing petroleum-based carbon with organic carbon from today can reduce the carbon footprint of the plastic material (Narayan 2011a). Biological carbon content analysis can establish the carbon footprint of bioplastics with life cycle assessment methods (Narayan 2011b). Biobased content of the plastic material can be established by tests that measure carbon isotopes of the material.

8.2.1 US Biobased Standard

8.2.1.1 ASTM D6866-10 Standard Test Methods for Determining the Biobased Content of Solid, Liquid, and Gaseous Samples Using Radiocarbon Analysis

8.2.1.1.1 Summary

This American test method establishes the procedures, equipment, materials, and conditions to measure the 14carbon content of the plastic sample through radiocarbon analysis. This test method is applicable to carbon-based plastic products that can undergo combustion in the presence of oxygen to produce carbon dioxide gas (CO_2). This test method can be useful for companies and the US federal agencies to promote the use of biobased products in commercial, nonfood products. The test method directly discriminates between 14C/12C content of plastic samples made from contemporary carbon and those plastic samples made from petroleum sources that do not have 14C. Measurement of a plastic product's 14C/12C content is determined relative a standard carbon-based reference material, namely, Standard Reference Material 4990c.

8.2.1.1.2 Procedures

The radiocarbon analysis can be performed with three methods, as listed in the following:

1) *Method B:* Accelerator mass spectrometry (AMS) and isotope ratio mass spectrometry (IRMS) techniques are used to quantify the biobased content of a

given product. Sample preparation methods are identical to Method A. Stored CO_2 is sent to an AMS facility for final processing and analysis. The maximum error is 1–2% for AMS and 0.1–0.5% for IRMS.

2) *Method C:* Liquid scintillation counting (LSC) radiocarbon techniques are used to quantify the biobased content of a product. Method C uses LSC analysis of sample carbon that has been converted to benzene rather than CO_2 solutions of Method A. The maximum error is $\pm3\%$.

8.2.1.1.3 Equipment

1) Liquid scintillation counting
2) Accelerator mass spectrometry
3) Isotope ratio mass spectrometry

8.2.2 International Biobased Standards

Currently, there are no ISO test methods that are equivalent to the biobased standard test method.

8.3 Industrial Compost Environment

Biodegradation standards for industrial compost environment require a biodegradation specification standard and a biodegradation test method standard. The specification standards in the United States are for compostable plastics and for coatings that can be used for packaging and containers. Both specification standards require the use of the same test method. Table 8.2 lists the biodegradation standards for industrial compost in several countries. The standards listed are the performance specification standard and the test method standards in the industrial compost environment. The standards are discussed in more detail in the following section. For complete information on the standards, consult the documentation from the standards agency.

8.3.1 US Biodegradation Standards for Industrial Compost Environment

8.3.1.1 Biodegradation Performance Specification Standard: ASTM D6400-04. Standard Specification for Compostable Plastics

8.3.1.1.1 Summary

This American specification standard establishes the performance requirements for biodegradation of compostable plastic materials that are designed to biodegrade into CO_2, water, and biomass in an industrial compost environment.

Table 8.2 Biodegradation standards for industrial composting environment.

Region	Performance specification	Test method	Measurement method	Disintegration test method
Australia	ISO 17088	ISO 14855-2	CO_2 measurement	ISO 16929, ISO 20200
Europe	ISO 17088/ EN13432	ISO 14855-2	CO_2 measurement	ISO 16929, ISO 20200
Japan	ISO 17088	ISO 14855-2	CO_2 measurement	ISO 16929, ISO 20200
USA	ASTM D6400, D6868	ASTM D5338	CO_2 measurement	ISO 16929

The industrial compost environment is one that maintains a temperature above 40 °C and results in thermophilic conditions. The performance specification standard requires the use of the ASTM D5338-05 test method to measure the amount of CO_2 that is emitted from the degrading plastic sample. ASTM D6400-04 performance specification standard requires the product must demonstrate each of the three characteristics as follows:

1) Sufficient disintegration during composting.
2) Adequate level of inherent biodegradation.
3) No adverse impacts on the ability of compost to support plant growth.

8.3.1.1.2 Procedures

Three test characteristics for the ASTM D6400-04 standard specify that three types of tests are performed on the plastic samples. The first characteristic measures the percentage of disintegration of the plastic samples while under hot and moist compost conditions. The plastic samples are weighed prior to exposure to test conditions. The samples are placed in compost soil with the use of a sack, bag, or screened container. The plastic samples are exposed to industrial composting conditions of at least 58 °C and 50% moisture for 12 weeks. The mass of the plastic sample is measured after 12 weeks by passing the plastic sample and compost through a 2.0-mm sieve. ASTM D-6400 specifies the test specimen must disintegrate at least 90% of the dry weight within 12 weeks. The second characteristic for ASTM D6400-04 standard specifies that a satisfactory rate of *biodegradation* of the plastic material is under industrial composting conditions of at least 58 °C and 50% moisture for 180 days, that is, more than 90% of the carbon in the original plastic sample is converted into CO_2 as measured by a CO_2 respirometer or wet chemistry methods. The details of the test procedure are listed in ASTM D5338-11 test method. The third characteristic for ASTM D6400-04 standard specifies a satisfactory ability of the compost soil at the end of the biodegradation testing to *support plant growth* through phytotoxicity testing and very low regulated heavy metal concentrations. Phytotoxicity testing is achieved through the planting of tomato, cucumber, radish, rye, barley, or cress grass seeds in the tested compost soil. The growth of the plants after 10 days indicates positive soil conditions for plant growth. Plant biomass tests can reveal quality differences between composts and can indicate potential plant stress induced by the compost at the given level used in the test. Regulated heavy metal testing measures the concentrations of regulated heavy metals in the compost soil at the end of the biodegradation testing. The level of regulated heavy metals can be measured with flame atomic absorption spectrophotometer using an air–acetylene flame and equipped with a Pb hollow-cathode lamp. The compost samples must have regulated metal concentrations less than 50% of the acceptable levels of regulated heavy metals as prescribed in 40 Code of Federal Regulations (CFR) Part 503.13, that is, lead (75 mg/kg), cadmium (8.5 mg/kg), chromium (not specified), copper (375 mg/kg), nickel (105 mg/kg), zinc (700 mg/kg), and mercury (4.25 mg/kg) (Title 40: Protection of the Environment 2013).

8.3.1.1.3 Specification

ASTM D6400-04 standard specifies that a plastic material is compostable if it meets the following specifications:

1) *Disintegration:* Greater than 90% disintegration of the original dry weight of the plastic material disintegrates after 12 weeks of exposure to industrial composting conditions specified in ISO 16929 test method standard for disintegration.
2) *Biodegradation:* Greater than 90% of the carbon in the original plastic sample is converted into CO_2 after 180 days of exposure to industrial composting conditions specified in ASTM D5338 test method standard.
3) *Nontoxic to plants:* No measurable phytotoxicity or regulated heavy metals greater than allowable levels as described in 40 CFR Part 503.13.

8.3.1.2 Biodegradation Performance Specification Standard: ASTM D6868-03. Standard Specification for Biodegradable Plastics Used as Coatings on Paper and Other Compostable Substrates

8.3.1.2.1 Summary

This American specification standard establishes the performance requirements for biodegradation of materials and products (including packaging, films, and coatings) that are attached (either through lamination or extrusion directly onto the paper) to compostable substrates plastic materials. The compostable plastics are designed to biodegrade in an industrial compost environment. The industrial compost environment is one that maintains a temperature above 40 °C and results in thermophilic conditions. The performance specification standard requires the use of ASTM D-5338-05 test method to measure the amount of CO_2 that is emitted from the degrading plastic sample. ASTM D-6868-03 performance specification standard requires the product must demonstrate each of the three characteristics as follows:

1) Sufficient disintegration during composting.
2) Adequate level of inherent biodegradation.
3) No adverse impacts on the ability of compost to support plant growth.

8.3.1.2.2 Procedures

The test procedures for ASTM D6868-03 standard specify that three types of tests are performed on the plastic samples. The first characteristic measures the percentage of disintegration of the plastic samples while under hot and moist compost conditions. The plastic samples are weighed prior to exposure to test conditions. The samples are placed in compost soil with the use of a sack, bag, or screened container. The plastic samples are exposed to industrial composting conditions of at least 58 °C and 50% moisture for 12 weeks. The mass of the plastic sample is measured after 12 weeks by passing the plastic sample and compost through a 2.0-mm sieve. ASTM D-6868 specifies that no more than 10% of the original dry weight of the plastic material remains. The second characteristic for ASTM D6868-03 standard specifies that a satisfactory rate of *biodegradation* of the plastic material is under industrial composting conditions of at least 58 °C and 50% moisture for 180 days, that is, more than 90% of the carbon in the original plastic sample is converted into CO_2 as measured by a CO_2 respirometer or wet chemistry methods. If the packaging or paper product is a combination of several materials, then each of the materials must individually pass the respirometry test. The details of the test procedure are listed in ASTM D5338-11 test method. The third characteristic for ASTM D6868-03 standard specifies a satisfactory ability of the compost soil at the end of the biodegradation testing to *support plant growth* through phytotoxicity testing and very low regulated heavy metal concentrations. Phytotoxicity testing is achieved through the planting of tomato, cucumber, radish, rye, barley, or cress grass seeds in the tested compost soil. The growth of the plants after 10 days indicates positive soil conditions for plant growth. Regulated heavy metal testing measures the concentrations of regulated metals in the compost soil at the end of the biodegradation testing.

8.3.1.2.3 Specification

ASTM D6868-03 standard specifies that a plastic coating material is compostable if it meets the following specifications:

1) *Disintegration:* Less than 10% of the original dry weight of the plastic material remains after 12 weeks of exposure to industrial composting conditions.
2) *Biodegradation:* Greater than 90% of the carbon in the original plastic sample is converted into CO_2 as measured by a CO_2 respirometer after 180 days of exposure to industrial composting conditions.

3) *Nontoxic to plants:* No measure able phytotoxicity or regulated heavy metals greater than allowable levels as prescribed in 40 CFR Part 503.13.

8.3.1.3 Biodegradation Test Method Standard: ASTM D5338-11. Standard Test Method for Determining Aerobic Biodegradation of Plastic Materials under Controlled Composting Conditions

8.3.1.3.1 Summary

This test method establishes the procedures, equipment, materials, and conditions to measure the degree and rate of biodegradation of plastic materials under aerobic thermophilic composting conditions. This test method is designed to produce repeatable and reproducible test results under controlled composting conditions that simulate industrial compost conditions. The plastic test samples are exposed to an inoculum that is derived from industrial compost. The aerobic thermophilic conditions of the test are provided in an environment where temperature, aeration, and humidity are closely monitored and controlled. The plastic samples are cut into small pieces and placed in a vessel with warm and moist compost soil. The test containers are at 58 °C (±2 °C) for 180 days. The biogas from the container is measured for CO_2 and O_2 over the testing period.

8.3.1.3.2 Procedures

The test procedures for ASTM D-5338-11 involve placing 500 g of industrial compost soil that has a maturity level of two to three months into a container vessel greater than 2 l and less than 5 l. The ASTM test procedure for industrial compost is explained in more detail in Appendix D. A minimum of 12 composting vessels are required for the test. The containers are provided for the test specimens that include plastics, blank, positive control, and negative control samples. The testing is completed in triplicate. The plastic samples are added to the container in quantities of 100 g. The container must have 25% free air space above the samples. The pH, moisture content, C/N ratio, and percentage solids are measured for the compost soil. The sample containers must have moist air supplied to the container that is free of CO_2. The biogas from the sample containers is measured periodically for CO_2 and O_2.

The O_2 level must be greater than 6% during the duration of the test. The test containers should be rotated every two or three days to reduce the packing of the soil and to mix fresh air with the compost soil. An example of a test apparatus is provided in Figure 8.1.

8.3.1.3.3 Materials

The materials used in this standard include the following:

- Plastic samples.
- Blank compost soil.

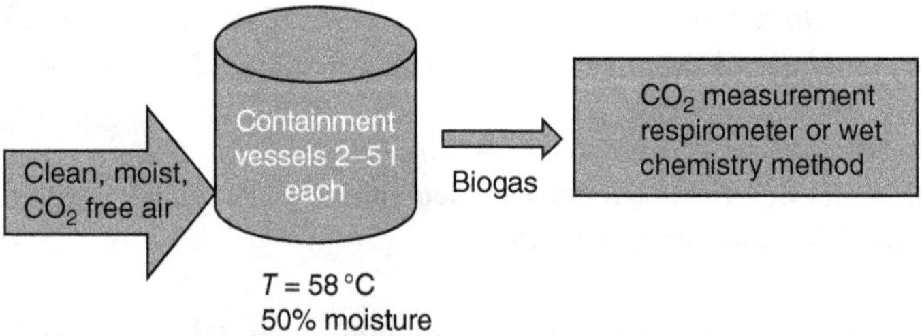

Figure 8.1 Experimental setup for laboratory environment.

- Positive control: cellulose powder or cellulose paper.
- Negative control: polyethylene film.
- Barium hydroxide solution.

8.3.1.3.4 Equipment

The equipment used in this standard includes the following:

- Composting vessels with a size greater than 2 l and less than 5 l.
- Water bath or other temperature control device.
- Pressurized air system with CO_2 free and H_2O saturated air.
- Carbon dioxide trapping apparatus with $Ba(OH)_2$ solution.
- Flexible tubing nonpermeable to CO_2.
- Stoppers equipped with sampling ports.
- Analytical balance (±0.1 mg).
- 100-ml Burette.
- 0.05 N HCL.
- pH meter.
- Respirometer devices for measuring CO_2 and O_2 periodically (optional).
- CO_2 trapping equipment with gas flow meter, gas chromatograph, or other suitable equipment.

8.3.2 International Biodegradation Standards for Industrial Compost Environment

The ISO is the world's largest developer of standards. ISO is a network of the national standards institutes of 157 countries, which agree on specifications and criteria to be applied consistently in the classification of materials, in the manufacture and supply of products, in testing and analysis, in terminology, and in the provision of services. The test method is designed to yield the percentage conversion of the carbon in the test material to evolve carbon dioxide as well as the rate of conversion. In Europe, compostable plastics are being used in several applications. Compostable plastics must comply with the International NormEN13432 that is the criteria for compostability. The EN13432 specifications standard is very similar to

ASTM D-6400 specifications standard. Certification with ISO compostability standard is compatible with certification for ASTM D6400 standard. EN13432 requires a compostable plastic material to break down to the extent of at least 90% to H_2O and CO_2 and biomass within a period of six months. ISO14855 standard specifies a testing method to evaluate the ultimate aerobic biodegradability of plastics, based on organic compounds, under controlled composting conditions by the measurement of the amount of carbon dioxide evolved and the degree of plastic disintegration at the end of test. ISO 14852 is the determination of the ultimate aerobic biodegradability of plastic materials in an aqueous medium. The test method measures the evolved carbon dioxide and is similar to ASTM D-5338 test method standard. DIN-Certco is a very well known and utilized certification system in Europe (Bastioli 2005). Sample materials are tested for regulated metals, organic contaminants, complete biodegradation, disintegration under compost conditions, and phytotoxicity (plant toxicity). The regulated metals and organic chemical tests ensure that neither organic contaminants nor heavy metals such as lead, mercury, and cadmium can enter the soil via the biodegradable materials. The procedures for testing complete biodegradation in the laboratory and disintegration under compost conditions ensure that materials are completely degraded during one process cycle of a standard composting plant. The DIN compostability certification is very similar to BPI certification, which meets ASTM D-6400 standards.

Table 8.3 illustrates the heavy metal limits in the International standard and the US standards (Rudnik 2008). Heavy metal concentrations in the EN13432 standard

Table 8.3 Regulated heavy metal allowable concentrations.

Element	ASTM D-6400-USA (mg/kg)	ASTM D-6400-Canada (mg/kg)	EN 13432 (mg/kg)	Japan (mg/kg)
Zn	1400	463	150	180
Cu	750	189	50	60
Ni	210	45	25	30
Cd	17	5	0.5	0.5
Pb	150	125	50	10
Hg	8.5	1	0.5	0.2
Cr	—	265	50	50
Mo	—	5	1	—
Se	50	4	0.75	—
As	20.5	19	5	5
F	—	—	100	—
Co	—	38	—	—

allow a limited amount of metal, that is, lead (30 mg/kg), cadmium (0.3 mg/kg), chromium (30 mg/kg), copper (22.5 mg/kg), nickel (15 mg/kg), zinc (100 mg/kg), and mercury (0.3 mg/kg). The US standard allows the following amounts: lead (150 mg/kg), cadmium (17 mg/kg), chromium (not specified), copper (750 mg/kg), nickel (210 mg/kg), zinc (1400 mg/kg), and mercury (8.5 mg/kg) (Rudnik 2008). Acceptable levels of heavy metals in sewer sludge are provided as per US Environmental Protection Agency (EPA) Subpart 503-13. Biodegradation standards for industrial compost environment require a biodegradation specification standard and a biodegradation test method standard. The specification standards in Europe are for plastics and for packaging. Both specification standards require the use of the same test method.

8.3.2.1 Biodegradation Performance Specification Standard: EN 13432-2000. Packaging Requirements for Packaging Recoverable through Composting and Biodegradation Test Scheme and Evaluation Criteria for the Final Acceptance of Packaging

8.3.2.1.1 Summary

This international specification establishes the performance requirements for biodegradation of plastic materials and products (including packaging, films, and other products). The compostable plastics are designed to biodegrade in an industrial compost environment. The industrial compost environment is one that maintains a temperature above 40 °C and results in thermophilic conditions. The performance specification standard requires the use of EN 14995 (ISO 17088) test method to measure the amount of CO_2 that is emitted from the degrading plastic sample. EN 13432-2000 performance specification standard requires the product must demonstrate each of the following characteristics:

1) Biodegradability.
2) Disintegration during biological treatment.
3) Minimal effect on the biological treatment process.
4) Minimal effect on the quality of the resulting compost.

8.3.2.1.2 Procedures

The test procedures for EN 13432 standard specify that three types of tests are performed on the plastic samples. The first test measures the percentage of disintegration of the plastic samples while under hot and moist compost conditions. The plastic samples are weighed prior to exposure to test conditions. The samples are placed in compost soil with the use of a sack, bag, or screened container. The plastic samples are exposed to industrial composting conditions of at least 58 °C and 50% moisture for 12 weeks. The mass of the plastic sample is measured after three months by passing the plastic sample and compost through a 2.0-mm sieve. EN 13432 specifies that no more than 10% of the original dry weight of the plastic material remains. The second test procedure for EN 13432 standard specifies that a satisfactory rate of *biodegradation* of the plastic material is under industrial composting conditions of at least 58 °C and 50% moisture for 26 weeks, that is, more than 90% of the carbon in the original plastic sample is converted into CO_2 as measured by a CO_2 respirometer or wet chemistry methods. The details of the test procedures are listed in EN 14046 or ISO 14855 test method. The third test procedure for EN 13432 standard specifies a satisfactory ability of the compost soil at the end of the biodegradation testing to *support plant growth* through phytotoxicity testing and very low regulated heavy metal concentrations. Phytotoxicity testing is achieved through the planting of two seed crops in the tested compost soil. The growth of the plants after 10 days indicates positive soil conditions for plant growth. Plant biomass tests can reveal quality differences between composts and can indicate potential plant stress induced by the compost at the given level used in the test. Regulated heavy metal testing measures the concentrations of regulated metals in the compost soil at the end of the biodegradation testing. The test procedures for EN 13432 standard specify that three types of tests are performed on the plastic samples. The first test measures the percentage of disintegration of the plastic samples while under hot and moist compost conditions. The plastic samples are weighed prior to exposure to test conditions. The samples are placed in compost soil with the use of a sack, bag, or screened container. The plastic samples are exposed to industrial composting conditions of at least 58 °C and 50% moisture for 12 weeks. The mass of the plastic sample is measured after three months by passing the plastic sample and compost through a 2.0-mm sieve. EN 13432 specifies that no more than 10% of the original dry weight of the plastic material remains. The second test procedure for EN 13432 standard specifies that a satisfactory rate of *biodegradation* of the plastic material is under industrial composting conditions of at least 58 °C and 50% moisture for 26 weeks, that is, more than 90% of the carbon in the original plastic sample is converted into CO_2 as measured by a CO_2 respirometer or wet chemistry methods. The details of the test procedures are listed in EN 14046 or ISO 14855 test method. The third test procedure for EN 13432 standard specifies a satisfactory ability of the compost soil at the end of the biodegradation testing to *support plant growth* through phytotoxicity testing and very low regulated heavy metal concentrations. Phytotoxicity testing is achieved through the planting of two seed crops in the tested compost soil. The growth of the plants after 10 days indicates positive soil conditions for plant growth. Plant biomass tests can reveal quality differences between composts and can indicate potential plant stress induced by the compost at the given level used in the test. Regulated heavy metal testing measures the concentrations of regulated metals in the compost soil at the end of the biodegradation testing.

8.3.2.1.3 Specifications

EN 13432 standard specifies that a plastic material is compostable if it meets the following specifications:

1) Satisfactory *disintegration* of the plastic material that is under the pilot-scale composting test of at least 58 °C and 50% moisture for 84 days that no more than 10% of the original dry weight of the plastic material remains after passing through a 2.0-mm sieve. Also, the biodegradation testing shall not have any adverse effects on the composting process. The standard requires the use of ISO 16929 or ISO 2020 test method standard.
2) Satisfactory rate of *biodegradation* of the plastic material that is under industrial composting conditions of at least 58 °C and 50% moisture in six months, that is, more than 90% of the carbon in the original plastic sample is converted into CO_2 as measured by measuring the amount of release of CO_2 through wet chemistry methods. The standard requires the use of ISO 148455 or ASTM D5338 test method standard.
3) Satisfactory ability of the compost soil at the end of the biodegradation testing to support plant growth through phytotoxicity testing with seedling emergence and seedling growth test (OECD 208).

8.3.2.2 Biodegradation Performance Specification Standard: ISO 17088 (EN 13432). Plastics – Evaluation of compostability – Test Scheme and Specification

This international standard is very similar to EN 13432-2000 and establishes the performance requirements for biodegradation of plastic materials and products (including packaging, films, and other products). The compostable plastics are designed to biodegrade in an industrial compost environment. The industrial compost environment is one that maintains a temperature above 40 °C and results in thermophilic conditions. The performance specification standard requires the use of EN 14995 (ISO 17088) test method to measure the amount of CO_2 that is emitted from the degrading plastic sample. EN 13432-2000 performance specification standard requires the product must demonstrate each of the four following characteristics:

1) Biodegradability.
2) Disintegration during biological treatment.
3) Minimal effect on the biological treatment process.
4) Minimal effect on the quality of the resulting compost.

ISO 17088 standard specification establishes the performance requirements for biodegradation of compostable plastic materials that are designed to biodegrade into CO_2, water, and biomass in an industrial compost environment according to test methods described in EN 14046 and ISO 14855. The industrial compost environment is one that maintains thermophilic temperatures. The performance specification standard requires the use of EN 14046 or ISO 14855 test method to measure the amount of CO_2 that is emitted from the degrading plastic sample.

8.3.2.2.1 Procedures

The test procedures for ISO 17088 standard specify that three types of tests are performed on the plastic samples. The first test measures the percentage of disintegration of the plastic samples while under hot and moist compost conditions. The plastic samples are weighed prior to exposure to test conditions. The samples are placed in compost soil with the use of a sack, bag, or screened container. The plastic samples are exposed to industrial composting conditions of at least 58 °C and 50% moisture for 84 days. The mass of the plastic sample is measured after three months by passing the plastic sample and compost through a 2.0-mm sieve. EN 13432 specifies that no more than 10% of the original dry weight of the plastic material remains.

The second test procedure for ISO 17088 standard specifies that a satisfactory rate of *biodegradation* of the plastic material is under industrial composting conditions of at least 58 °C and 50% moisture for 26 weeks, that is, more than 90% of the carbon in the original plastic sample is converted into CO_2 as measured by a CO_2 respirometer. The details of the test procedures are listed in ISO 14855 or ASTM D5338 test method. The third test procedure for ISO 17088 standard specifies a satisfactory ability of the compost soil at the end of the biodegradation testing to *support plant growth* through phytotoxicity testing and very low regulated heavy metal concentrations. Phytotoxicity testing is achieved through the planting of tomato, cucumber, radish, rye, barley, or cress grass seeds in the tested compost soil. The growth of the plants after 10 days indicates positive soil conditions for plant growth. Regulated heavy metal testing measures the concentrations of regulated metals in the compost soil at the end of the biodegradation testing.

The compost samples must have regulated metal concentrations less than prescribed levels, listed in Table 8.3.

8.3.2.2.2 Specifications

ISO 17088 standard specifies that a plastic material is compostable if it meets the following specifications:

1) Satisfactory *disintegration* of the plastic material that is under the pilot-scale composting test of at least 58 °C and 50% moisture for three months that no more than 10% of the original dry weight of the plastic material remains after passing through a 2.0-mm sieve. Also, the biodegradation testing shall not have any adverse effects on the composting process.

Satisfactory rate of *biodegradation* of the plastic material that is under industrial composting conditions of at least 58 °C and 50% moisture for six months, that is, more than 90% of the carbon in the original plastic sample is converted into CO_2 as measured by a CO_2 respirometer or wet chemistry method.

2) Satisfactory ability of the compost soil at the end of the biodegradation testing to *support plant growth* through phytotoxicity testing and very low regulated heavy metal concentrations. Plant growth test (OECD 208).

8.3.2.3 Biodegradation Test Method Standard: ISO 14855-2 (EN 14046) Packaging. Evaluation of the Ultimate Aerobic Biodegradability and Disintegration of Packaging Materials under Controlled Composting Conditions. Method by Analysis of Released Carbon Dioxide

8.3.2.3.1 Summary

This international standard specifies a test method for the evaluation of the ultimate aerobic biodegradability of packaging materials based on organic compounds under controlled composting conditions by measurement of released carbon dioxide at the end of the test. This method is designed to resemble typical aerobic composting conditions for the organic fraction of mixed municipal solid waste.

The packaging material is exposed in a laboratory test to an inoculum that is derived from compost. The aerobic composting takes place in an environment where especially temperature, aeration, and humidity are closely monitored and controlled. The test method is designed to yield a percentage and rate of conversion of carbon of the test material to released carbon dioxide. The conditions described in this standard do not necessarily always correspond to the optimal conditions allowing the maximum degree of biodegradation to occur.

This test method establishes the procedures, equipment, materials, and conditions to measure the degree and rate of biodegradation of plastic materials under aerobic thermophilic composting conditions. This test method is designed to produce repeatable and reproducible test results under controlled composting conditions that simulate industrial compost conditions. The testing is completed in duplicate. The plastic test samples are exposed to an inoculum that is derived from industrial compost. The aerobic thermophilic conditions of the test are provided in an environment where temperature, aeration, and humidity are closely monitored and controlled. The plastic samples are cut into small pieces and placed in a vessel with warm and moist compost soil. The test containers are at 58 °C (±2 °C) for 180 days. The biogas from the container is measured for CO_2 and O_2 over the testing period.

8.3.2.3.2 Procedures

The test procedure for ISO 14855 involves placing 120 g of moist (~50% moisture) industrial compost soil that has a maturity level of six months and 50% total solids into a container vessel of approximately 500 ml. The containers will be provided for the plastic sample and the following samples, blank, positive control, and negative control. The testing is done in duplicate. The plastic samples are added to the container in quantities of 10 g. The pH, moisture content, C/N ratio, and percentage solids are measured for the compost soil. The sample containers must have moist air supplied to the container that is free of CO_2. The biogas from the sample containers is measured periodically for CO_2 and O_2. The test containers should be rotated every two or three days to reduce the packing of the soil and to mix fresh air with the compost soil. The tests are run in duplicate.

8.3.2.3.3 Materials

- Plastic sample
- Blank compost soil
- Positive control reference: cellulose powder
- Negative control: polyethylene film

8.3.2.3.4 Equipment

- Composting vessels with size approximately 500 ml.
- Water bath or other temperature control device.
- Pressurized air system with CO_2 free and H_2O saturated air.
- Carbon dioxide trapping apparatus.
- Flexible tubing nonpermeable to CO_2.
- Stoppers equipped with sampling ports.
- Analytical balance (±0.1 mg).
- 100-ml Burette.
- pH meter.
- Respirometer devices for measuring CO_2 and O_2 periodically (optional).
- CO_2 trapping equipment with gas flow meter, gas chromatograph, or other suitable equipment.

8.3.2.4 ISO 16929 (EN14045:2003) Plastics – Determination of the Degree of Disintegration of Plastic Materials under Simulated Composting Conditions in a Pilot-Scale Test

8.3.2.4.1 Summary

This international standard specifies a method for the evaluation of the disintegration of plastic materials while under industrial composting conditions. The testing environment is pilot scale at an industrial compost facility. The mass of plastic sample is measured prior to the test. The samples are placed in a sack or aerated bag with compost soil. The sack is placed in an aerated pile of at least 1 m deep at an industrial compost facility. The compost vessel should be a minimum of 140 l and a maximum of 250 l. The concentration is 1% by weight plastic sample and 99% compost. The temperature of the compost pile should be between 40 and 75 °C. The aerated compost pile should be turned twice per week.

8.3.2.4.2 Procedures

The test procedure for ISO 16929:

1) Measure the mass of plastic samples.
2) Place in sack or aerated container between 140 and 250 l.
3) Add samples and compost to sack or container.
4) Bury container in compost pile at least 1 m from the surface.
5) Rotate the compost soil around container at least twice per week.
6) Measure the temperature of the compost.
7) After 12 weeks, remove compost and plastic.
8) Sieve in 2-mm screen.
9) Wash and dry plastic to remove compost soil.
10) Measure the mass of the plastic samples.
11) Measure the quality of the compost soil for maturity, percentage of solid, pH, moisture content, carbon/nitrogen ratio, and temperature.

8.3.2.4.3 Materials

- Plastic
- Mature compost
- Container minimum 140 l

8.3.2.4.4 Equipment

- Industrial compost facility

8.3.2.5 ISO 20200 (EN14806:2005) Plastics – Determination of the Degree of Disintegration of Plastic Materials under Simulated Composting Conditions in a Laboratory-Scale Test

8.3.2.5.1 Summary

This international standard specifies a method for the evaluation of the disintegration of plastic materials while under industrial composting conditions. The testing environment is pilot scale at an industrial compost facility. The mass of plastic sample is measured prior to the test. The samples are placed in a sack or aerated bag with compost soil. The sack is placed in an aerated pile at least 1 m deep at an industrial compost facility. The compost vessel should be a minimum of 5 l and a maximum of 20 l. The concentration is 1% by weight plastic sample and 99% compost. The temperature of the compost pile should be between 40 and 75 °C. The aerated compost pile should be turned twice per week.

8.3.2.5.2 Procedures

The test procedure for ISO 20200:

1) Measure the mass of plastic samples.
2) Place in sack or aerated container between 5 and 20 l.
3) Add samples and compost to sack or container at concentrations of 1% (weight) plastic and 99% compost.
4) Bury container in compost pile at least 1 m from the surface.
5) Rotate the compost soil twice per week.
6) Measure the temperature of the compost.
7) After 12 weeks, remove compost and plastic.
8) Sieve in 2-mm screen.
9) Wash and dry plastic to remove compost soil.
10) Measure the mass of the plastic samples.
11) Measure the quality of the compost soil for maturity, percentage of solid, pH, moisture content, carbon/nitrogen ratio, temperature.

8.3.2.5.3 Materials

- Plastic
- Mature compost
- Container minimum 5 l

8.3.2.5.4 Equipment

- Industrial compost facility

8.3.2.6 Australian Biodegradation Standards for Industrial Compost

The Australian standard for degradable plastics includes test methods that enable validation of biodegradation of degradable plastics and designation as "compostable" in Australia. The Australian standard for industrial compost is AS4736-2006 (AS4736-2013 2013). The AS4736 standard is based on EN 13432-2000 standard for the determination of the ultimate aerobic biodegradability and disintegration of plastic materials under controlled composting conditions. AS4736-2006 requires that plastic materials meet the following requirements:

1) A minimum of 90% biodegradation of plastic material within 180 days while under industrial composting conditions of 58 °C and 50% moisture.
2) A minimum of 90% disintegration of plastic material within 12 weeks while under industrial composting conditions of 58 °C and 50% moisture.
3) No toxic effects on the resulting compost for plants and earthworms.
4) Toxic levels of regulated metals are less than the maximum allowable levels.
5) Minimum of 50% organic material in the plastic.

8.3.2.7 Japanese Biodegradation Standards for Industrial Compost

The Japanese JIS standards are met with a GreenPLA certification system. The GreenPLA system has very similar testing requirements as per the US and International certification methods. In particular, the GreenPLA certification assures biodegradability by measuring carbon dioxide evolution after microbial biodegradation, mineralization by the ability to disintegrate and not have visible fragments after composting, and organic compatibility by the ability of the compost to support plant growth. The same amount of carbon dioxide evolution (60%) in 45 days is required for certification. The same 11 regulated metals are monitored in GreenPLA as EN 13432. However, several aspects of the certification are different than the US BPI and International Din-Certco certifications.

8.4 Marine Environment

Plastic debris is accumulating in the oceans around the world endangering animal life, releasing toxic chemicals, and collecting floating toxins that can enter the food stream through fish. The fate of plastics in the oceans can lead to fragmentation and result in slurry of plastic particles floating or suspended in the water column. The plastic fragments can degrade and release toxic chemicals into the ocean waters.

Biodegradation of biodegradable plastics in the marine environment is based upon two levels of standards: the first for a test method standard and the second for a performance specification standard. Biodegradation standards for the marine environment are listed in Table 8.4. The marine biodegradation standards cover nonfloating products made from plastics that are designed to biodegrade in the aerobic marine environment. The standard applies to deep seawater, shallow seawater, and brackish inland waters. In the standards, plastic materials must demonstrate disintegration and inherent biodegradation during marine water

Table 8.4 Biodegradation standards for marine environment.

Region	Performance specification	Test method	Measurement method	Other requirements
Australia	None	ISO 14851, ISO 14852	CO_2 measurement	None
Europe	None	ISO 14851, ISO 14852	CO_2 measurement	None
Japan	None	ISO 14851, ISO 14852	CO_2 measurement	None
USA	ASTM D7081	ASTM D6691	CO_2 measurement	ASTM D6400

exposure and not exhibit any adverse environmental impacts on the survival of marine organisms while in the marine environment.

The plastic sample also must pass several marine toxicity tests, including Polytox (microbial oxygen absorption), Microtox (microbial bioluminescence) test, fish Acute Toxicity (static conditions) OPPTS 850.1075, Daphnia Acute Toxicity (static conditions) OPPTS 850.1010, or Static Algal Toxicity Test OPPTS 850.5400. The plastic samples must also have less than 25% of the maximum allowable concentrations of regulated heavy metals.

Marine biodegradation standards require that the plastic samples should also pass the ASTM D-6400 standard for biodegradation under industrial aerobic compost conditions. The ASTM D-6400 standard requires plastic samples to convert 90% of the carbon in the plastic sample to CO_2 after 180 days while at 58 °C.

8.4.1 US Biodegradation Standards for Marine Environment

8.4.1.1 Biodegradation Performance Specification Standard: ASTM D-7081-05. Nonfloating Biodegradable Plastic in the Marine Environment

8.4.1.1.1 Summary

This American specification establishes the performance requirements for biodegradation of plastic materials and products, including packaging, films, and coatings. The marine environment includes conditions of aerobic marine waters or anaerobic marine sediments or both. This specification establishes the requirements for biodegradation of plastic materials that have rates that are similar to known compostable materials. The specification also specifies that the degradation of the plastic materials will not cause any harm to sea life or habitat.

The performance specification standard requires the use of ASTM D6691-08 test method to measure the amount of CO_2 that is emitted from the degrading plastic sample while in the cool marine water. ASTM D-7081-05 performance specification standard requires the product must demonstrate each of the three characteristics as follows:

1) Disintegration during marine biodegradation.
2) Inherent biodegradation of the plastic material.
3) Minimal adverse effect on the marine environment.

8.4.1.1.2 Procedures

The test procedures for ASTM D-7081-05 standard specify that three types of tests are performed on the plastic samples. The first test measures the percentage of disintegration of the plastic samples while exposed to cool marine water conditions, described in ASTM D-6691-08 test conditions. The plastic samples are weighed prior to exposure to test conditions. The samples are placed in containers with ocean water. The plastic samples are exposed to aerobic marine water conditions of 30 °C for 12 weeks. The mass of the plastic sample is measured after 12 weeks by passing the plastic sample and marine water through a 2.0-mm sieve. ASTM D-7081-05 specifies that no more than 30% of the original dry weight of the plastic material remains.

The second test procedure for ASTM D-7081-05 standard specifies that a satisfactory rate of *biodegradation* of the plastic material is under marine conditions of 30 °C for 180 days, that is, more than 30% of the carbon in the original plastic sample is converted into CO_2 as measured by a CO_2 respirometer or another measurement device. The details of the test procedures are listed in ASTM D6691-08 test method.

The third test procedure for ASTM D-7081-05 standard specifies that the plastic product shall undergo toxicity testing with one of the following tests: Polytox, Microtox, fish acute toxicity, Daphnia acute toxicity, or static algal toxicity. The fourth test procedure for ASTM D-7081-05 standard specifies that the plastic sample shall have less than 25% of the regulated heavy metal concentrations that is specified in the country where the product is sold.

8.4.1.1.3 Specifications

ASTM D-7081-05 standard specifies that a plastic coating material is compostable if it meets the following specifications:

1) Satisfactory *disintegration* of the plastic material that is under aerobic marine water test of 30 °C for 12 weeks that no more than 30% of the original dry weight of the plastic material remains after passing through a 2.0-mm sieve.
2) Satisfactory rate of *biodegradation* of the plastic material that is under aerobic marine water test of 30 °C in six months, that is, more than 30% of the carbon in the original plastic sample is converted into CO_2 as measured by measuring the amount of release CO_2.
3) Satisfactory passing of one of the several marine toxicity tests.
4) Less than 25% of the regulated heavy metal concentrations that is specified in the country where the product is sold.
5) Satisfactory rate of *biodegradation* specified in ASTM D6400 standard for compostability is a necessary but not sufficient prerequisite.

8.4.1.2 Biodegradation Test Method Standard: ASTM D6691-09. Standard Test Method for Determining Aerobic Biodegradation of Plastic Materials in the Marine Environment by a Defined Microbial Consortium or Natural Seawater Inoculum

8.4.1.2.1 Summary

ASTM D6691-09 test method provides a description of the testing procedures that best simulate the marine environment and a method by which to measure biodegradation. ASTM D6691-09 test method is used to determine the degree and rate of aerobic biodegradation of plastic materials exposed to the indigenous population of existing seawater or synthesized seawater with pre-grown population of at least 10 aerobic marine microorganisms of known genera. The ASTM test procedure for marine biodegradation is explained in more detail in Appendix E. The test method consists of preparing a uniform inoculum of marine water, exposing the plastic samples to the marine water, measuring biodegradation with a carbon dioxide respirometer or equivalent measurement method, and assessing the percentage of carbon conversion in the plastic to carbon dioxide. The testing is carried out at 30 ± 2 °C under controlled laboratory conditions for 180 days. The standard recommends the use of 125-ml autoclave bottles, a recipe of marine organisms and nutrients, and seawater. The standard recommends adding 20 mg of plastic sample with 75 ml of marine water inoculum in a constant temperature environment of 30 ± 2 °C. The amount of biodegradation of the plastic sample is compared to the biodegradation of a positive control, for example, cellulose powder. This test method establishes the procedures, equipment, materials, and conditions to measure the degree and rate of biodegradation of plastic materials under aerobic mesophilic marine water conditions. This test method is designed to produce repeatable and reproducible test results under controlled test conditions that simulate the marine environment.

8.4.1.2.2 Procedures

The test procedures for ASTM D6691-08 involve placing 20 mg of plastic sample with 75 ml of marine stock solution or natural seawater with inorganic nutrients are placed in 25-ml bottles in a sterile environment as shown in Figure 8.2. A minimum of 12 composting vessels are required for the test.

8.4.1.2.3 Materials

- Plastic samples.
- Blank marine water.
- Positive control: cellulose powder or paper.
- Negative control: polyethylene film.

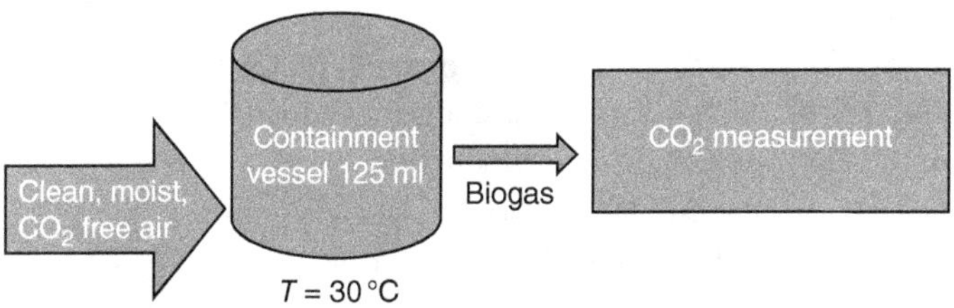

Figure 8.2 Experimental setup for laboratory environment of marine biodegradation test.

8.4.1.2.4 Equipment

- Sterile jar with a volume of 125 ml.
- Water bath or other temperature control device.
- Carbon dioxide trapping apparatus with $Ba(OH)_2$ solution.
- Flexible tubing nonpermeable to CO_2.
- Stoppers equipped with sampling ports.
- Analytical balance (±0.1 mg).
- pH meter.
- Respirometer devices for measuring CO_2 and O_2 periodically (optional).
- CO_2 trapping equipment with gas flow meter, gas chromatograph, or other suitable equipment.

8.4.2 International Aqueous Biodegradation Standards

8.4.2.1 Biodegradation Test Method Standard: ISO 14852-1999 (EN14047). Determination of Ultimate Aerobic Biodegradability of Plastic Materials in an Aqueous Medium – Method by Analysis of Evolved Carbon

8.4.2.1.1 Summary

This international standard specifies a method for the evaluation of the biodegradation of plastic materials while exposed to an aqueous mineral medium. Biodegradation is determined by measuring the evolved carbon dioxide from the biodegrading sample. The test is used to measure the degree and rate of aerobic biodegradation of plastic materials on exposure to activated sludge biomass or a suspension of active soil under laboratory conditions at solid concentrations between 100 and 2000 mg/l. The temperature of the test environment should be between 20 and 25 °C. The tests are run in duplicate.

8.4.2.1.2 Procedures

The test procedure for ISO 14852 is:

1) Measure the mass of plastic samples.
2) Measure the quality of the inoculum soil for a percentage of solid, pH, moisture content, carbon/nitrogen ratio, and temperature. Carbon/nitrogen ratio should be approximately 40 : 1.
3) Add the plastic sample in the form of powder or small pieces to a concentration of 100–2000 mg/l of TOC with activated sludge or active soil in a test flask.
4) Connect the sample flasks to CO_2-free air production.
5) Maintain the test flasks at a temperature between 20 and 25 °C.
6) Bubble CO_2-free air through flasks at a rate of 50–100 ml/min.
7) Measure CO_2 from the sample flasks.

8.4.2.1.3 *Materials*

- Plastic samples.
- Blank activated sludge or active soil.
- Positive control: cellulose powder or cellulose paper.
- Negative control: polyethylene film.
- Barium hydroxide solution.

8.4.2.1.4 *Equipment*

- Flasks.
- Water bath or other temperature control device.
- Pressurized air system with CO_2 free and H_2O saturated air.
- Carbon dioxide trapping apparatus.
- Flexible tubing nonpermeable to CO_2.
- Stoppers equipped with sampling ports.
- Analytical balance (±0.1 mg).
- 100-ml Burette.
- 0.05 N HCL.
- pH meter.
- CO_2 trapping equipment with gas flow meter, gas chromatograph, or other suitable equipment.

8.4.2.2 Biodegradation Test Method Standard: ISO 14851 (EN14048). Determination of Ultimate Aerobic Biodegradability of Plastic Materials in an Aqueous Medium – Method by Measuring the Oxygen Demand in a Closed Respirometer

8.4.2.2.1 *Summary*

This International standard specifies a method for the evaluation of the biodegradation of plastic materials while exposed to an aqueous mineral medium. Biodegradation is determined by measuring oxygen demand produced from the biodegrading sample. The percentage of biodegradation is calculated by comparing the biological oxygen demand to the chemical oxygen demand, determined by the chemical oxidation of the test item or by stoichiometric formula. The principle and test procedures are very similar to ISO 14852 (Bastioli 2005).

The test is used to measure the degree and rate of aerobic biodegradation of plastic materials on exposure to activated-sludge biomass or a suspension of active soil under laboratory conditions at solid concentrations between 100 and 2000 mg/l. The temperature of the test environment should be between 20 and 25 °C. The tests are run in duplicate.

8.4.2.2.2 Procedures

The test procedure for ISO 14852 is:

1) Measure the mass of plastic samples.
2) Add the plastic sample in the form of powder or small pieces to a concentration of 100–2000 mg/l of TOC with activated sludge or active soil in a test flask.
3) Connect the sample flasks to CO_2-free air production.
4) Maintain the test flasks at a temperature between 20 and 25 °C.
5) Bubble CO_2-free air through flasks at a rate of 50–100 ml/min.
6) Measure CO_2 from the sample flasks.

8.4.2.2.3 Materials

- Plastic samples.
- Blank activated sludge or active soil.
- Positive control: cellulose powder or cellulose paper.
- Negative control: polyethylene film.

8.4.2.2.4 Equipment

- Flasks.
- Water bath or other temperature control device.
- Pressurized air system with CO_2 free and H_2O saturated air.
- Flexible tubing nonpermeable to CO_2.
- Stoppers equipped with sampling ports.
- Analytical balance (±0.1 mg).
- 100-ml Burette.
- pH meter.
- O_2 trapping equipment with gas flow meter, gas chromatograph, or other suitable equipment.

8.5 Anaerobic Digestion

Anaerobic digestion is a process where food and agricultural waste can be converted to methane and carbon dioxide in a thermophilic reactor. Anaerobic digestion is a series of processes in which microorganisms break down biodegradable material in the absence of oxygen (NNFCC Renewable Fuels and Energy Factsheet: Anaerobic Digestion 2013). It is used for industrial or domestic purposes to manage waste and/or to release energy. The digestion process begins with bacterial hydrolysis of the input materials to break down insoluble organic polymers, such as carbohydrates, and make them available for other bacteria. Acidogenic bacteria then convert the sugars and amino acids into carbon dioxide,

Table 8.5 Biodegradation standards for anaerobic-digestion environment.

Region	Performance specification	Test method	Measurement method
Australia	None	ISO 14853	CO_2 measurement
Europe	None	ISO 14853	CO_2 measurement
Japan	None	ISO 14853	None
USA	None	ASTM D5511	CO_2 measurement

hydrogen, ammonia, and organic acids. Acetogenic bacteria then convert these resulting organic acids into acetic acid, along with additional ammonia, hydrogen, and carbon dioxide. Finally, methanogens convert these products to methane and carbon dioxide. Anaerobic digestion is used as part of the process to treat organic and agricultural waste and sewage sludge. Anaerobic digestion can also be used to remove organic waste at waste water treatment operations. Biodegradable plastics can hold organic and agricultural waste with the use of biodegradable plastic bags. The biodegradable plastic can be a food source for the anaerobes in the digester. The biodegradable plastic and organic waste can be converted into carbon dioxide, methane, and other trace gases. Standards are necessary to monitor the performance of biodegradable plastics in an aerobic digestion reactor. The anaerobic decomposition takes place under high solids (more than 30% total solids) and static non-mixed conditions. The biodegradation process usually requires one to two weeks of dwell time. Table 8.5 lists the biodegradation standards for anaerobic digestion.

8.5.1 US Biodegradation Standards for Anaerobic Digestion

8.5.1.1 Biodegradation Test Method Standard: ASTM D5511-02. Standard Test Method for Determining Anaerobic Biodegradation of Plastic Materials under High Solids Anaerobic-Digestion Conditions

8.5.1.1.1 Summary

This test method establishes the procedures, equipment, materials, and conditions to measure the degree and rate of biodegradation of plastic materials under anaerobic thermophilic to produce repeatable and reproducible test results under controlled anaerobic-digestion conditions that simulate industrial conditions. The plastic test samples are exposed to an inoculum that is derived from an aerobic digester or waste water treatment operation. The anaerobic thermophilic conditions of the test are provided in an environment where temperature, methane, carbon dioxide, and oxygen levels are closely monitored and controlled. The plastic samples can be in the form of films, powders, pellets, or molded pieces and are placed in a vessel with warm inoculum with proper anaerobic bacteria. The test containers are at 50 °C (±2 °C) for 30 days. The biogas from the container is measured for CH_4, CO_2, and O_2 over the testing period. The standard for anaerobic digestion comprises a biodegradation test method and not a biodegradation specification.

8.5.1.1.2 Procedures

The test procedures for ASTM D-5511-02 involve placing 1000 g of inoculum derived from a properly operating anaerobic digester that is made from pretreated household waste. The inoculum should be derived from a digester operating under greater than 20% total solids conditions. A minimum of 12 test vessels are required for the test. The containers will be provided for the plastic sample and the following plastic samples, blank, positive control, and negative control. The testing is completed in triplicate. The plastic samples are added to each test container in quantities up to 100 g. The positive control must obtain greater than 70% biodegradation in 30 days. The inoculum is tested for pH, volatile fatty acids, and NH_4^+–N/NO_3 ratio. The test apparatus can include an inverted graduated cylinder or plastic column. The graduated cylinder or plastic column is inverted in a low pH fluid to avoid CO_2 loss through the dissolution in the fluid. The biogas is calculated through a pressure measurement of the inverted tubes. The biogas can also be measured with a gas chromatograph to determine the percentage of CH_4 and CO_2 in the biogas. An example of a test apparatus is provided in Figure 8.3. The testing apparatus will provide measurements of pressure over 30 days. Through ideal gas law, the pressure can be converted to grams of biogas. The

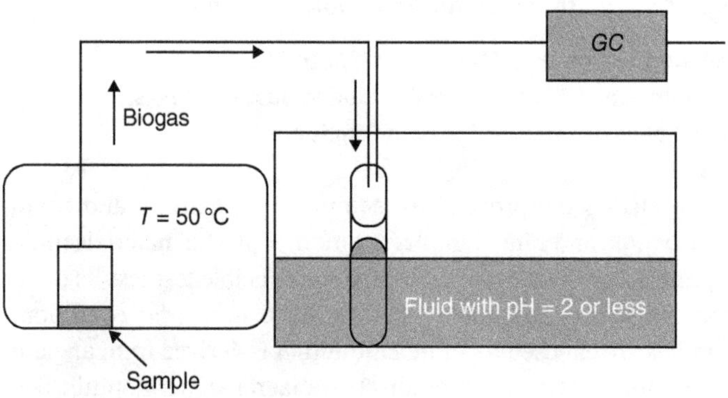

Figure 8.3 Experimental setup for laboratory environment for anaerobic digester test per ASTM standards.

concentration of biogas can be converted to concentrations of CO_2 and CH_4. The conversion of carbon from the plastic sample to CO_2 and CH_4 can be determined. This will result in the carbon biodegradation percentage over 30 days in a high solids anaerobic digester.

8.5.1.1.3 Materials

- Plastic samples.
- Blank anaerobic digester inoculum.
- Positive control: cellulose powder or cellulose paper.
- Negative control: polyethylene film.

8.5.1.1.4 Equipment

- Test jars.
- Low pH fluid bath or other temperature control device.
- Flexible tubing nonpermeable to CH_4, CO_2, O_2.
- Stoppers equipped with sampling ports.
- Graduated cylinder or plastic tube.
- Analytical balance ($\pm$0.1 mg).
- pH meter.
- Gas chromatograph.

8.5.2 International Biodegradation Standards for Anaerobic Digestion

8.5.2.1 Biodegradation Test Method Standard: ISO 14853:2005 Plastics. Determination of Ultimate Anaerobic Biodegradation of Plastic Materials in an Aqueous System. Method of Biogas Production

8.5.2.1.1 Summary

This test method specifies a method for determining the ultimate biodegradability of plastics by anaerobic microorganisms. The conditions described in this test method simulate the conditions of some anaerobic digesters. The plastic test samples are exposed to an inoculum that is derived from digested sludge from waste water treatment operations. The anaerobic thermophilic conditions of the test are provided in an environment where temperature, methane, carbon dioxide, and oxygen levels are closely monitored and controlled.

The plastic samples can be in the form of films, powders, pellets, or molded pieces and are placed in a vessel with warm inoculum with proper anaerobic bacteria. The test containers are at 35 °C ($\pm$2 °C) for 60 days. The biogas from the container is measured for CH_4, CO_2, and O_2 over the testing period. The standard for anaerobic digestion comprises a biodegradation test method and not a biodegradation specification.

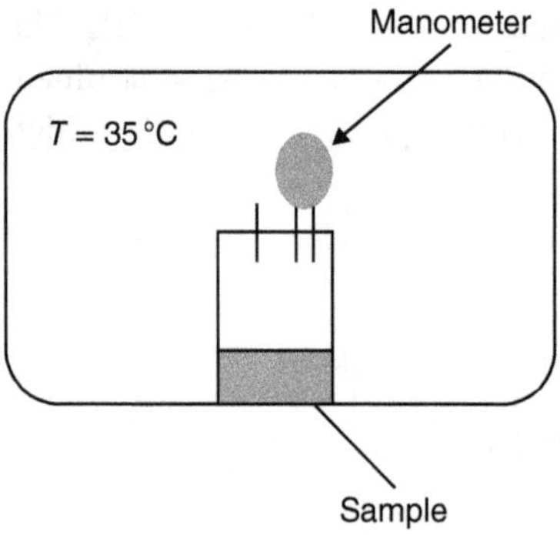

Figure 8.4 Experimental setup for laboratory environment for anaerobic digester test per ISO standards.

8.5.2.1.2 Procedures

The test procedures for ISO 14853:2005 involve placing 1000 ml of inoculum derived from properly operating anaerobic digester that is made from digested sludge. The concentration of test materials is 20–200 mg/l of organic carbon. Approximately, 350 ml of inoculum is added to the test sample. A head space of 20% is recommended. A reference material or positive control can be cellulose, poly (hydroxybutyrate), or poly (ethylene glycol). The glass vessels have sizes of 0.1–1 l with gas-tight septum rated for 2000 kPa. Plastic samples are added to glass containers that include plastic samples, blank, positive control, and negative control. The testing is completed in triplicate. The positive control must obtain greater than 70% biodegradation in 60 days. The inoculum is tested for pH, volatile fatty acids, and NH_4^+–N/NO_3 ratio. Biogas measurement is achieved with a monometer or volumetric device. An example of a test apparatus is provided in Figure 8.4. The testing apparatus will provide measurements of pressure over 60 days. Through ideal gas law, the pressure can be converted to grams of biogas. The concentration of biogas can be converted to concentrations of CO_2 and CH_4. The conversion of carbon from the plastic sample to CO_2 and CH_4 is determined. This will result in the carbon biodegradation percentage over 30 days in a high solids anaerobic digester.

8.5.2.1.3 Materials

- Plastic samples.
- Blank anaerobic digested sludge inoculum.
- Positive control: cellulose powder or poly(hydroxybutyrate), or poly(ethylene glycol).
- Negative control: polyethylene film.

8.5.2.1.4 Equipment

- Test jars: 0.1–1 l.
- Low pH fluid bath or other temperature control device.
- Flexible tubing nonpermeable to CH_4, CO_2, O_2.
- Stoppers equipped with sampling ports.
- Graduated cylinder or plastic tube.
- Manometer.

- Analytical balance (±0.1 mg).
- pH meter.
- Gas chromatograph.

8.6 Active Landfill

Landfills in the United States are typically built with the EPA guidelines with the use of clay linings and a landfill cap (Criteria for Solid Waste Disposal Facilities 2013). The most common material for landfill caps is made from asphalt or concrete (Remediation Technologies Screening Matrix and Reference Guide 2013). Landfills can operate with creation of biogas that is composed of methane, carbon dioxide, and other trace gases. Methane gas can be vented and burned or can be captured and stored for energy purposes. The carbon dioxide and other gases must be scrubbed to provide a clean methane gas without carbon dioxide or other gases. Some landfills are considered active and provide clean methane gas for energy consumption. Biodegradable plastics can hold the waste as trash bags for disposal and provide a food source for the aerobic and anaerobic bacteria that are in the landfill. Standards are needed to evaluate the biodegradation of biodegradable plastics in landfills. Table 8.6 lists the biodegradation standards for an active landfill environment. The standard for active landfill comprises a biodegradation test method and not a biodegradation specification.

8.6.1 US Biodegradation Standards for Active Landfill

8.6.1.1 Biodegradation Test Method Standard: ASTM D5526-11. Determining Anaerobic Biodegradation of Plastic Materials under Accelerated Landfill Conditions

8.6.1.1.1 Summary

This test method establishes the degree and rate of anaerobic biodegradation of plastic materials under anaerobic mesophilic conditions in an accelerated landfill

Table 8.6 Biodegradation standards for active landfill environment.

Region	Performance specification	Test method	Measurement method
Australia	None	None	—
Europe	None	None	—
Japan	None	None	—
USA	None	D5526, D7575	CO_2 measurement

test environment. This test method is designed to yield a percentage of carbon conversion from the plastic sample to methane or carbon dioxide biogas. This test method is designed to provide repeatable and reproducible test results under controlled anaerobic conditions that simulate high solids (>30%) active landfill conditions. The plastic test samples are exposed to an inoculum that is derived from an aerobic digester or waste water treatment operation.

The anaerobic mesophilic conditions of the test are provided in an environment where temperature, methane, carbon dioxide, and oxygen levels are closely monitored and controlled. The plastic samples can be in the form of films, powders, pellets, or molded pieces and are placed in a vessel with cool inoculum with proper anaerobic bacteria. The test containers are at 35 °C (±2 °C) for up to 365 days. The biogas from the container is measured for CH_4, CO_2, and O_2 over the testing period.

8.6.1.1.2 Procedures

The test procedures for ASTM D5526 involve placing 600 g of household waste with 100 g of inoculum derived from a properly operating anaerobic digester that is made from pretreated household waste. The inoculum should be derived from a digester operating under greater than 30% total solids conditions. A minimum of 27 test vessels are required for the test with a volume between 4 and 6 l. The containers will be provided for the plastic sample and the following plastic samples, blank, positive control, and negative control. The testing is completed in triplicate. The plastic samples are added to each test container in quantities between 60 and 100 g of dry weight. The mixture should contain approximately 60% dry weight of solids. The positive control must obtain greater than 70% biodegradation at the end of the test. The inoculum is tested for pH, volatile fatty acids, and NH_4^+–N/NO_3 ratio. The test apparatus can include a pressure-resistant glass vessel. The vessel should withstand a pressure of two atmospheres. The graduated cylinder or plastic column is inverted in a low pH fluid to avoid CO_2 loss through the dissolution in the fluid. The biogas is calculated through a pressure measurement of the inverted tubes. The biogas can also be measured with a gas chromatograph to determine the percentage of CH_4 and CO_2 in the biogas. An example of a test apparatus is provided in Figure **8.5**. The testing apparatus will provide measurements of pressure over 30 days. Through ideal gas law,

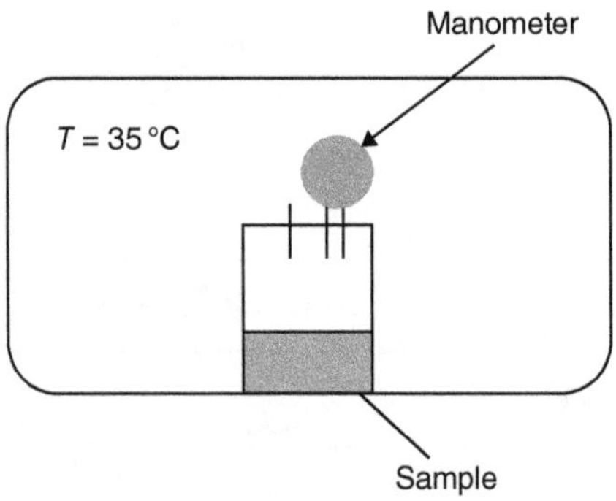

Figure 8.5 Experimental setup for laboratory environment for active landfill test.

the pressure can be converted to grams of biogas. The concentration of biogas can be converted to concentrations of CO_2 and CH_4. The conversion of carbon from the plastic sample to CO_2 and CH_4 can be determined. This will result in the carbon biodegradation percentage over 30 days in a high solids anaerobic digester.

8.6.1.1.3 Materials

- Plastic samples.
- Anaerobic digester inoculum.
- Pretreated household solid organic waste.
- Positive control: cellulose powder or cellulose paper.
- Negative control: polyethylene film.

8.6.1.1.4 Equipment

- Test jars.
- Flexible tubing nonpermeable to CH_4, CO_2, O_2.
- Stoppers equipped with sampling ports.
- Incubator.
- Pressure transducer.
- Analytical balance ($\pm$0.1 mg).
- pH meter.
- Gas chromatograph.

8.6.1.2 Biodegradation Test Method Standard: ASTM D7475-11. Determining Aerobic Degradation and Anaerobic Biodegradation of Plastic Materials under Accelerated Landfill Conditions

8.6.1.2.1 Summary

This two-tiered test method modifies ASTMD5526 by adding an aerobic degradation test method to the anaerobic biodegradation test method. ASTM D7475 establishes the degree of aerobic degradation and rate of anaerobic biodegradation of plastic materials under anaerobic mesophilic conditions in an accelerated landfill test environment. This test method is designed to establish mechanical tests or materials tests in addition to a percentage of carbon conversion from the plastic sample to methane or carbon dioxide biogas. This test method is designed to provide repeatable and reproducible test results under controlled aerobic and anaerobic conditions that simulate high solids (>30%) active landfill conditions. The plastic test samples are exposed to an inoculum that is derived from an aerobic digester or waste water treatment operation. The first tier of the test method is designed to estimate aerobic biodegradation through the measurement of loss of mechanical properties, for example, tensile strength or modulus, molecular weight, or other characteristics. In the second tier of the test method, anaerobic is designed to provide mesophilic conditions of the test in an environment where temperature, methane, carbon dioxide, and oxygen levels are closely monitored and controlled. The plastic samples can be in the form of films, powders, pellets, or molded pieces and are placed in a vessel with cool inoculum with proper anaerobic bacteria. The test containers are at 35 °C (±2 °C) for up to 365 days. The biogas from the container is measured for CH_4, CO_2, and O_2 over the testing period.

8.6.1.2.2 Procedures

The test procedures for the first tier of ASTM D7475-11 involve measuring baseline mechanical or chemical properties and then placing test specimens in an aerobic degradation environment. The mechanical or chemical properties are measured after a specified time. Differences in mechanical or chemical properties are recorded. The test procedures for the second tier of ASTM D7475-11 involve placing 600 g of household waste with 100 g of inoculum derived from properly operating anaerobic digester that is made from pretreated household waste. The inoculum should be derived from a digester operating under greater than 30% total solids conditions. A minimum of 27 test vessels are required for the test with a volume between 4 and 6 l. The containers will be provided for the plastic sample and the following plastic samples, blank, positive control, and negative control. The testing is completed in triplicate. The plastic samples are added to each test container in quantities between 60 and 100 g of dry weight. The mixture should contain approximately 60% dry weight of solids. The positive control must obtain greater than 70% biodegradation at the end of the test. The inoculum is tested for pH, volatile fatty acids, and NH_4^+–N/NO_3 ratio. The test apparatus can include a pressure-resistant glass vessel. The vessel should withstand a pressure of two atmospheres. The graduated cylinder or plastic column is inverted in a low pH fluid to avoid CO_2 loss through the dissolution in the fluid. The biogas is calculated through a pressure measurement of the inverted tubes. The biogas can also be measured with a gas chromatograph to determine the percentage of CH_4 and CO_2 in the biogas. An example of a test apparatus is provided in Figure **8.5**. The testing apparatus will provide measurements of pressure over 30 days. Through ideal gas law, the pressure can be converted to grams of biogas. The concentration of biogas can be converted to concentrations of CO_2 and CH_4. The conversion of carbon from the plastic sample to CO_2 and CH_4 can be determined. This will result in the carbon biodegradation percentage over 30 days in a high solids anaerobic digester.

8.6.1.2.3 Materials

- Plastic samples.
- Anaerobic digester inoculum.
- Pretreated household solid organic waste.
- Positive control: cellulose powder or cellulose paper.
- Negative control: polyethylene film.

8.6.1.2.4 Equipment

- Test jars.
- Flexible tubing nonpermeable to CH_4, CO_2, O_2.
- Stoppers equipped with sampling ports.
- Incubator.
- Pressure transducer.
- Analytical balance (±0.1 mg).
- pH meter.
- Gas chromatograph.
- Mechanical tests, for example, tensile, impact, etc.
- Chemical tests, for example, Fourier transform infrared spectroscopy (FTIR), differential scanning calorimetry (DSC), gas chromatograph/mass spectroscopy (GC/MS), and molecular weight.

8.6.2 International Biodegradation Standards for Active Landfill

Currently, there are no ISO test methods that are equivalent to the active landfill standard test method.

8.7 Home Compost

8.7.1 European Home Compost Certification

AIB-Vincotte International has a biodegradable home compost test certification for plastic products that references EN 13432 performance specification and other ISO biodegradable test methods. Biodegradation in soil environments is based on International OK Biodegradable Soil standards. Table **8.7** lists the biodegradation standards for home composting.

Table 8.7 Biodegradation standards for home composting environment.

Region	Performance specification	Test method	Measurement method
Australia	None	None	—
Europe	EN13432 modified for longer times	ISO 14855-2 modified for lower temperatures	CO_2 measurement
Japan	None	None	—
USA	None	None	—

8.7.1.1 Summary

This European certification establishes the performance requirements for biodegradation of plastic materials and products (including packaging, films, and other products) while under cool home composting conditions. The certification specifies modifications to EN 13432. The home compost environment is one that maintains a temperature below 30 °C and results in mesophilic conditions. The certification requires the use of EN 14855 (ISO 17088) test method to measure the amount of CO_2 that is emitted from the degrading plastic sample. Vincotte home compost certification requires the product must demonstrate each of the four characteristics as follows:

1) Biodegradability.
2) Disintegration during biological treatment.
3) Minimal effect on the organic matter with the limit on heavy metals and other hazardous substances.
4) Minimal effect on the quality of the resulting compost.

8.7.1.2 Procedures

The Vincotte home compost certification specifies that three types of tests are performed on the plastic samples. The first test measures the percentage of disintegration of the plastic samples while under cool home compost conditions. The plastic samples are weighed prior to exposure to test conditions. The samples are placed in home compost soil with the use of a sack, bag, or screened container. The plastic samples are exposed to home composting conditions between 20 and 30 °C for 12 months. The mass of the plastic sample is measured after six months by passing the plastic sample and compost through a 2.0-mm sieve. EN 13432 standard specifies that no more than 10% of the original dry weight of the plastic material remains. The second test procedure for EN 13432 standard specifies that a satisfactory rate of *biodegradation* of the plastic material that is cool home composting conditions between 20 and 30 °C for 12 months, that is, more than 90% of the carbon in the original plastic sample is converted into CO_2 as measured by a CO_2 respirometer or wet chemistry methods. The details of the test procedures are listed in ISO 14855, ISO 14851, ISO 14852, or ISO 17566 test methods. Testing to *support plant growth* through phytotoxicity testing and very low regulated heavy metal concentrations. Phytotoxicity testing is achieved through the planting of two seed crops in the tested compost soil. The growth of the plants after 10 days indicates positive soil conditions for plant growth. Plant biomass tests can reveal quality differences between composts and can indicate potential plant stress induced by the compost at the given level used in the test.

Regulated heavy metal testing measures the concentrations of regulated heavy metals or other toxic and hazardous substances in the compost soil at the end of the biodegradation testing. The concentrations of regulated heavy metals or other toxic and hazardous substances must be less than 0.1% of the dry weight of the plastic material.

8.7.1.3 Specifications

Home compost specification specifies that a plastic material is compostable if it meets the following specifications:

1) Satisfactory *disintegration* of the plastic material that is under home composting conditions between 20 and 30 °C for 12 months that no more than 10% of the original dry weight of the plastic material remains after passing through a 2.0-mm sieve. Also, the biodegradation testing shall not have any adverse effects on the composting process.
2) Satisfactory rate of *biodegradation* of the plastic material that is under industrial composting conditions between 20 and 30 °C for 12 months, that is, more than 90% of the carbon in the original plastic sample is converted into CO_2 as measured by measuring the amount of release CO_2 through wet chemistry methods.
3) Satisfactory ability of the compost soil at the end of the biodegradation testing to *support plant growth* through phytotoxicity testing and less than 0.1% of dry weight of regulated heavy metals, toxic, or hazardous substances.

8.7.2 US Home Composting Standards

Currently, ASTM does not have a test method or specification for home compost biodegradation.

8.8 Soil Biodegradation

8.8.1 European Soil Biodegradation Certification

AIB-Vincotte International has a certification for biodegrade in soil environments based on International OK Biodegradable Soil standards. The certification for plastic products references EN 13432 performance specification and other ISO biodegradable test methods. Table **8.8** lists the biodegradation standards for soil biodegradation.

8.8.1.1 Summary

This European certification establishes the performance requirements for biodegradation of plastic materials and products (including packaging, films, and other products) while under cool soil burial conditions. The certification specifies

Table 8.8 Biodegradation standards for soil biodegradation environment.

Country	Performance specification	Test method	Measurement method
Australia	None	None	—
Europe	EN13432 modified for longer times	ISO 14855-2 modified for lower temperatures	CO_2 measurement
Japan	None	None	–
USA	None	None	—

modifications to EN13432. The soil biodegradation environment is one that maintains a temperature below 30 °C and results in mesophilic conditions. The certification requires the use of EN 14855 (ISO 17088) test method to measure the amount of CO_2 that is emitted from the degrading plastic sample. Vincotte soil biodegradation certification requires the product must demonstrate each of the following characteristics:

1) Biodegradability.
2) Minimal ecotoxicity on the organic matter with limit on heavy metals and other hazardous substances.

8.8.1.2 Procedures

The Vincotte soil biodegradation certification specifies that two types of tests are performed on the plastic samples. The first test procedure for modified EN 13432 standard specifies that a satisfactory rate of *biodegradation* of the plastic material is under soil burial conditions between 20 and 30 °C for 24 months, that is, more than 90% of the carbon in the original plastic sample is converted into CO_2 as measured by a CO_2 respirometer or wet chemistry methods. The details of the test procedures are listed in ISO 14855, ISO 14851, ISO 14852, or ISO 17566 test methods.

The third test procedure for the modified EN 13432 standard specifies a satisfactory ability of the soil at the end of the biodegradation testing to *support plant growth* through ecotoxicity testing to identify detrimental environmental effects. Testing is conducted according to methods listed in EN 13432 on ecotoxicity.

8.8.1.3 Specifications

Soil biodegradation specification specifies that a plastic material is compostable if it meets the following specifications:

- Satisfactory rate of *biodegradation* of the plastic material that is under industrial composting conditions between 20 and 30 °C for 24 months, that is, more than 90% of the carbon in the original plastic sample is converted into CO_2 as

measured by measuring the amount of release CO_2 through wet chemistry methods.
- Satisfactory ability of the compost soil at the end of the biodegradation testing to *support plant growth* through ecotoxicity testing.

8.8.2 US Soil Biodegradation Standards

Currently, ASTM does not have a test method or specification for soil biodegradation.

8.9 Summary

Biodegradation standards of plastic materials require that plastic materials must convert to CO_2, water, and biomass in a disposal environment in a defined timespan. Biodegradation standards are established from several worldwide standards organizations for disposal environments, including industrial compost, marine environment, anaerobic digestion, landfill, and home compost. Biodegradation standards for plastic materials in industrial compost and marine environments are established based on two necessary categories for biodegradation: biodegradation testing method and biodegradation performance specifications. For industrial compost, ISO and ASTM composting standards require that plastic materials under prescribed testing conditions disintegrate sufficiently after 12 weeks during composting, biodegrade sufficiently after 180 days, and result in no adverse impacts on the ability of compost to support plant growth. Certification is available for meeting ISO or ASTM standards for biodegradation under industrial compost. For home compost, Vincotte has a certification for biodegradion in soil environments based on International OK Biodegradable Soil standards. The home composting certificate requires that plastic materials under prescribed testing conditions biodegrade sufficiently after 24 months, and result in no adverse impacts on the ability of compost to support plant growth. ASTM does not, currently, provide standards for home composting. For the marine environment, ISO and ASTM composting standards require that plastic materials under prescribed testing conditions disintegrate sufficiently after 12 weeks during composting, biodegrade sufficiently after 180 days, and result in no adverse effect on the marine environment. Certification is available for meeting ISO or ASTM standards for marine biodegradation.

Anaerobic digestion has ISO and ASTM standards for testing methods but not for biodegradation performance. Anaerobic-digestion test method establishes the procedures, equipment, materials, and conditions to measure the degree and rate of biodegradation of plastic materials under anaerobic thermophilic conditions. Certification is not available for meeting performance standards for biodegradation under anaerobic digestion conditions.

Test method standards for plastic materials in active landfill conditions are available with ASTM standards but not with ISO standards. This test method establishes the degree and rate of anaerobic biodegradation of plastic materials under anaerobic mesophilic conditions in an accelerated landfill test environment. Certification is not available for meeting performance standards for biodegradation under active landfill conditions.

References

AS4736-2013 (2013). http://www.biofiba.com/documents/AS4736-2006.pdf (accessed December 2013).

Australasian Bioplastics. https://bioplastics.org.au/. (accessed September 2022).

Bastioli, C. (2005). *Handbook of Biodegradable Polymers*, 170. Shropshire: Smithers Rapra Press.

BPI Certified Compostable (2013). http://www.bpiworld.org/ (accessed March 2013).

Criteria for SolidWaste Disposal Facilities (2013). Environmental Protection Agency. http://www.epa.gov/wastes/nonhaz/municipal/landfill/criteria/landbig.pdf (accessed March 2013).

Narayan, R. (2011a). Rationale, drivers, standards, and technology for biobased materials. In: *Renewable Resources and Renewable Energy: A Global Challenge*, 2nde (ed. M. Graziani and P. Fornasiero), 10–16. Boca Raton, FL: CRC Press.

Narayan, R. (2011b). Carbon footprint of bioplastics using biocarbon content analysis and life-cycle assessment. *Mater. Res. Soc. Bull.* 36 (9): 716–721.

Narayan, R. and Pettigrew, C. (1999). ASTM standards help define and grow a new biodegradable, plastic industry. *ASTM Standardization News* 36.

National Service Center for Environment Publications, United States Environmental Protection Agency (2022). https://www.epa.gov/nscep (accessed September 2022).

NNFCC Renewable Fuels and Energy Factsheet: Anaerobic Digestion (2013). National Non-Food Crops Centre. https://www.scribd.com/document/252403062/NNFCC-Renewable-Fuels-and-Energy-Factsheet-Anaerobic-Digestion-Nov-11 (accessed March 2010).

Remediation Technologies Screening Matrix and Reference Guide (2013). http://www.frtr.gov/matrix2/top_page.html (accessed June 2014).

Rudnik, E. (2008). *Compostable Polymer Materials*, 101. Oxford: Elsevier.

Title 40: Protection of the Environment (2013). Electronic Code of Federal Regulations, U.S. Government Printing Office. http://www.ecfr.gov/cgi-bin/text-idx?tpl=/ecfrbrowse/Title40/40tab_02.tpl (accessed March 2013).

9

Commodity Plastics

This chapter will discuss the plastics that are considered commodity plastics and the reasons for the classification. The plastics will include low density polyethylene (LDPE), linear low density polyethylene (LLDPE), high density polyethylene (HDPE), and ultra high molecular weight density polyethylene (UHMWPE), polypropylene (PP), thermoplastic polyolefin (TPO), polystyrene (PS), poly vinyl chloride (PVC), vinyl, poly methyl methacrylate (PMMA), polyethylene copolymers, and blends. The discussion will include the manufacturing process to make the plastic pellets, their mechanical and thermal properties, shrinkage, and many typical applications of these plastics.

Plastics are versatile, durable, and inexpensive to produce. Plastic production has increased dramatically in the last 10 years. Figure 9.1 shows worldwide plastic production from 1950 to 2015.

Olefins are unsaturated, aliphatic hydrocarbons made from ethylene gas. Many commodity plastics have a simple chemical structure that is very similar to that of polyethylene and are called olefins. Ethylene gas is produced by cracking higher hydrocarbons of natural gas or petroleum. Olefin means oil forming and it has been given to ethylene because oil was formed when ethylene was treated with Cl. It now applies to all hydrocarbons with linear C::C double bonds (not aromatic C::C double bonds). Polyethylene was discovered around 1900, though using an expensive process. LDPE was commercialized in 1939. HDPE was commercialized in 1957.

9.1 Definition of Commodity Plastics

Commodity thermoplastics are plastics that have lower mechanical and thermal properties than engineering plastics. They can be made in packaging products. They are cheaper per pound than engineering plastics. Conventional plastic

Sustainable Plastics: Environmental Assessments of Biobased, Biodegradable, and Recycled Plastics, Second Edition. Joseph P. Greene.

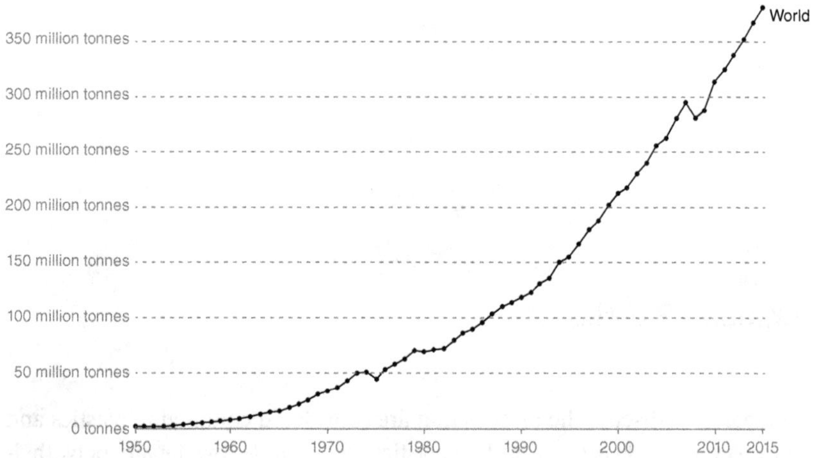

Figure 9.1 Worldwide plastic production from 1950 to 2015. Source: Plastic annual production (2020).

molding operations can be used to produce inexpensive parts. They cannot usually be painted, but paint adhesion promoters can be added.

9.2 Commodity Plastics

Polyolefins are the biggest group of commodity plastics. They can include the following:

- Polypropylene
- Polyethylene
- Polystyrene
- Polyvinyl chloride (PVC)
- TPO
- SAN
- PMMA
- Acrylics

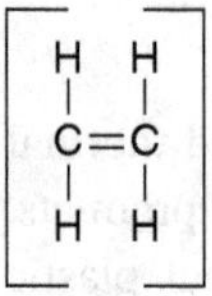

Figure 9.2 Chemical structure of ethylene monomer.

The chemical structures of polyolefins are listed in Figures 9.2–9.5. The chemical structure of ethylene gas involves two carbon atoms and four hydrogen atoms with an unsaturation chemical bond of C=C, double bonds. This monomer is polymerized to form polyethylene.

The unsaturated chemicals can be polymerized with several methods to produce polyethylene, polypropylene, poly butene, and poly 4-methyl pentene as shown in

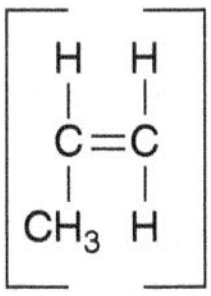

Figure 9.3 Chemical structure of propylene monomer

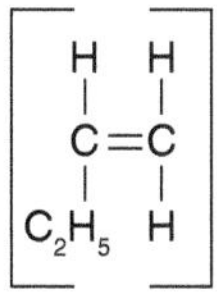

Figure 9.4 Chemical structure of butene monomer.

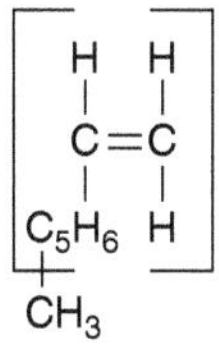

Figure 9.5 Chemical structure of 4-methyl pentene monomer.

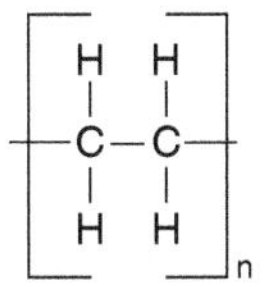

Figure 9.6 Chemical structure of polyethylene.

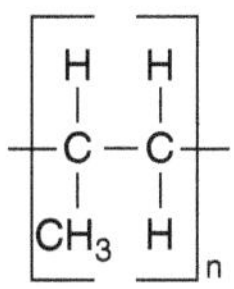

Figure 9.7 Chemical structure of polypropylene.

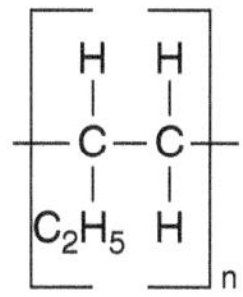

Figure 9.8 Chemical structure of poly isobutene.

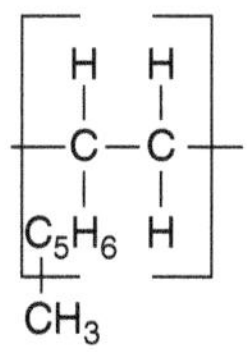

Figure 9.9 Chemical structure of poly methylpentene.

Figures 9.6–9.9. The chemical structure of polyethylene involves two carbon atoms and four hydrogen atoms that repeat over and over again in a polymer chain. The repeating monomer can be over 1000 units in some polyethylene plastics.

Polyethylene can be made by purifying a quantity of ethylene, known as the feedstock, before adding a catalyst (Polyethylene 2020). This will begin a reaction that causes the ethylene molecules to form the polymer polyethylene. The first step in the polymerization process is to purify the ethylene feedstock. During the manufacture of ethylene, this monomer can pick up many impurities, such as moisture, sulfur, and ammonia. Further impurities can occur during storage and transportation. Run the ethylene through a purity regulator. This device will mix several gases with the feedstock ethylene under high pressure and draw off the impurities and foreign matter. At the end of the purification process, pass the ethylene into the reaction tank.

The second step is to add a catalyst to the purified ethylene. A typical catalyst used in this process is benzyl peroxide. The molecules in benzyl peroxide have the ability to fall apart. A benzyl peroxide molecule breaks down into two pieces, each with an unpaired electron. This type of molecule fragment is called a free radical. The unpaired electron in a free radical will now search for electrons to pair with in the ethylene feedstock.

The third step is to allow the benzyl peroxide to react with the ethylene. Since the fragments of the catalyst molecules take electrons from the previously stable ethylene molecules, the latter now try to replace their missing electrons by taking electrons from other ethylene molecules and forming a bond with them. Each time this happens, an electron gap occurs and must be filled by bonding with yet another ethylene molecule. Provided there are few impurities in the feedstock, this process can go on for a long time.

Typically, an additional catalyst is added whenever the reaction begins to slow. The growing chains of molecules can find each other and join, ending the search for electrons to fill the gaps created by the initial reaction. More catalysts will restart the reaction.

The fourth step is to pour the polyethylene into a pelletizer. This machine will shape small amounts of polyethylene into pellets for storage and transportation. These tiny pellets can then be reheated and formed into whatever shape is needed.

Well over 75 million tonnes of poly(ethene), often known as polyethylene, is manufactured each year, making it an important plastic. This accounts for over 60% of the ethene manufactured each year (Polyethylene basics 2020). Poly(ethene) is made by several methods which use the addition polymerization of ethene, which is principally produced by the cracking of ethane and propane, naphtha, and gas oil.

There are several polyethylene types that are used today. The two simplest ones are low-density polyethylene and high-density polyethylene. They are produced using different processes that influence their chemical structure. Polyethylene is produced using the low-pressure process (Ziegler) or the high-pressure process (ICI). Polyethylene can be produced with linear (HDPE) or branched (LDPE) chains. Over 100 million tonnes of polyethylene was produced in 2019 (Ibid).

Low density has lower density is caused by a branching process of the polymer during polymerization. Figure 9.10 shows the configuration of the branched low-density polyethylene.

It can be represented by Figure 9.11:

There are about 20 branches per 1000 carbon atoms. The relative molecular mass and the branching influence the physical properties of LDPE. The branching affects the degree of crystallinity, which in turn affects the density of the material. LDPE is generally amorphous and transparent with about 50% crystallinity. The branches prevent the molecules from fitting closely together and so it has low density.

Branching can affect mechanical and physical properties in low-density polyethylene. For example, branching of the polymer causes the plastic to have lower:

- Density
- Crystallinity

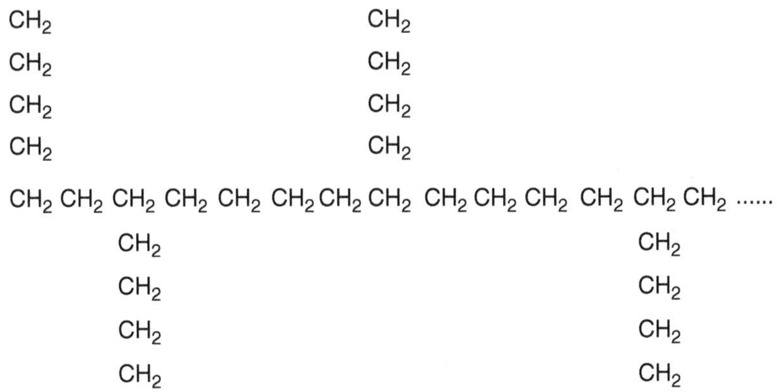

Figure 9.10 Branched low density polyethylene configuration.

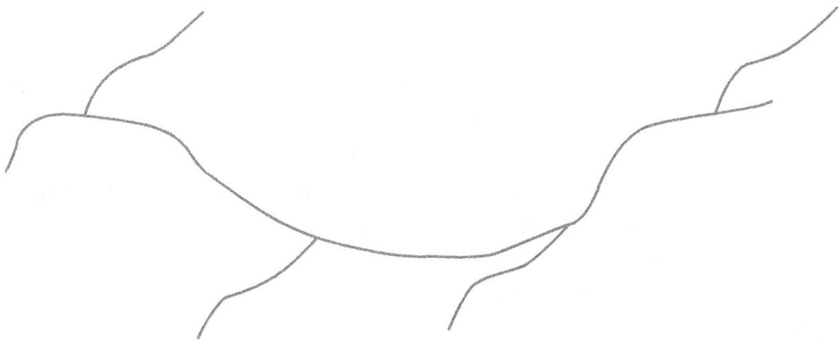

Figure 9.11 Branched LDPE schematic.

- Melting point
- Creep resistance
- Tensile strength
- Tensile modulus
- Stiffness
- Hardness
- Shrinkage

Branching can also increase impact toughness, elongation, transparency, and permeability of non-polar molecules, like oxygen.

Polyethylene can be produced in Brazil for the production that is made from sugar cane by way of bioethanol. This new bio-polyethylene will have the same mechanical properties as petroleum-based polyethylene. This bio-based polyethylene can be made with a lower carbon footprint (BioPE 2022).

Polyethylene and other polymers can be made with addition polymerization and the following methods:

- Bulk Polymerization
- Solution Polymerization
- Suspension Polymerization
- Emulsion Polymerization

9.2.1 Low-Density Poly(ethylene) (LDPE)

LDPE produced with a higher-pressure process using free radical polymerization with "Addition" polymerization, as shown in Figure 9.12.

Free radical polymerization is a method of polymerization by which a polymer forms by the successive addition of free-radical building blocks. Free radicals can be formed via a number of different mechanisms, usually involving separate initiator molecules. Following its generation, the initiating free radical adds monomer units, thereby growing the polymer chain.

The process is operated under very high pressure of 14 000–28 000 psi at moderate temperatures of 150–300 °C. This is a radical polymerization process and an initiator, such as a small amount of oxygen, and/or an organic peroxide, is used. Ethene gas with purity in excess of 99.9% is compressed and passed into a reactor together with the initiator. The molten polyethylene is removed, extruded, and cut into granules. Unreacted ethene is recycled. The average polymer molecule contains 4000–40 000 carbon atoms, with many short branches.

There are four types of polyethylene. Each has a different density.

- Type 1: (Branched) Low density of 0.910–0.925 g/cc
- Type 2: Medium density of 0.926–0.940 g/cc
- Type 3: High density of 0.941–0.959 g/cc
- Type 4: (Linear) High density to ultra-high density > 0.959 g/cc

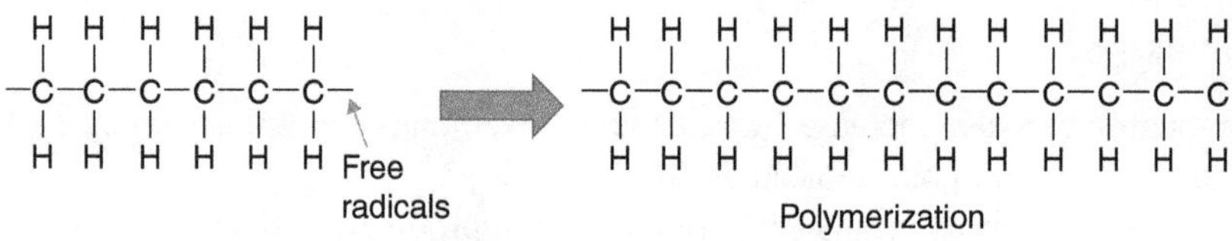

Figure 9.12 Low density polyethylene polymerization.

9.2.1.1 High-Density Poly(ethene) (HDPE)

HDPE is expected to produce 360 million metric tonnes in 2020 globally (Plastic-production 2022).

HDPE is produced using a low-pressure and -temperature process with Ziegler- and metallocene-based catalysts. HDPE has a density of greater than or equal to 0.941 g/cc. HDPE has a low degree of branching. The mostly linear molecules pack together well, so intermolecular forces are stronger than in highly branched polymers. HDPE can be produced using Cr/silica catalysts, and Ziegler- and metallocene-based catalysts. By choosing catalysts and reaction conditions, the small amount of branching that does occur can be controlled. They cause new ethylene monomers to add to the ends of the molecules, rather than along the middle, causing the growth of a linear chain. HDPE has high tensile strength. Products include packaging such as milk jugs, detergent bottles, butter tubs, garbage containers, and water pipes. One-third of all toys are manufactured from HDPE.

Two types of catalyst are used principally in the manufacture of HDPE:

- A Ziegler-Natta organometallic catalyst. This can be titanium compounds with an aluminum alkyl.
- An inorganic compound, known as a Phillips-type catalyst. A well-known example is chromium(VI) oxide on silica, which is prepared by roasting a chromium(III) compound at *ca* 1000 K in oxygen and then storing prior to use, under nitrogen.

HDPE is produced using three types of processes. All operate at relatively low pressures (10-80 atm) in the presence of a Ziegler-Natta or inorganic catalyst. Typical temperatures range between 350 K and 420 K. In all three processes, hydrogen is mixed with ethene to control the chain length of the polymer.

1) Slurry process (using either a CSTR [continuous stirred tank reactor] or a loop reactor). The Ziegler-Natta catalyst, as granules, is mixed with a liquid hydrocarbon (for example, 2-methylpropane [isobutane] or hexane), which simply acts as a diluent (Essential chemistry 2020).
   a) A mixture of hydrogen and ethene is passed under pressure into the slurry and ethene is polymerized to HDPE.
   b) The reaction takes place in a large loop reactor with the mixture constantly stirred (Figure 9.13).
   c) On opening a valve, the product is released and the solvent is evaporated to leave the polymer, still containing the catalyst.
   d) Water vapor, on flowing with nitrogen through the polymer, reacts with the catalytic sites, destroying their activity.
   e) The residue of the catalyst, titanium(IV) and aluminum oxides, remains mixed, in minute amounts, in the polymer (Figure 9.14).

Figure 9.13 The slurry process using a loop reactor (Ibid).

9.2.1.1.1 Solution Process

The second method involves passing ethene and hydrogen under pressure into a solution of the Ziegler-Natta catalyst in a hydrocarbon (a C_{10} or C_{12} alkane). The polymer is obtained in a way similar to the slurry method.

9.2.1.1.2 Gas Phase Process

A mixture of ethene and hydrogen is passed over a Phillips catalyst in a fixed-bed reactor (Figure 9.15).

Ethene polymerizes to form grains of HDPE, suspended in the flowing gas, which passes out of the reactor when the valve is released. Modern plants sometimes use two or more of the individual reactors in series (for example, two or more slurry reactors or two gas phase reactors), each of which is under slightly

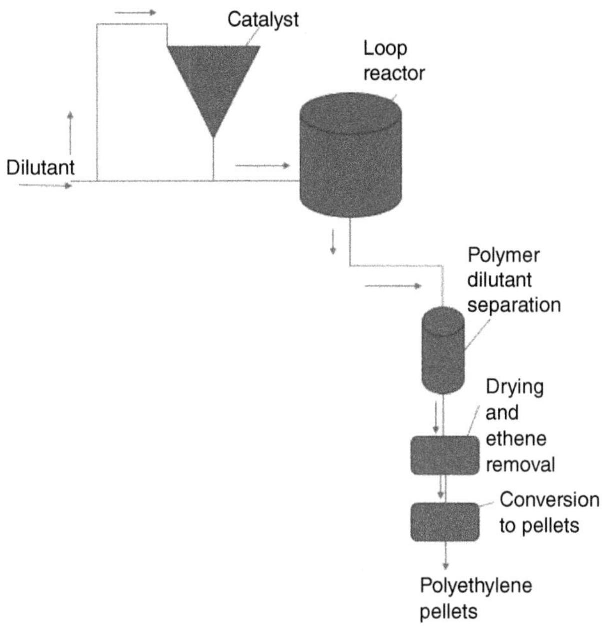

Figure 9.14 The manufacture of poly(ethene) using the slurry process in a loop reactor (Ibid).

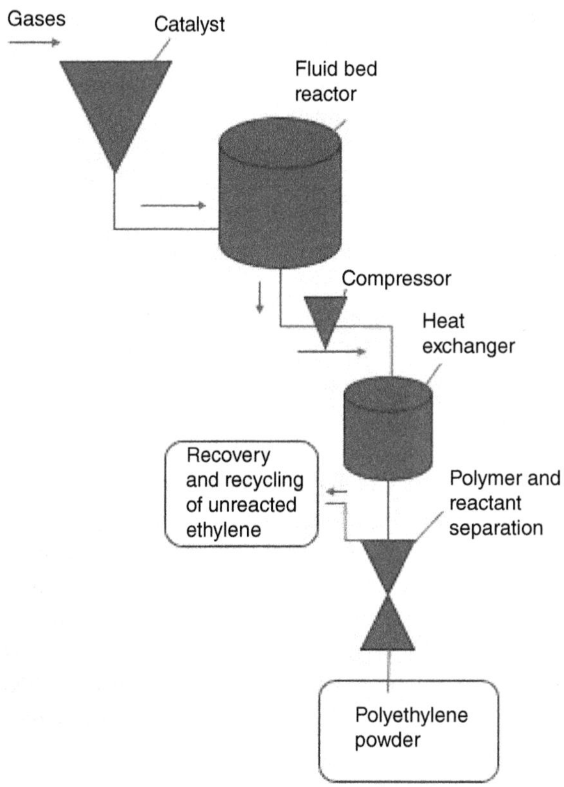

Figure 9.15 Low-pressure gas-phase process for HDPE (Ibid).

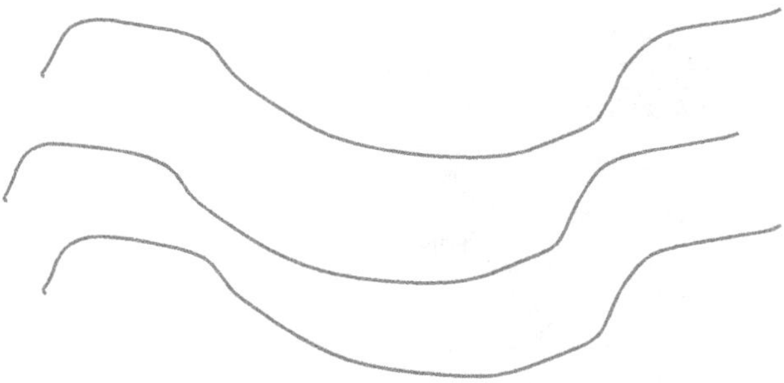

Figure 9.16 HDPE plastic schematic.

different conditions, so that the properties of different products from the reactors are present in the resulting polymer mixture, leading to a broad or bimodal molecular mass distribution. This provides improved mechanical properties such as stiffness and toughness (Ibid).

The HDPE powder coming out of any of the reactors discussed above is separated from the diluent or solvent (if used) and is extruded and cut up into granules. This method gives linear polymer chains with few branches. The poly(ethene) molecules can fit closer together. The polymer chains can be represented as shown in Figure 9.16.

This leads to strong intermolecular bonds, making the material stronger, denser, and more rigid than LDPE. The polymer is not transparent.

9.2.2 Linear Low-Density Poly(ethene) (LLDPE)

Low-density polyethylene has many uses but the high-pressure method of manufacture by which it is produced has high capital costs. However, an elegant technique has been developed based on both Ziegler-Natta and inorganic catalysts to produce linear low-density poly(ethene) (LLDPE), which has better properties than LDPE. Any of the three processes, slurry, solution, and gas phase, can be used when a Ziegler-Natta catalyst is chosen. The gas phase process is used when an inorganic catalyst is employed.

Linear low-density polyethylene (LLDPE) can have densities between 0.916 and 0.930 g/cc. It contains little, if any, branching. The properties include good flex life, low warpage, and improved stress-crack resistance. The applications include films for ice, trash, garment, and produce bags. It is made with Ziegler- and metallocene-based catalysts and low pressure, similar to the HDPE process. Linear low-density polyethylene (LLDPE) produces longer branches. LLDPE has 16 to 25 branches per 1000 backbone carbons versus one to two branches per 1000 backbone carbons in HDPE. LLDPE is made with olefin copolymer, which makes

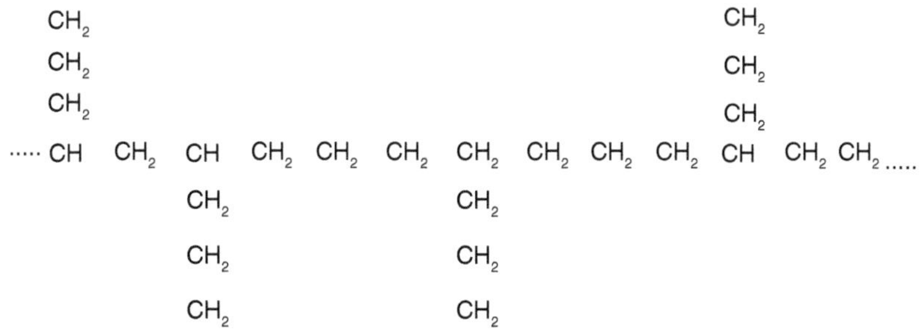

Figure 9.17 Branched LLDPE plastic schematic.

Figure 9.18 LLDPE plastic chains.

up the side chains that are 2-, 4-, 6-, 8--carbon long depending on the comonomer used (butane, hexane, or octane). LLDPE is referred to as a homopolymer due to the chemical similarity of copolymer.

Small amounts of a co-monomer such as but-1-ene or hex-1-ene are added to the feedstock. The monomers are randomly polymerized and there are small branches made up of a few carbon atoms along the linear chains (Figures 9.17 and 9.18).

For example, with but-1-ene, $CH_3CH_2CH{=}CH_2$, the structure of the polymer is:

The side chains are known as pendant groups or short chain branching. The molecule can be represented as:

The structure is essentially linear but because of the short chain branching, it has low density. The structure gives the material much better resilience, tear strength, and flexibility without the use of additives and plasticizers. This makes linear low-density poly(ethene) an ideal material for the manufacture of film products, such as those used in wrappings. LLDPE is used in many plastic bags that are made from recycled polyethylene and used for reusable plastic bags. The properties of the polymer, and hence its uses, can be changed by changing the proportion of ethene and co-monomer, and by using different co-monomers.

LLDPE is cheaper to produce than LDPE. It is also produced with low pressure of 1 psi (6 KPa), low temperature of 212 °F (100 °C), lower construction costs, and lower capital, and are easy to maintain and have minimal environmental impacts.

Table 9.1 Properties of polyethylene products.

	LDPE	MDPE	HDPE	Linear HDPE
Density	0.91–0.925	0.926–094	0.941–0.95	0.959–0.965
Cryallinity	30–50%	50–70%	70–80%	80–91%
Molecular weight	10K–30K	30K–50K	50K–250K	250K–1.5M
Tensile strength, psi	600–2300	1200–3000	3100–5500	5000–6000
Tensile modulus, psi	25K–41K	35K–75K	150K–158K	150K–158K
Tensile elongation	100–650%	100–965%	10–1300%	10–1300%
Impact strength, ft-lbs/in	5–no break	1–no break	0.4–4.0	0.4–4.0
Hardness	D44–D50	D50–D60	D60–D70	D66–D73

Source: PE properties (2020).

The molecular weight of polyethylene is related to crystallinity and density. The soft materials have lower molecular weight and have lower crystallinity.

Some mechanical properties of polyethylene products is provided in Table 9.1.

9.2.3 Metallocene Linear Low-Density Poly(ethene) (mLLDPE)

This polyethylene, known as mLLDPE, is produced by a new family of catalysts, the metallocenes. Another name for this family is *single site catalyst*. The benefit is that the mLLDPE is much more homogenous in terms of molecular structure than classical LLDPE produced by Ziegler-Natta catalysts. The structure has a transition metal (often zirconium or titanium) "filling" a hole between layers of organic compounds.

The catalysts are even more specific than the original Ziegler-Natta and it is possible to control the polymer's molecular mass as well as its configuration. Either the slurry or the solution processes are usually used (Figure 9.19).

Polyethylene produced using a metallocene can be used as a very thin film which has excellent optical properties and sealing performance, thus making it very effective for wrapping food. Mechanical properties are a real plus for the metallocene catalysts to make mLLDPE. This can be used to make a stronger polyethylene blown film.

9.2.3.1 Ultra-High Molecular Weight Polyethylene (UHMWPE)

UHMWPE is more expensive that LDPE or HDPE. UHMWPE densities are 0.96 g/cc or greater. The molecular weight ranges from 3 million to 6 million. UHMWPE properties include improved high wear resistance, chemical inertness, and low coefficient of friction. The yield strength is 2.4 GPa (350 000 psi). Strength/density is greater than steel and aramid fibers. The high viscosities result in the

Figure 9.19 Polyethylene film is used extensively for wrapping foods. Source: Plastic bag (2020).

material not flowing or melting. It is usually machined with CNC or pressed into shape. It is processed similar to PTFE (Teflon). Ram extrusion and compression molding, and not injection molding, are used. Applications for UHMWPE include pump parts, seals, surgical implants, pen tips, and butcher-block cutting surfaces.

9.2.3.2 Cross-Linkable Polyethylene (XLPE)

XLPE is crosslinked by electron irradiation or chemical means. The high energy breaks the C–H and C–C bonds creating free radicals. It can then react with other nearby C–H or C–C free radicals causing crosslinking of the polyethylene. LDPE is the most common plastic used with this method. The chemical processing uses peroxide and a bridge molecule as a crosslinker that forms free radicals. This is effective with LDPE and HDPE. Cross-linkable polyethylene cannot be melted due to crosslinking. It forms a thermoset plastic that is part of the thermoplastic. It can be molded into rotational molded into hollow parts. The applications include rotational molded tanks, hot water pipes, and power cable insulation.

9.2.3.3 Copolymers of Polyethylene

A common copolymer of polyethylene is ethylene vinyl acetate. Ethylene-Vinyl Acetate (EVA), shown in Figure 9.20.

The chemistry of EPDM is shown in Figure 9.21.

Repeating groups in the EPDM polymer are ethylene with an acetate functional or vinyl. That part of the vinyl pendent group are highly polar. Vinyl acetate

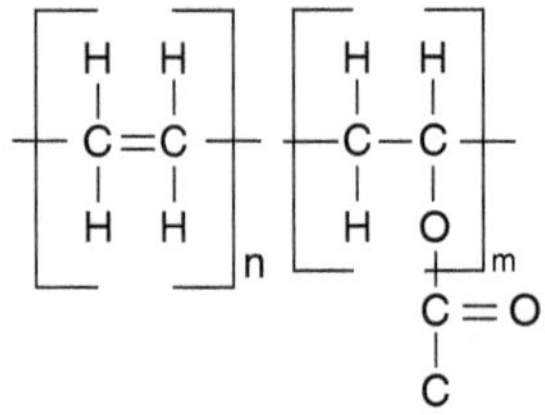

Figure 9.20 Copolymers of polyethylene, EVA.

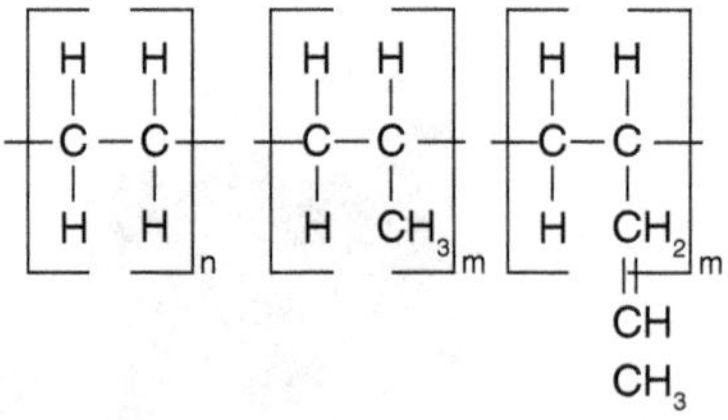

Figure 9.21 Copolymers of polyethylene, EPDM.

reduces crystallinity and increases chemical reactivity because of high regions of polarity. This results in a flexible polymer that bonds well to other materials. Polyvinyl acetate (PVA) is the main ingredient in "Elmers Glue" and makes an excellent adhesive to wood. Other applications include flexible packaging, shrink wrap, auto bumper pads, flexible toys, and tubing.

Another copolymer of ethylene is ethylene-propylene (EPM). Ethylene and propylene are copolymerized in a random manner and it causes a delay in the crystallization. Thus, the copolymer is rubbery at room temperature because the Tg is between HDPE (−110C) and PP (−20C). Ethylene and propylene can be copolymerized with small amounts of a monomer containing 2 C=C double bonds (dienes). This results in a ter polymer, EPDM, or thermoplastic rubber, TPR. If polypropylene is added, then the copolymer is called Thermoplastic Olefin (TPO). The copolymers of ethylene and propylene (EPR) or ter polymer of ethylene-propylene diene (EPDM) can be blended with PP in ratios that determine the stiffness of the elastomer. An 80/20 EPDM/PP ratio gives a soft elastomer (TPO).

9.2.4 Polypropylene (PP)

Polypropylene is a very popular plastic for many applications. Much of it has to do with its low density and good tensile strength. Polypropylene (PP) is a tough, rigid plastic produced in a variety of molecular weights and crystallinities (Polypropylene 2020).

Over 35 million metric tonnes of polypropylene was produced in 2020 (Plastic-production 2020).

Polypropylene is made from the polymerization of propylene gas in the presence of a catalyst system, usually a Ziegler-Natta or metallocene catalyst. Polymerization conditions (temperature, pressure, and reactant concentrations) are set by the polymer grade to be produced.

Various production processes exist with some general similarities. They take place either in a gas-phase (fluidized bed or stirred reactor) or a liquid-phase process (slurry or solution). An example of flow diagram corresponding to each of the

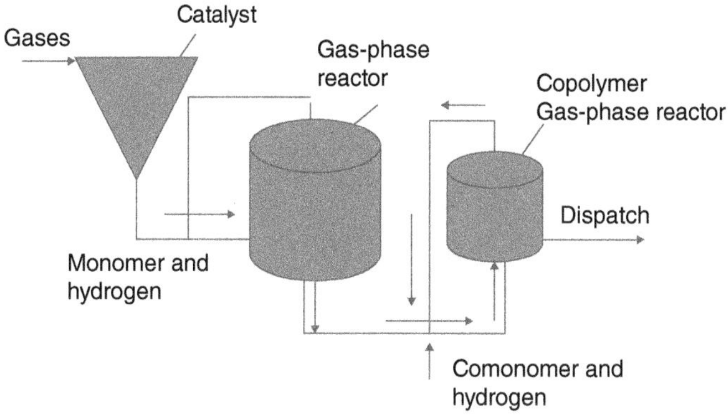

Figure 9.22 Polypropylene reactor.

two types of processes is illustrated in Figure 9.22. The gas-phase polymerization is economical and flexible and can accommodate a large variety of catalysts. It is the most common technology in modern polypropylene production plants.

The powder is conveyed to powder silos and then converted into pellets with a twin screw that includes additives, like thermal stabilizers, processing aids, ultraviolet (UV) stabilizers, antioxidants, impact modifiers, and others.

Bio-based PP is being produced by Braskem in Brazil from sugar cane (Bio-based PP 2022). Bio-based PP will have a lower carbon footprint as the petroleum-based PP with very similar mechanical properties. It will not be biodegradable but can be current recycled plastics. Some mechanical properties of polypropylene products are provided in Table 9.2.

Table 9.2 Properties of polypropylene.

	PP	LDPE	HDPE
Density	0.9	0.91–0.925	0.941–0.95
Cryallinity	30–50%	30–50%	70–80%
Molecular weight	200K–600K	10K–30K	50K–250K
Tensile strength, psi	4500–5500	600–2300	3100–5500
Tensile modulus, psi	165K–225K	25K–41K	150K–158K
Tensile elongation	100–600%	100–650%	10–1300%
Impact strength, ft-lbs/in	0.4–1.2	5–no break	0.4–4.0
Hardness	R80–R102	D44–D50	D60–D70

Source: PP properties (2020).

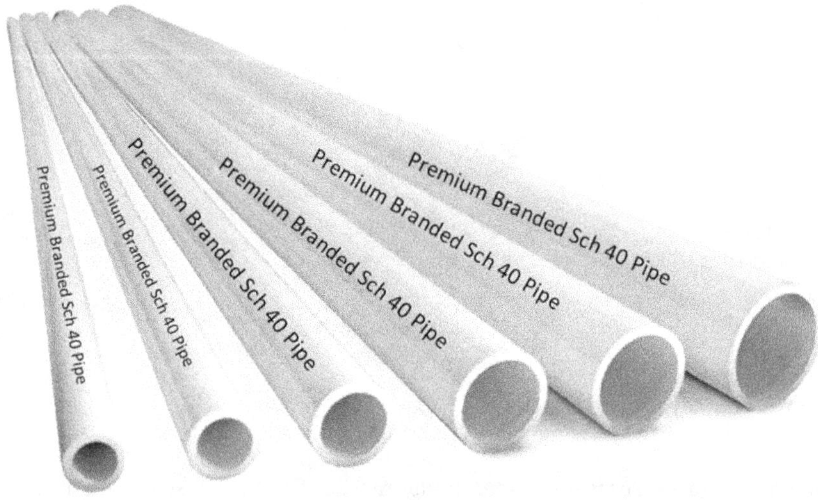

Figure 9.23 PVC pipe. Source: PVC pipe (2020).

9.2.4.1 Polyvinyl Chloride (PVC)

Polyvinyls were invented in 1835 by German chemist Eugen Baumann, when he discovered that a white residue could be synthesized from ethylene dichloride in an alcohol solution (Sunlight was catalyst) (PVC Wikipedia 2020). PVC was patented in 1933 by BF Goodrich Company in a process that combined a plasticizer, tritolyl phosphate, with PVC compounds making it easily moldable and processed.

PVC is the world's third-most widely produced plastic after polyethylene and polypropylene. About 40 million tonnes of PVC is produced each year worldwide (Ibid). PVC comes in two basic forms: rigid, as used in PVC white pipes, and flexible, as used in latex gloves. The rigid form of PVC is used in the construction of pipe and in profile applications such as doors and windows. Figure 9.23 shows an example of PVC pipe. It is also used in making bottles, packaging, and rigid cards. It can be made softer and more flexible with the addition of plasticizers, usually phthalates. It can be used to make flexible coverings and tents.

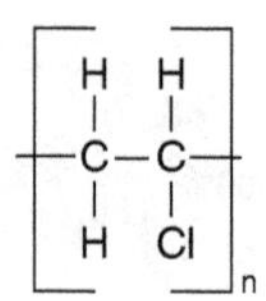

Figure 9.24 Chemical structure of PVC.

Polyvinyl chloride is produced by polymerization of the vinyl chloride monomer. The chemistry of PVC is similar to that of polyethylene but the Cl atom is substituted for the hydrogen atom as shown in Figure 9.24.

PVC can be made with suspension polymerization, emulsion polymerization, or bulk polymerization accounts. Suspension polymerization affords particles with average diameters of 100–180 μm, whereas

Table 9.3 Properties of PVC products.

	PVC (rigid)	PVC (flexible)	PVB	PVDC
Density	1.3–1.58	1.16–1.35	1.05	1.65–1.72
Tensile strength, psi	6000–7500	1500–3500	500–3500	3500–5000
Tensile modulus, psi	350K–600 K	N/A	N/A	160K–240K
Tensile elongation	40–80%	200–400%	150–450%	160–240%
Impact strength, ft-lbs/in	0.4–22	1–50	1–50	0.4–1
Hardness	Shore D 65–85	Shore D 50–100	N/A	M 60–65
CLTE, 10-6 in/in/F	23–50	35–75	N/A	85
HDT @ 264 psi	140F–170F	N/A	N/A	130F–150F

Source: PVC properties (2020).

emulsion polymerization gives much smaller particles of average size around 0.2 μm. VCM and water are introduced into the reactor along with a polymerization initiator and other additives. The contents of the reaction vessel are pressurized and continually mixed to maintain the suspension and ensure a uniform particle size of the PVC resin. The reaction is exothermic and thus requires cooling. All PVC is produced by addition polymerization from the vinyl chloride monomer in a head-to-tail alignment. PVC is amorphous with partially crystalline (syndiotactic) because structural irregularity increases with the reaction temperature. PVC (rigid) decomposes at 100 °C, leading to the release of dangerous HCl gas.

Some mechanical properties of PVC products are provided in Table 9.3.

9.2.4.2 PVC Plasticizers

Plasticizers can be added to PVC to make a flexible product. Plasticizers are chemicals that have strong solvent effects on plastics in low to moderate concentrations. The plasticizers causes the polymer to swell, which makes the plastic more flexible. Ideal plasticizers would have the following properties:

- Low cost
- Nonvolatile and stays in the plastic
- Low flammability of plastic
- Nontoxic
- Not harmful to the environment
- Not water extractable
- Stable through processing and use
- Low migration to the surface
- Stable to UV light

Plasticizers are mixed using the calendaring process. Phthalates are an ester type. Plasticizers for plastics are additives and can be called Phthalate ester.

Most of the plasticizers are used with PVC, giving this material improved flexibility and durability. The majority are used in films and cables. For plastics such as PVC, the more the plasticizer added, the lower their cold flex temperature will be. Plastic items containing plasticizers exhibit improved flexibility and durability. Plasticizers can become available for exposure due to migration and abrasion of the plastic since they are not bound to the polymer matrix. The "new car smell" is often attributed to plasticizers or their degradation products.

Plasticizers make it possible to achieve improved compound-processing characteristics while also providing flexibility in the end-use product. Ester plasticizers are selected based upon cost–performance evaluation. The rubber compounder must evaluate ester plasticizers for compatibility, processing, permanence, and other performance properties. The wide variety of ester chemistries that are in production include sebacates, adipates, terephthalates, dibenzoates, gluterates, phthalates, azelates, and other specialty blends. This broad product line provides an array of performance benefits required for the many elastomer applications such as tubing and hose products, flooring, wall-coverings, seals and gaskets, belts, wire and cable, and print rolls.

In Western Europe, about one million tonnes of phthalates are produced each year, of which approximately 900 000 tonnes are used to plasticize PVC. The most common are: diisononyl phthalate (DINP), diisodecyl phthalate (DIDP), and di-2-ethylhexyl phthalate (DEHP, sometimes also referred to as DOP), which have the well-balanced properties described above. These plasticizers account for about 75% of all plasticizers used for PVC. Apart from phthalates, several other kinds of plasticizers are used to meet specific requirements, including adipates for low temperature resistance and trimellitates for heat resistance. Phthalates can be restricted to reduce emissions. The European Union prohibited the use of four phthalates in consumer products. The four phthalates are butylbenzyl phthalate (BBP), di(2-ethylhexyl) phthalate (DEHP), dibutyl phthalate (DBP), and diisobutyl phthalate (DIBP). Restrictions or bans have been placed on phthalates in PVC toys in the entire European Union (Ban 2020).

Phthalates were banned in toys in the United States in 2008 (US ban phthalates 2020). Similar bans have been enacted by the states of California, Washington, and Vermont. In 1998, the U.S. Consumer Product Safety Commission (CPSC) reached a voluntary agreement with manufacturers to remove two phthalates from PVC rattles, teethers, pacifiers, and baby bottle nipples. Sweden first proposed restrictions on PVC use in 1995 and is working toward discontinuing all PVC uses. In Spain, over 60 cities have been declared PVC-free. Germany has banned the disposal of PVC in landfills as of 2005, is minimizing the incineration of PVC, and is encouraging the phasing out of PVC products that cannot easily be recycled. Since 1986, at least 274 communities in Germany have enacted restrictions against PVC. Research suggests that there is an association between PVC flooring and the development of bronchial obstruction in children.

Also, increased incidence of asthma could be partially due to the increasing household use of plastics containing phthalates. Butlybenzylphthalate (BBP) and Dibutylphthalate (DBP) are estrogenic to breast cancer cell lines. BBP can also act as an antiandrogen, blocking the action of dihydrotestosterone in a yeast screen containing the androgen receptor. DBP can damage the reproductive system of male rats at low exposures, Action disrupts the androgen system, not imitiating the estrogen. DBP can be an endocrine disrupter. DEHP is toxic to the developing fetus per large studies involving hundreds of animals, to smaller ones with few animals, as well as cell culture studies, and case reports in humans.

9.2.4.3 Polystyrene (PS)

Polystyrene was discovered in 1839 by Eduard Simon, an apothecary from Berlin (Styrene Wikipedia 2022). Dr. Simon distilled an oily substance, a monomer that he named styrol, now called styrene. Several days later, Simon found that the styrol had thickened into a jelly he dubbed styrol oxide ("Styroloxyd") because he presumed an oxidation. They called the product "metastyrol"; analysis showed that it was chemically identical to Simon's Styroloxyd (Blyth and Hofmann (1845)).

This eventually led to the substance receiving its present name, polystyrene. PS was produced in 1851 by French chemist M. Berthelot by passing benzene and ethylene through a red-hot-tube. A similar process is used today to produce PS.

PS is an amorphous polymer made from addition polymerization of styrene. It is a homopolymer of which 25 million metric tonnes was produced in 2020 (Styrene-volume 2020).

PS is clear and colorless with excellent optical properties and high stiffness. It is brittle until biaxially oriented when it becomes flexible and durable. Polystyrene (PS) is a graft copolymer which means that it is the polymer has one or more blocks of homopolymer are grafted as branches on the main chain. PS can be blended with elastomers, like butadiene, to create a more ductile PS. The properties are dependent upon the elastomer %, but are grouped into:

- medium impact (Izod < 1.5 ft-lb),
- high impact (Izod between 1.5 and 2.4 ft-lb), and
- super-high impact (Izod between 2.6 and 5 ft-lb)

The chemistry of PS is similar to that of polyethylene with a substitution of a benzene ring for the hydrogen. The chemical structure is shown in Figure 9.25.

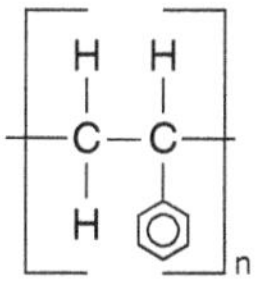

Figure 9.25 Chemical structure of polystyrene.

PS has number "6" as the recycling code from SPI. Less than 1% of the PS is recycled in the US. This can be discussed more in a future chapter. PS is made with addition polymerization. Initiation of the polymer is done with free radicals. The mechanism is chain propagation, where each monomer unit adds to the growing chain in a manner that generates the most stable radicals. PS forms head-to-tail regioselectivity. The molecular arrangement is atactic with a random benzene group. This helps create low crystallinity. The different forms of PS are shown in Figure 9.26.

Isotactic PS and syndiotactic PS are not commercially available. Only atactic is available. PS degrades in chlorinated solvents and swells. This can help with bonding. PS is subject to environmental stress cracking, where cracks appear in the plastic part. PS is good for packaging since it can be easily foamed. PS is resistant to water absorption due to its non-polar nature.

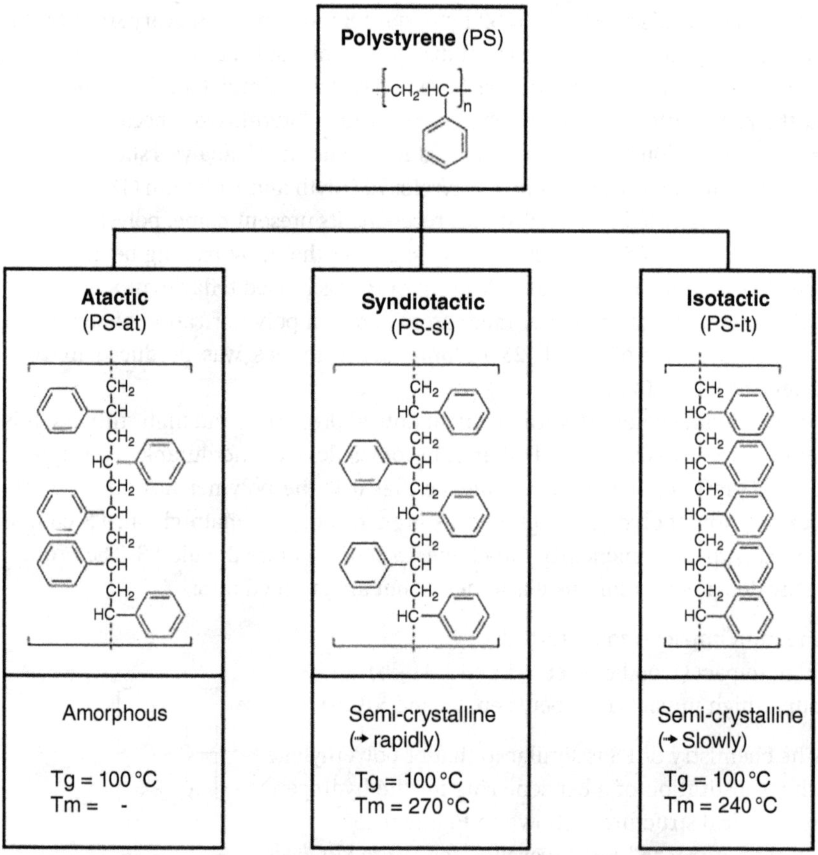

Figure 9.26 Forms of polystyrene chemical arrangement. Source: Styrene-atactic (2020).

PS has one large benzene pendant group that is much larger than the C and H atoms. The large benzene ring creates steric effects that

- reduce the flexibility of the polymer chain
- prevent crystallization (PS is ~100% amorphous)
- provide high tensile strength
- provide higher melting temperature
- inhibit disentanglements and sliding polymer chain
- result in low impact strength and brittleness

Amorphous PS allows light to pass through it without significant refraction. Crystal PS refers to its glitter and sparkle and not amorphous nature. The large benzene ring affects its chemical resistance as well. The aromatic portion of molecule is benzene. PS has poor resistance to aromatic and chlorinated solvents, for example, HCL, methylene chloride, toluene, and ethylbenzene. PS can dissolve in these solvents. Solvent adhesives can cause PS to swell and improves bonding.

PS burns readily with a yellow flame and dark sooty smoke. It is similar dark smoke for plastics with benzene rings. Flame retardants are added to reduce smoke generation. PS clarity is hurt by exposure to oxygen and UV light. PS can turn yellow and become brittle in the outdoors. PS is not recommended for outdoor applications

The mechanical properties are listed in Table 9.4.

The high tensile strength is due to high molecular weight. PS has a high HDT softening temperature of 100 °C. The decomposition temperature is 250 °C.

Polystyrene can be made into foam structure. The process includes the following:

- Ethylene is added to benzene with $AlCl_3$ catalyst to form ethyl benzene, C_8H_8.
- Ethyl benzene, C_8H_8, is dehydrated at 650 °C to form the styrene monomer.

Table 9.4 Mechanical properties of polystyrene.

Property	PS	LDPE	HDPE	PP
Density, g/cc	1.04	0.91–0.925	0.959–0.965	0.9
Tensile strength, psi	5000	1400	5500	5000
Tensile Modulus, psi	80 000	30 000	155 000	200 000
Tensile elongation %	1	500	300	200
Impact strength, ft-lbs/in	0.5	10	2	1
Hardness, rockwell	R53–R120	D44–D50	D60–D70	R80–R102
CLTE $\times 10^{-6}$ mm/mm/°C	70	150	80	55
HDT °F at 264 psi	85	45	75	125

Source: HDT CLTE (2020) and PS Props (2020).

- Through suspension polymerization (addition) at 100 °C, free radicals are formed and then head-to-tail polymerization occurs to form polystyrene.

Expanded polystyrene (EPS) is made by:

- Heating PS with steam or hot air in expansion tank.
- Aging the EPS beads for 24 hours with cool air and getting 5X expansion.
- Molding in tool
  - Add EPS beads and low-pressure steam to expand them.
  - Cool the molds (20X expansion)

This forms a closed-cell foam. EPS has a closed cell like all hard foams. Expanded polystyrene (EPS) is a rigid and tough, closed-cell foam with a normal density range of 11–32 kg/m^3. (EPS foam. *Expanded Polystyrene (EPS) Technical Data (PDF). Australia: Australian Urethane & Styrene. 2010)* EPS is usually white and made of pre-expanded polystyrene beads.

Figure 9.27 shows closed- and open-cell foams.

PS and PP bead foams are closed cell. Polyurethanes can produce open-cell foams. The problem with open-cell foams is that the thermoset resin can flow into the foam structure during thermoset molding of polymer composites. The closed-cell foam will not have resin penetration. With open-cell foams, the tiny cells of the foam are not completely closed. They are broken and air fills the "open" space inside the material. This makes the foam weaker or softer than closed-cell foam.

EPS is used mostly for foamed parts, for example, drinking cups, fast food containers, plates, wall insulation, packaging pellets, bumper absorbers, and so on. Expanded PS (EPS) is very popular for cups and insulation foam. EPS is made

(a) (b)

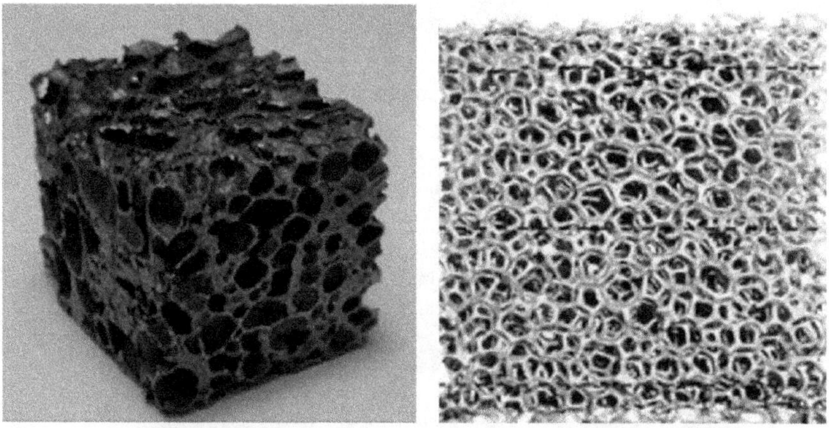

Figure 9.27 Open (b) and closed (a) cell foam structures. Source: Foam (2020).

with blowing agents, such as pentane and isopentane. The properties are dependent upon cell size and cell-size distribution.

The advantages are low thermal conductivity, which leads to good insulation properties, high buoyancy for marine applications, good energy absorption for many bumper absorbers, a high stiffness-to-weight ratio, and low cost per volume. The biggest disadvantage is low impact strength.

9.2.4.4 Blends and Alloys

Styrene is very commonly used in producing blends and alloys. A blend is a mixture of two polymers that results in the polymer having two or more phases, one for each material. Mechanical and physical properties are averages of the individual properties. PS alloy can be formed by combining the two polymers after they are formed, or of a polymer and a monomer into a single phase. The resulting polymer is a homogeneous polymer material. Chemical attraction between the polymers is usually required to form an alloy. An alloy is formed when two different materials combine to form a new polymer. Mechanical and physical properties are unique in the new PS-based polymer and has unique individual properties. Copolymerization is the combination of two or more types of monomers such that the new polymer has some of each monomer. Mechanical and physical properties are unique in the new polymer to the individual properties.

The advantages of alloying or blending is that it is less expensive than copolymerization, requires less time and investment, and can obtain a variety of properties in the new polymer.

The disadvantages are that the properties are lower for alloys and blends versus copolymerization. It is very common for PP, PS, HDPE, and PVC alloys and blends. If polymers are miscible, then alloys usually form. Otherwise a blend is formed. Compatibility agents are added to improve properties of immiscible polymers. Physical and mechanical properties of alloys can be 10X more than blends.

A very common alloy for many applications is high impact polystyrene (HIPS). Several elastomers can be added to PS to improve impact strength. Most common rubbers can be made with 30% SBR-Styrene butadiene rubber or Butadiene rubber. Figure 9.28 shows the chemical structure of SBR rubber.

The process to produce HIKPS is to dissolve the rubber polymer in styrene monomer and then polymerize PS by regular means. The properties show a 7X improvement in impact strength. There is, however, a drop in tensile strength, hardness, and HDT due to the presence of the elastomer.

9.2.4.5 Copolymers

Copolymers include SAN (poly styrene-acrylonitrile), SMA (styrene maleic anhydride), SBS (styrene butadiene styrene), and acrylic copolymers. Figure 9.29 shows the chemical structure of SBR plastic.

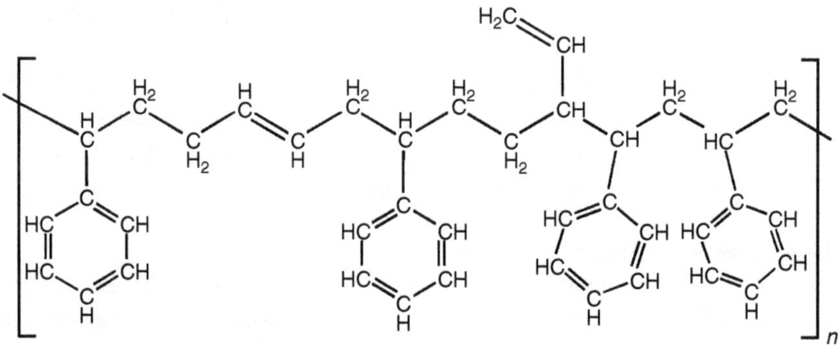

Figure 9.28 Styrene-butadiene rubber chemical structure. Source: Rubber chemical structure (2020).

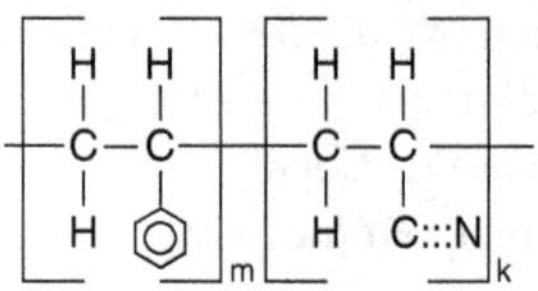

Figure 9.29 Alternating (ABABABAB) copolymer arrangement of SMA. Source: SMA structure (2020).

SAN is similar in use to polystyrene. Like polystyrene, it is optically transparent and brittle in mechanical behavior. The copolymer has a Tg greater than 100 °C owing to the acrylonitrile units in the chain, thus making the material resistant to boiling water. Uses include food containers, kitchenware, computer products, packaging material, battery cases, and plastic optical fibers. It is structurally related to ABS plastic, where polybutadiane is copolymerized with SAN to give a much tougher material. The rubber chains form separate phases which are 10–20 micrometers in diameter. When the product is stressed, crazing from the particles helps to increase the strength of the polymer. The method of rubber toughening has been used to strengthen other polymers such as PMMA and nylon.

Styrene acrylonitrile (SAN) resin is a copolymer plastic made from styrene and acrylonitrile. SAN has higher thermal resistance than PS. The chains of the polymer show alternating repeat units of styrene and acrylonitrile. The relative composition is typically between 70 and 80% by weight styrene and 20–30% acrylonitrile. The higher acrylonitrile content improves mechanical properties and chemical resistance, but also adds a yellow tint to the normally transparent plastic.

Styrene maleic anhydride (SMA) is a synthetic polymer that is produced from styrene and maleic anhydride monomers. The monomers can be almost perfectly alternating, making it an alternating copolymer, but (random) co-polymerization with less than 50% maleic anhydride content is also possible. The polymer is formed by radical polymerization, which a polymer forms by the successive addition of free-radical building blocks using an organic peroxide as the initiator. The main characteristics of SMA copolymer are its transparent appearance, high heat

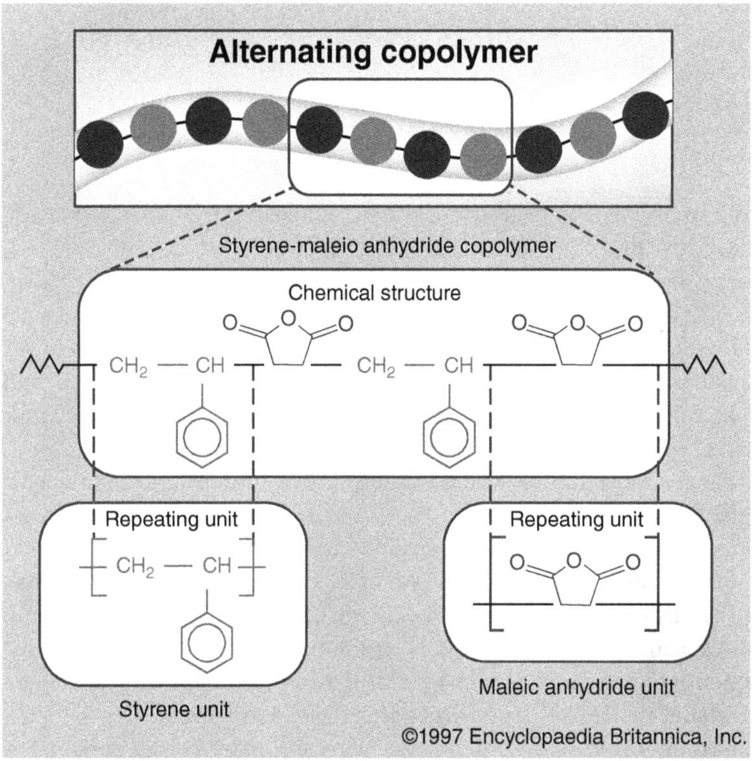

Figure 9.30 Chemical structure of SAN plastic.

resistance, high dimensional stability, and the specific reactivity of the anhydride groups. The latter feature results in the solubility of SMA in alkaline (water-based) solutions and dispersions. SMA has high molecular weight.

SMA has an alternating (ABABABAB) copolymer arrangement of styrene-maleic anhydride copolymer. This is shown in Figure 9.30.

Each colored ball in the molecular structure diagram represents a styrene or maleic-anhydride repeating unit as shown in the chemical structure formula. Styrene is a clear liquid obtained by the dehydrogenation of ethyl benzene. Maleic anhydride is a white solid obtained by the oxidation of benzene or butane. SMA is a rigid, heat-resistant, and chemical-resistant thermoplastic that is used in automobile parts, small appliances, and food-service trays. The molecular weights range from 5000 to 250 000 g/mol. SMA has a high Tg of 177 °C.

9.2.4.6 Acrylics

Poly(methyl methacrylate) (PMMA), also known as acrylic or plexiglass. PMMA can be transparent plastic often used in sheet form as a lightweight or

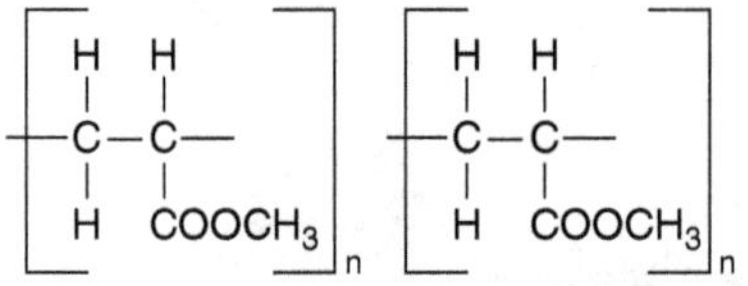

Figure 9.31 Chemical structure of polymethacrylate and polymethyl methacrylate (PMMA).

shatter-resistant plastic. PMMA can also be used as a casting resin (Acrylics 2022). The chemical structure of acrylics is shown in Figure 9.31.

PMMA was developed in 1928 in several different laboratories by many chemists and was first brought to market in 1933 in Germany. The reaction between methacrylic acid and methanol results in the ester methyl methacrylate PMMA was used extensively in World War II as for submarine periscopes and aircraft windshields. Today acrylics can be used in homes and in recreational vehicles for interior parts.

PMMA was created in the 1800s and was formulated in 1865. PMMA is an economical alternative to polycarbonate, PC. It has higher mechanical properties than PC, including tensile modulus, flexural strength, and clarity. It also has higher UV tolerance but lower impact strength, chemical resistance, and heat resistance than PC. Additionally, PMMA does not contain the potentially harmful bisphenol A chemical that is found in polycarbonate. It, also, is easier to mold with injection molding and has a lower cost than PC. PMMA can have higher scratch resistance than PC. The properties are provided in Table 9.5.

PMMA has low moisture absorption and has a shrinkage rate of about 0.5%. Acrylics, or PMMA, are inexpensive and cost $1.30 per pound (Plastics prices 2020). Acrylics includes acrylic and methacrylic esters, acids, and derivatives. PMMA

Table 9.5 Mechanical properties of PMMA, PC, and ABS.

	Acrylic	PC	ABS	PC/ABS
Density, g/cc	1.16–1.19	1.2	1.16–1.21	1.07–1.15
Tensile strength, psi	5000–9000	9500	3300–8000	5800–9300
Tensile modulus, psi	200K–500K	350K	320K–400K	350K–450K
Tensile elongation, %	20–70%	110%	1.5–25%	50–60%
Impact strength ft-lb/in	0.65–2.5	16	1.4–12	6.4–11
Hardness	M38–M68	M70	R100–120	R95–R120
CLTE 10^{-6}mm/m m/C	48–80	68	65–95	67
HDT 264 psi	165–209F	270	190F–225F	225F

Source: PMMA properties (2020).

Figure 9.32 Helicopter cockpit with PMMA.(Ibid).

can be used singularly or in combination with other polymers to produce products ranging from soft, flexible elastomers to hard, stiff thermoplastics and thermosets.

PMMA is an economical alternative to PC when extreme strength is not necessary. Acrylics or plexiglass can be used with window frames, (glass filled): tubs, counters, vanities. Optical applications can include outdoor advertising signs, aircraft windshields, cockpit covers, bubble bodies for helicopters. PMMA glazing provides a panoramic view in Airbus Helicopters as shown in Figure 9.32 (PMMA 2020).

Acrylics can be used in motorhomes on the inside of the cabin. These can be used in custom vans on the interior parts.

Additionally, PMMA does not contain the potentially harmful bisphenol-A chemicals found in polycarbonate. It is often preferred because of its moderate properties, easy handling and processing, and low cost. Non-modified PMMA is brittle when under load, especially under an impact force, and is more prone to scratching than conventional inorganic glass, but modified PMMA is sometimes able to achieve high scratch and impact resistance.

Cellulose acetate is the acetate ester of cellulose (Cellulose acetate 2020). It was first prepared in 1865. Cellulose acetate was developed in 1927. Cellulose acetate had reduced limitations of flammability. Cellulose acetate can be used as a film base in photography. Cellulose acetate can be used as a frame for eyeglasses. It is also used as a filter paper fiber for cigarettes. In 1923, CA became the first material to be injection molded. Cellulose acetate butyrate (CAB) in 1938 and cellulose acetate propionate (CAP) in 1945 found applications for hair brushes, toothbrushes, combs, cosmetic cases, hand tool handles, steering wheels, knobs, armrests, speakers, grilles, and so on.

9.2.4.7 Additives for Plastics

Additives for plastics help reduce environmental exposure of the plastics, especially by protecting the plastic from UV sunlight and oxygen. Most plastics readily biodegrade when exposed to sunlight or oxygen. The ultraviolet light stabilizer is TiO_2. Anti-oxidants include phenolic and acryl amines (Additives 2022).

Plastics are readily degraded by ultraviolet light and oxygen. These processes degrade the polymer on a chemical level and cause chain scission. This can greatly reduce the mechanical properties. Antioxidants and antioxidants can be used to help the plastics be used for outdoor use. The degradation can occur at room temperature or elevated temperatures. Once initiated, autooxidation proceeds vis free radical chain reaction which can be autocatalytic. (Antioxidants 2020)

As such, even though efforts are usually made to reduce oxygen levels, total exclusion is often not achievable and even exceedingly low concentrations of oxygen can be sufficient to initiate degradation. Sensitivity to oxidation varies significantly depending on the polymer in question; without stabilizers, polypropylene and unsaturated polymers such as rubber will slowly degrade at room temperature, whereas polystyrene can be stable even at high temperatures (Geuskens et al. 1981).

Antioxidants are of great importance during the process stage. They can provide plastics with long-term stability at ambient temperature. They can be used with hindered amine light stabilizers (HALs). Antioxidants are often referred to as being primary or secondary depending on their mechanism of action.

Hindered amine light stabilizers (HALS) are chemical compounds containing an amine functional group that contains nitrogen. They are used as stabilizers in plastics and polymers. Stabilizers are a class of chemical additives added to plastics to reduce degradation. Polymeric plastics can be subject to various degrading processes like oxidation, photodegradation, thermal degradation, and other environmental exposure. These processes all degrade the polymer on a chemical level, leading to a chain scission that can adversely affect its mechanical properties such as strength, modulus, and appearance. These compounds are primarily used to protect the polymers from the effects of photo-oxidation.

Natural and synthetic plastics and polymers can be attacked by ultraviolet radiation, and products using these materials may turn yellow, crack, or disintegrate. This is called problem *UV degradation*, and is a common problem in products exposed to sunlight. Continuous exposure is a more serious problem than intermittent exposure due to the extent and degree of exposure.

Thermal stability is achieved with the addition of Sb_2O_3, which is a fire retardant. Fire retardants can include zinc borate and chlorinated paraffins. Flame retardants include a diverse group of chemicals which are added to plastic materials. Flame retardants are activated by the presence of an ignition source and are intended to prevent or slow any further development of ignition by a variety of different physical and chemical methods.

Brominated flame retardants (BFRs) are organo-bromine compounds that have an inhibitory effect on combustion chemistry and tend to reduce the flammability of products containing them. The brominated variety of commercialized chemical flame retardants comprise approximately 19.7% of the market. They are effective in plastics and textile applications like electronics, clothes, and furniture (Flame retardants Wikipedia 2022).

Many brominated chemicals are under increasing criticism in their use in household furnishings and for causing harm to some people. Some believe PFR's could have harmful effects on humans and animals. Flame-retardant additives are put into various materials, such as foams for furniture seating and mattresses, plastics used for electronic and electrical applications, and aircraft and vehicle interiors. It became so important for people to be protected from the ravages of clothing and furniture fires, and particularly in the transportation industry, that the federal government began mandating the use of these chemicals in countless products.

The flip side is that advocacy groups such as Safer Chemicals, Healthy Families and about 21 states are calling for the elimination of flame retardants, calling them "toxic." Recently, Safer Chemicals, Healthy Families released a study on TV and flame retardants (TV flame retardants 2020). The study found that "TVs could be bad for your health in an unexpected way: TVs contain toxic flame-retardant chemicals that can contaminate homes." Testing done by Toxic-Free Future and Clean Production Action found that TV manufacturers continue to use toxic flame-retardant chemicals in their products despite evidence that the chemicals are harmful to health. Two TVs were found to contain the banned chemical flame retardant deca-BDE in apparent violation of Washington state law, where the TVs were purchased, said the press release. The big problem that the Toxic-Free Future group has with flame retardants is that the chemicals "can escape products and end up in household dust, exposing adults and children to the chemicals through ingestion, such as through hand-to-mouth activity." (Flame retardants-toxicity 2020).

An antiozonant is an organic compound that prevents or retards the degradation of material caused by ozone. Antiozonants are added to polymers to reduce cracking. Cracks can be formed in many different rubbers and elastomers when they are exposed to ozone gas. The problem is very common in rubber tires. Tiny traces of ozone in the air will attack unsaturated, or double bonds of carbon, in rubber polymers of natural rubber, polybutadiene, styrene butadiene rubber, and nitrile rubber. Nitrile rubber is a copolymer of butadiene and acrylonitrile, shown in Figure 9.33. This will be discussed more in Chapter 10.

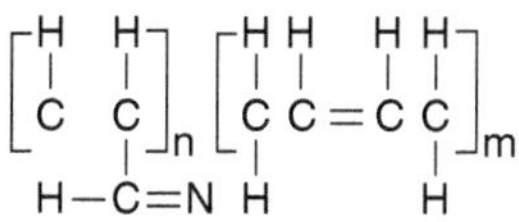

Figure 9.33 Nitrile rubber: copolymer of butadiene and acrylonitrile.

Butyl rubber is more resistant but still has a small number of double bonds in its chains, so attack is possible. Exposed surfaces are attacked first, with the density of cracks varying with ozone gas concentration. The higher the concentration, the greater the number of cracks formed (Ozone cracking Wikipedia 2022). Common antiozonants include:

- *p*-Phenylenediamines such as 6PPP (*N*-(1,3-dimethylbutyl)-*N*′-phenyl-*p*-phenylenediamine) or IPPD (*N*-isopropyl-*N*′-phenyl-*p*-phenylenediamine),
- 6-ethoxy-2,2,4-trimethyl-1,2-dihydroquinoline (ETMQ, EMQ, or Ethoxyquin)
- Ethylene diurea (EDU), and
- Paraffin waxes that form a surface barrier.

Other additives can include scratch- and mark-resistance plastics that are made from alkyl modified siloxane. Other processing aids can be made from butyl stearate. Dispersing additives can be made from siloxane and polyolefin compounds. Antistatic agents can be made from glycerol monostearate and butyl stearate.

Colorants are pigments of dyes for masterbatch of colorants that are added to the plastic to give it some color. Those compounds come in the form of dyes and pigments. Dyes are usually used with polycarbonates, polystyrene, and acrylic plastics. Pigments are better suited for use with polyethylene and polypropylene. The colorant must be chemically compatible with the base resin, be a suitable match with a color standard (International Color Consortium 2020), and be chemically stable in the environment that the plastic is exposed to for temperature, chemicals, moisture, and others. The colorant must also survive the processing temperatures and processing conditions and then last the life duration of the product.

Typically colorants are added to master batches in the plastic. This method involves colorants that are a concentrate being separated into resin type and color concentrations greater than 40%. Then, the master batch is blended with virgin plastic to get 2 or 3% of color in the plastic.

Blowing agents are used for foaming of polystyrene (PS) and polypropylene (PP). The blowing agents can be made from azodicarbonamide. A blowing agent is a substance which is capable of producing a cellular structure via the foaming process in either open-cell or closed-cell foams in plastics. The open cell has a structure that is connected, whereas a closed structure does not have interconnecting cells. Thus, resin can flow into the open-cell foam structure causing an increase in resin impregnation in polymer composites. PP and PS foams are closed-cell foams. EPS is closed cell as are all hard foams. The closed-cell foams have higher R value, and are denser and more expensive. Polyurethane creates open-cell foams.

The cellular structure in a matrix reduces density, increasing thermal and acoustic insulation, while increasing relative stiffness of the original polymer.

Blowing agents create holes in a matrix, producing cellular materials, and have been classified as follows: Physical blowing agents usually are added to the monomer and then polymerized. For PS and PP foams, the plastic is added to the mold and then steam or hot air is added to the mold, causing the PS or PP to expand. The EPS beads are set in the mold for 24 hours and then cooled with cold air to achieve a 5X to 20X expansion.

Chemical blowing agents used in polyurethanes (PU) include isocyanate and water. Gaseous products and other byproducts are formed by a chemical reaction of the chemical blowing agent, promoted by the heat of the foam production process or a reacting polymer's exothermic heat. Mixed physical/chemical blowing agents are used to produce flexible PU foams with very low densities. Here both chemical and physical blowing are used in tandem to balance each other out with respect to thermal energy released and absorbed, minimizing temperature rise. For instance, polyurethane systems combine isocyanate and water with liquid carbon dioxide (which boils to give gaseous form) in the production of very low-density flexible PU foams for mattresses.

Filler materials are particles added to plastics that can improve specific properties, make the product cheaper, or a combination of both. The two largest segments for filler material use are elastomers and plastics. The most common filler materials used are ground calcium carbonate (GCC), precipitated calcium carbonate (PCC), kaolin, talc, and carbon black. Filler materials can increase, slightly, tensile strength, tensile modulus, toughness, heat resistance, color, and clarity. Additives typically make the plastic thicker and increase viscosity. They decrease the impact strength and the coefficient of linear thermal expansion (CLTE). As an example, sheet molding compound is a composite material used for doors and hoods for the Corvette, and has about 30% calcium carbonate. This reduces the cost of the composite and also reduces its CLTE.

Fibers can also be added to increase the tensile strength, tensile modulus, impact strength, and thermal properties of plastics and composites. The fibers can be glass, carbon, mineral, aramid, and mica. This will be discussed in more detail in the Polymer Composites chapter.

References

Acrylics (2022). https://www.acmeplastics.com/acrylic (accessed February 2022).

Additives (2022). http://polymerdatabase.com/Additives/Polymer%20Additives.html (accessed February 2022).

Blyth, J. and Hofmann, A.W. (1845). On styrole, and some of the products of its decomposition. *Memoirs and Proceedings of the Chemical Society of London* 2: 334–358.

https://doi.org/10.1039/mp8430200334. Archived from the original on 1 May 2018.; see p. 339.

Geuskens, G.; Bastin, P.; Lu Vinh, Q.; Rens, M. (1981). Photo-oxidation of polymers: part IV—influence of the processing conditions on the photo-oxidative stability of polystyrene. *Polymer Degradation and Stability* 3 (4): 295–306. doi:https://doi.org/10.1016/0141-3910(81)90025-2.

Websites

Antioxidants (2020). https://en.wikipedia.org/wiki/Polymer_stabilizers (accessed September 2020).

Antioxonants (2020). http://www.nocil.com/Downloadfile/ETechnicalNote-Antioxidants-Dec2010.pdf (accessed September 2020).

Ban (2020). https://cen.acs.org/articles/95/i26/European-Union-further-restricts-four.html (accessed April 2020).

Bio-based PP (2022). https://www.braskem.com.br/imgreen/bio-based-en (accessed February 2022).

Essential chemistry (2020). https://www.essentialchemicalindustry.org/polymers/polyethene.html (accessed September 2020).

Flame retardants-toxicity (2020). https://www.plasticstoday.com/materials/toxic-flame-retardants-are-still-being-used-plastics-even-though-there-are-alternatives/199932598957570 (accessed September 2020).

Flame retardants Wikipedia (2022). https://en.wikipedia.org/wiki/Flame_retardant (accessed February 2022).

Foam (2020). https://www.researchgate.net/publication/310488467_Numerical_and_experimental_study_of_the_mechanical_response_of_aluminum_foams_under_compressive_loading_using_CT_data/figures?lo=1 (accessed September 2020).

HDT CLTE (2020). https://omnexus.specialchem.com/polymer-properties/properties/CLTE and hdt-0-46-mpa-67-psi (accessed September 2020).

I'm Greene Bio-Based Polyethylene. https://www.braskem.com.br/imgreen/bio-based-en (accessed Septemebr 2022).

International Color Consortium (2020). http://www.color.org/index.xalter (accessed September 2020).

Ozone cracking Wikipedia (2022). https://en.wikipedia.org/wiki/Ozone_cracking (accessed February 2022).

Plastic bag (2020). https://www.thriftyfun.com/Uses-for-Plastic-Produce-Bags.html (accessed September 2020).

Plastics prices (2020). https://www.plasticsnews.com/resin/currentPricing/commodity-thermoplastics (accessed September 2020).

Plastic-production (2022). https://www.statista.com/statistics/282732/global-production-of-plastics-since-1950/ (accessed February 2022).
Plastic annual production (2020). https://ourworldindata.org/faq-on-plastics#how-much-plastic-and-waste-do-we-produce (accessed September 2020).
PMMA (2020). https://www.plastics.gl/automotive/through-a-plexiglas-clearly/ (accessed September 2020).
Polyethylene (2020). https://sciencing.com/make-polyethylene-10029798.html (accessed September 2020).
Polyethylene basics (2020). https://www.essentialchemicalindustry.org/polymers/polyethene.html (accessed September 2020).
PE properties (2020). www.matweb.com/HDPE (accessed September 2020).
PMMA properties (2020). http://www.matweb.com/search/DataSheet.aspx?MatGUID=3cb08da2a0054447a3790015b7214d07 (accessed September 2020).
Polypropylene (2020). https://guichon-valves.com/faqs/pp-polypropylene-manufacturing-process-of-pp-polypropylene/ (accessed September 2020).
PP properties (2020). www.matweb.com/PP (accessed September 2020).
PS Props (2020). www.matweb.com (accessed September 2020).
PVC pipe (2020). www.amazon.com (accessed September 2020).
PVC properties (2020). www.matweb.com/pvc (accessed September 2020).
PVC Wikipedia (2020). https://en.wikipedia.org/wiki/Polyvinyl_chloride (accessed September 2020).
Rubber chemical structure (2020). https://en.wikipedia.org/wiki/Styrene-butadiene#/media/File:SBRwithexplicitC.png (accessed September 2020).
SMA structure (2020). https://www.britannica.com/science/styrene-maleic-anhydride-copolymer (accessed September 2020).
Styrene-atactic (2020). https://en.wikipedia.org/wiki/Polystyrene (accessed April 2020).
Styrene-volume (2020). https://www.prnewswire.com/news-releases/global-polystyrene-market-to-reach-29-01-billion-by-2023--says-beroe-inc-300942912.html (accessed September 2020).
Styrene Wikipedia (2022). https://en.wikipedia.org/wiki/Styrene (accessed February 2022).
US ban phthalates (2020). https://www.cnn.com/2017/10/20/health/phthalates-banned-in-childrens-toys/index.html (accessed April 2020).

10

Engineering Plastics

The chapter will discuss the plastics that are considered engineering plastics and the reasons for the classification. The plastics will include nylon, polycarbonate (PC), polyethylene terephthalate (PET), polybutylene terephthalate (PBT), thermoplastic polyurethane (TPU), acrylonitrile butadiene styrene (ABS), Styrene acrylonitrile (SAN), acetal, polytetrafluoroethylene (PTFE), polyphenylene oxide (PPO), Polyphylene sulfide (PPS), polyether-ether-ketone (PEEK), liquid crystal polymers (LCPs), copolymers, and blends. The discussion will include manufacturing processing to make the plastic pellet, mechanical and thermal properties, shrinkage, and typical applications with these plastics.

10.1 Engineering Plastics Definition

Engineering thermoplastics are plastics that have higher mechanical and thermal properties than commodity plastics. They can sometimes replace metal parts. They can use conventional plastic molding operations. They usually can be painted. They, typically, have higher costs per pound than commodity plastics. The costs of some engineering plastic are listed in Table 10.1.

Engineering plastics can use conventional plastic molding operations to produce parts. They usually can be painted. They typically are polar in chemical nature. Common engineering thermoplastics are ABS, LCPs, nylon (polyamide), PC, PET, PBT, polyurethane, PEEK, polyether ketone (PEK), PPO, polyphenylene sulfide, styrene acrylonitrile, styrene maleic anhydride, thermoplastic elastomer, tetrafluoroethylene (Teflon), and ultrahigh molecular weight polyethylene.

Sustainable Plastics: Environmental Assessments of Biobased, Biodegradable, and Recycled Plastics, Second Edition. Joseph P. Greene.

Table 10.1 Plastic costs per pound per May 2020.

ABS	$1.10
Acetal	$1.44
LCP (glass)	$9.80
Nylon 6	$1.45
Polyarylate	$2.10
PC	$1.50
PEEK	$45.00
PBT	$1.24
PET	$1.24
Polyurethane	$2.06
PPO	$1.87
PPS (glass)	$3.90
Polysulfone	$6.50
PTFE	$7.65
SAN	$1.36
SMA	$1.78
TPE (olefinic)	$1.15
UHMWPE	$1.24

Source: Data from Plastic cost 2021.

10.2 Acrylonitrile Butadiene Styrene

ABS (chemical formula $(C_3H_3N)_n)\cdot(C_4H_6)_m\cdot(C_8H_8)_k$ is a common thermoplastic polymer. The chemical structure of ABS is shown in Figure 10.1.

ABS lists a glass transition temperature (TG) of approximately 105 °C (221 °F). (TG-ABS 2020) ABS is amorphous and therefore has no true melting point. ABS is a terpolymer consisting of acrylonitrile for chemical resistance, butadiene for impact resistance, and styrene for rigidity and processing ease. ABS is made with graft polymerization techniques. ABS is a family of materials that vary from high gloss to low matte finish, and from low to high impact resistance. Additives are

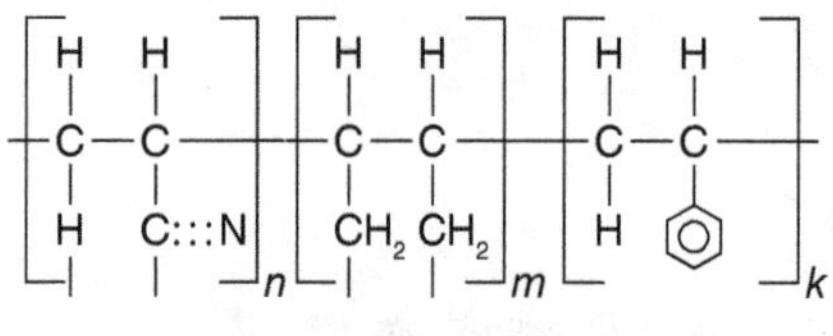

Figure 10.1 Chemical structure of ABS.

used to enable ABS grades that are flame retardant, transparent, high heat resistance, foamable, or Ultra-violet (UV) light stabilized.

ABS is a terpolymer made by polymerizing styrene and acrylonitrile in the presence of polybutadiene. The proportions can vary from 15 to 35% acrylonitrile, 5 to 30% polybutadiene, and 40 to 60% styrene. (ABS 2020). The nitrile groups from neighboring chains, being polar, attract each other and bind the chains together, making ABS stronger than pure polystyrene. The styrene gives the plastic a shiny, impervious surface. The polybutadiene, a rubbery substance, provides toughness even at low temperatures. For the majority of applications, ABS can be used between −20 and 80 °C (−4 and 176 °F) as its mechanical properties vary with temperature. Polyacrylonitrile (PAN) provides weather resistance and barrier properties, good tensile strength, and high-temperature resistance. Polystyrene provides rigidity, stiffness, and transparency. Polybutadiene provides impact strength, elongation, low-temperature properties. ABS has excellent impact strength and is higher than all other plastics besides PC.

PC and ABS can be blended to form an excellent polymer for doors and fenders. They were used on the Camaro and Saturn cars in the 1990s as shown in Figures 10.2 and 10.3.

A compounded blend of PC and ABS: The PC contributes impact and heat-distortion resistance, while the ABS contributes processability and chemical stress

Figure 10.2 Plastic use on the door panels for the Camaro in the 1990s. Source: Camaro 2020.

Figure 10.3 Plastic use on the door panels for the Saturn in the 1990s. Source: Saturn 2020.

resistance, and cost reduction below PC. Properties of PC/ABS include the following:

- Impact resistance between ABS and PC, (10–14 ft.-lb./in.).
- Strength and stiffness between PC and ABS.
- Heat resistance superior to ABS, less than PC.
- Flame retardant system more stable in processing than ABS.
- Very good indoor UV light color stability.
- Low-temperature impact and ductility.
- Superior processability to PC.

The mechanical properties of PC and ABS are shown in Table 10.2. The table shows that ABS has equivalent tensile strength and modulus as acrylics and PC. PC has the highest elongation and impact strength. ABS has higher impact strength and thermal properties, heat deflection temperature (HDT) than acrylic but slightly lower than PC.

PC/ABS can be used for appearance housings and structural parts, which need stiffness, gloss, impact, and heat resistance which is higher than ABS, but requires costs below PC.

Housings requiring UL V0, but not 5VA-that is, portable electronic devices, not permanently fixed electric devices are also good applications for PC/ABS. Appearance parts require colorfastness under interior UV light, interior parts where temperature environment is between 180 and 230 F – "above the beltline" applications, and where nonshattering ductility is required like instrument panel retainers. Applications can be painted exterior parts that require

Table 10.2 Properties of ABS, PC, acrylic, and PC/ABS blends.

	Acrylic	PC	ABS	PC/ABS
Density, g/cc	1.16–1.19	1.2	1.16–1.21	1.07–1.15
Tensile strength, psi	5000–9000	4000–5000	3300–8000	5800–9300
Tensile modulus, psi	200k–500k	350k	320k–400k	350k–450k
Tensile elongation, %	20–70%	110%	1.5–25%	50–60%
Impact strength, ft-lb/in	0.65–2.5	16	1.4–12	6.4–11
Coefficient of Linear Thermal Expansion (CLTE), 10^{-6} mm/mm/C	48–80	68	65–95	67
HDT, @264 psi	165–209F	270F	190–225F	225F

Source: Data from ABS 2020.

low-temperature impact strength greater than 10 ft.-lb./in. is another place PC/ABS should be used. Any outdoor application should be painted.

10.3 Acetal (Polyoxymethylene)

Polyoxymethylene (POM), also known as acetal, Delrin, polyacetal, and polyformaldehyde, is an engineering thermoplastic used in precision parts requiring high stiffness, low friction, and excellent dimensional stability (POM 2020). As with many other synthetic polymers, it is produced by different chemical firms with slightly different formulas. POM is characterized by its high strength, hardness, and rigidity to −40 °C.

To make POM homopolymer, anhydrous formaldehyde must be generated. The principal method is by reaction of the aqueous formaldehyde with alcohol to create a hemiformal, dehydration of the hemiformal/water mixture (either by extraction or vacuum distillation), and release of the formaldehyde by heating the hemiformal. The formaldehyde is then polymerized by anionic catalysis and the resulting polymer is stabilized by reaction with acetic anhydride. Due to the manufacturing process, large diameter cross-sections may have pronounced centerline porosity. A typical example is DuPont's Delrin. The chemical structure of POM is shown in Figure 10.4.

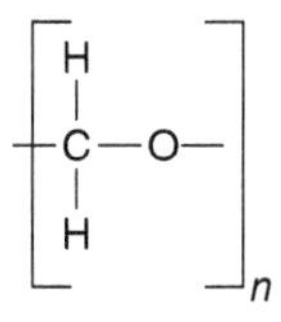

Figure 10.4 Chemical structure of POM.

POM was discovered by Hermann Staudinger, a German chemist who received the 1953 Nobel Prize in Chemistry (Ibid). He had studied the polymerization and structure of POM in the 1920s while researching macromolecules, which he characterized as polymers. Due to problems with thermal stability, POM was not commercialized at that time. Around 1952, research chemists at DuPont synthesized a version of POM, and in 1956 the company filed for patent protection of the homopolymer. The inventor of a heat-stable POM homopolymer was Dal Nagore, who discovered that reacting the hemiacetal ends with acetic anhydride converts the readily depolymerizable hemiacetal into a thermally stable, melt-processable plastic.

POM is intrinsically opaque white due to its high crystalline composition, but it is available in all colors. POM has a high density of approximately 1.410 g/cm^3. Typical applications for injection-molded POM include high-performance engineering components such as small gear wheels, eyeglass frames, ball bearings, ski bindings, fasteners, guns, knife handles, and lock systems. POM can be easily machined with CNC. The material is widely used in the consumer electronics industry. It has the same shrinkage (2.5%) as PP and be molded with the same injection molds.

The mechanical properties of POM are compared to Polypropylene (PP) and Nylon 6,6 in Table 10.3. The table shows that POM has a higher density than PP and nylon 6,6. POM has lower tensile strength and modulus than nylon 6,6 but higher than PP, higher CLTE than PP and nylon 6,6, and higher thermal properties (HDT) and cost than nylon 6,6 and PP.

POM can be made into the following parts through injection molding:

- Mechanical gears, sliding and guiding elements, housing parts, springs, chains, screws, nuts, fan wheels, pump parts, and valve bodies.
- Electrical engineering: insulators, bobbins, connectors, parts for electronic devices such as televisions, telephones, etc.

Table 10.3 Mechanical properties of POM.

	POM	**PP**	**Nylon 6,6**
Density, g/cc	1.38–1.44	0.9	1.13–1.15
Tensile strength, psi	5000–9000	4000–5000	14 000
Tensile modulus, psi	447k	125k—225k	230k—550k
Tensile elongation, %	10–76%	100–600%	15–80%
Impact strength, ft-lb/in	1.6–4.0	0.4–1.2	0.55–1.0
CLTE, 10^{-6} mm/mm/C	100	60–80	80
HDT, @264 psi	241F	150F	180F

Source: Data from POM 2020.

- Automotive: fuel sender unit, light/control stalk/combination switch (including shifter for light, turn signal), power windows, door lock systems, and articulated shells.
- Model: model railway parts, such as trucks (bogies) and handrails (handlebars).
- Hobbies: radiocontrolled helicopter main gear, landing skid, yo-yos, and vaping drip tips.
- Furniture: hardware, locks, handles, and hinges.
- Construction: structural glass — pod holder for point
- Sports: paintball accessories, like paintball markers.
- Airsoft guns with reduced piston noise.
- Music: picks, Irish flutes, bagpipes, practice chanters, harpsichord plectra, instrument mouthpieces, and tips of some drumsticks.
- Home uses: fully automatic coffee brewers; knife handles.

POM can be machined with manual or Computer Numerical Control (CNC) methods. POM should not be painted. POM has similar shrinkage as PP of about 2% and can replace the PP in application that would like to increase the impact performance and thermal resistance. POM is tougher than ABS, comes in bright translucent colors, and is not paintable.

POM can be reinforced with fibers for increased strength and mechanical properties. The reinforcement fibers can be glass, carbon, aramid, metal, mineral, PTFE, and stainless steel.

10.4 Liquid Crystal Polymer

LCPs are excellent thermoplastic polymers that have excellent mechanical and thermal properties. Liquid crystallinity in polymers may occur either by dissolving a polymer in a solvent or by heating a polymer above its glass or melting transition point (LCP 2020). Liquid-crystal polymers are present in melted/liquid or solid form. In solid form, the main example of lyotropic LCPs is the commercial aromatic polyamides known as Kevlar. The chemical structure of this aramid consists of linearly substituted aromatic rings linked by amide groups. In a similar way, several series of thermotropic LCPs have been commercially produced by several companies (e.g.,Vectra and Ticona).

LCPs can provide high precision and high-heat properties (HDT) (LCP-Celanese 2020). LCPs are highly crystalline, inherently flame retardant, thermotropic (melt-orienting) thermoplastics. They are similar to conventional semicrystalline polymers but have higher mechanical and thermal properties. When melted, conventional semicrystalline polymers have a random amorphous structure, which, as they cool, form crystalline regions surrounded by an amorphous matrix. LCP molecules have very little amorphous regions and can be over 90% crystalline. Conventional plastics have between 30 and 60% crystalline regions with the remaining amorphous regions. LCP molecules remain well ordered even in the melt and slide past each other easily under shear. The result is very low melt viscosity making it easy to fill extremely thin walls and to replicate intricate features. They exhibit little or no shrinkage in the flow direction and require very little time to set up or solidify. Quick fills and shorter cycle times mean increased productivity.

Injection-moldable LCP products have proven successful in a variety of applications, including sockets, bobbins, switches, connectors, chip carriers, sterilizable trays, dental tools, surgical instruments, pumps, meters, and valve liners. Unfilled polymers can be extruded for film or sheet, coextruded for barrier packaging, or melt spun for applications needing high-performance fibers or yarn.

A high number of LCPs have order in the melt phase analogous to that exhibited by nonpolymeric liquid crystals (LCPs). Processing of LCPs from liquid-crystal phases (or mesophases) gives rise to fibers and injected materials having high mechanical properties as a consequence of the self-reinforcing properties due to the macromolecular orientation in the mesophase. The highly ordered polymers are highly crystalline, which can cause high mechanical properties.

Celanese Vectra can be made into several plastic parts shown in Figure 10.5.

LCPs can be melt processed on conventional equipment to obtain excellent mechanical properties. A unique class of partially crystalline aromatic polyesters are based on p-hydroxybenzoic acid and related monomers. Liquid-crystal polymers are capable of forming regions of a highly ordered structure while in the liquid phase. However, the degree of order is somewhat less than that of a regular solid crystal. LCPs have a high tensile strength and modulus at high temperatures, excellent chemical resistance, and flame retardancy. The mechanical properties of

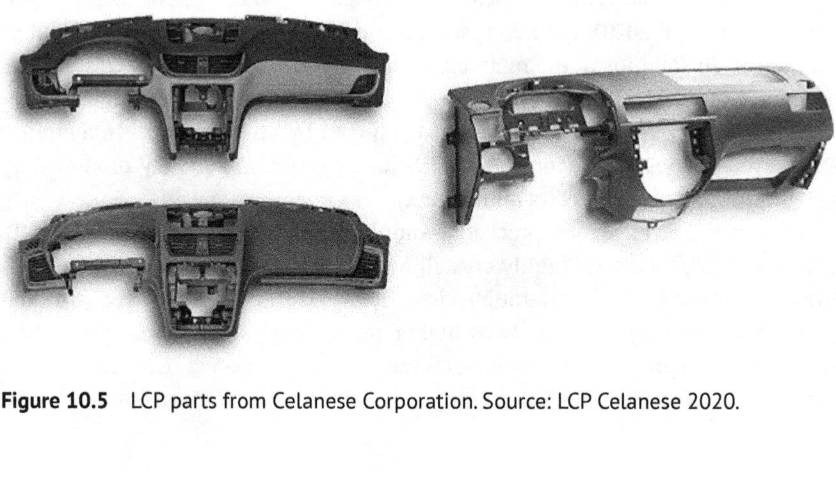

Figure 10.5 LCP parts from Celanese Corporation. Source: LCP Celanese 2020.

Table 10.4 Mechanical properties of LCP thermoplastics.

	PET	PBT	LCP
Density, g/cc	1.29–1.40	1.30–1.38	1.35–1.40
Tensile strength, psi	7000–10 500	8200	16 000–27 000
Tensile modulus, psi	400–600k	280–435k	1400–2800k
Tensile elongation, %	30–300%	50–300%	1.3–4.5%
Impact strength, ft-lb/in	0.25–0.70	0.7–1.0	2.4–10
CLTE, 10^{-6} mm/mm/C	65	60–95	25–30
HDT, @264 psi	70–100F	122–185F	356–671F

Source: Data from LCP 2020.

some LCPs are listed in Table 10.4. The properties of LCP polyester are higher in tensile modulus and strength, impact strength, and thermal properties (HDT) than PBT or PET. LCPs have lower elongation and CLTE than PBT or PET.

Liquid crystallinity in polymers may occur either by dissolving a polymer in a solvent (lyotropic liquid-crystal polymers) or by heating a polymer above its glass or melting transition point (thermotropic liquid-crystal polymers) (Shibaev and Lam (1994). The chemical structures are shown in Figures 10.6 and 10.7. Two common LCP plastics are Kevlar and Vectra.

Most thermoplastic LCP appears to be aromatic copolyesters. Wholly aromatic copolyesters (LCP). High melting sintered: Oxybenzoyl (does not melt below its

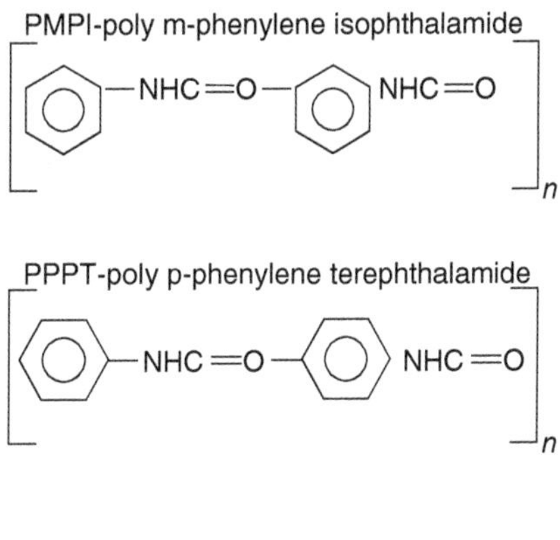

Figure 10.6 Kevlar chemical structure from Celanese Corporation.

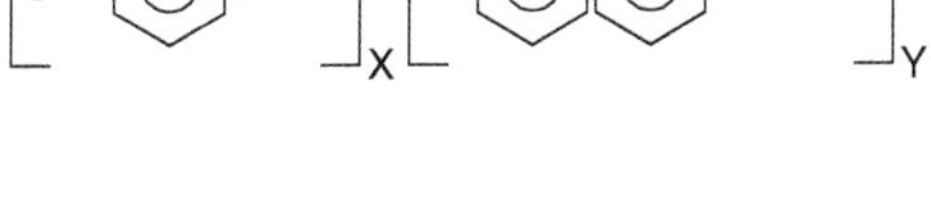

Figure 10.7 VECTRA chemical structure Celanese Corporation.

Table 10.5 Properties of LCP, PET, and PBT.

	PET	PBT	LCP
Density, g/cc	1.29–1.40	1.30–1.38	1.35–1.40
Tensile strength, psi	7000–10 500	8200	16 000–27 000
Tensile modulus, psi	400–600k	280–435k	1.400–2800k
Tensile elongation, %	30–300%	50–300%	1.3–4.5%
Impact strength, ft-lb/in	0.25–0.70	0.7–1.0	2.4–10
CLTE, 10^{-6} mm/mm/C	65	60–95	25–30
HDT, @264 psi	70–100F	122–185F	356–671F

Source: LCP-Properties 2020.

decomposition temperature. Must be compression molded). Injection moldable grades are Xydar and Vectra.

Xydar from Amoco Performance Products is made with terephthalic acid, *p*,*p*′-dihydroxybiphenyl, and *p*-hydroxybenzoic acid. Grade 1 has an HDT of 610 F. Grade 2 has an HDT of 480 F. Vectra from Hoechst Celanese Corporation is made from para-hydroxybenzoic acid and hydroxynaphthoic acid. It contains rigid chains of long, flat monomer units, which are thought to undergo parallel ordering in the melt and form tightly packed fibrous chains in molded parts. Properties of LCP are compared with other polyesters, PET and PBT and are shown in Table 10.5. The table shows that LCP has higher density, tensile strength, tensile modulus, impact strength, and thermal property of HDT than PET and PBT. LCP has lower elongation and CLTE than PET and PBT. LCP can be used as an actuator pin for parts.

10.5 PBT (Polybutylene Terephthalate)

PBT is a thermoplastic engineering polymer that is used in electronics industries. It is a thermoplastic semicrystalline polymer and a type of polyester. PBT is resistant to solvents, shrinks very little during forming, is mechanically strong, heat-resistant up to 150 °C (or 200 °C with glass–fiber reinforcement). It was developed by Britain's Imperial Chemical Industries (ICI) (ICI 2020).

PBT is closely related to other thermoplastic polyesters, e.g., PET. Compared to PET, PBT has slightly lower strength and rigidity, slightly better impact resistance, and a slightly lower glass transition temperature. PBT and PET are sensitive to hot water above 60 °C (140 °F). PBT and PET need UV protection, as with most plastics, if used outdoors, and most grades of these polyesters are flammable, although additives can be used to improve both UV and flammability properties. The chemical structure of PBT is shown in Figure 10.8.

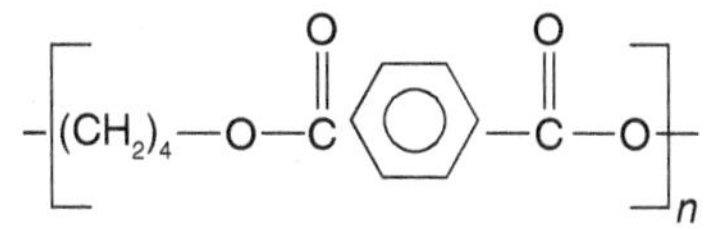

Figure 10.8 Chemical structure of PBT.

PBT is produced by polycondensation of terephthalic acid or dimethyl terephthalate with 1,4–butanediol using special catalysts (PBT 2020).

PBT produces popular products that consume millions of pounds every year. The molded products of PBT are Valox, Xenoy, Vandar, and Pocan. It can be molded into the following parts, including, distributers, door panels, fenders, bumper fascia covers, cables, connectors, terminal blocks, fuse holders and motor parts, distributor caps, door, and window hardware. The extruded applications include extrusion-coat wire.

PBT is used for housings in electrical engineering, but also in construction as plug connectors and in households, for example, in showerheads or irons. It is also found processed into fibers in toothbrushes, false eyelashes, and in some high-end computer keyboards because the texture is highly resistant to wear and discoloration due to UV radiation. PBT can also be made into yarn for sportswear. Due to its chlorine resistance, it is commonly found in swimwear. Further, recent studies have shown that PBT has superior UV properties to PET-based fabrics. PBT especially glass fiber reinforced grades can be effectively fire retarded with halogen-free products.

The longer, more flexible $(CH_2)_4$ groups allow for more rapid crystallization than PET. PBT is not as conveniently oriented as PET and is normally injection molded. PBT has a sharp melting transition with a rather low melt viscosity. PBT has rapid crystallization and a high degree of crystallization causing warpage concerns. It can be combined with PC to form Zenoy plastic for fenders or door skins.

The mechanical properties of PBT are shown in Table 10.6. The table shows that PBT has lower density, tensile strength, tensile modulus than PET and LCP. It, also, shows that PBT has higher thermal property of HDT, CLTE, and impact

Table 10.6 Mechanical Properties of PBT, PET, and other plastics.

Property	PBT	PET
Tensile strength psi (MPa)	17400 (120)	23100 (159.3)
Flexural modulus kpsi (MPa)	1100 (7590)	1300 (8965)
Glass transition temp, F (C)	154 (68)	203 (95)
Melting point, F (C)	433 (223)	486 (252)

Source: Data from PBT-properties- PBT-PET properties 2020.

strength than PET. The advantages of PBT include its resistance to solvents and low shrinkage rate when forming. It also has good electrical resistance and because of its fast crystallization is easy to mold. It has excellent heat resistance up to 150 °C and a melting point reaching 225 °C. PBT can be used for lighting applications in the industry.

10.6 PET (Polyethylene Terephthalate)

PET polyesters were invented by Dr. W. H. Carothers in 1929. He invented the classification of polymers into two groups, *condensation* and *addition* polymers. Dr. Carothers found, though, if he produced an aromatic polyester, the properties significantly improved. He added a *p*-phenylene group for stiffening and leads to polymers with high melting points and good fiber-forming properties, e.g., PET. PET polymers are used for films and fibers. Polyesters are one of many heterochain thermoplastics, which has atoms other than C in the backbone chain. Polyesters include unsaturated for thermosets with C=C bonds and saturated for aromatic thermoplastic polyesters with C–C bonds.

Thermoplastic polyesters are produced with an ester, –C–O, repeating link. Linear polyesters are PET and PBT. The chemical structures of PET and PBT are shown in Figure 10.9.

The properties of PET and PBT are shown in Table 10.6. The table shows that PET has higher density, tensile strength, tensile modulus than PBT. It, also, shows that PET has the lower thermal property of HDT, CLTE, and impact strength than PBT.

PET has flexible, but short, $(CH_2)_2$ groups tend to leave the chains relatively stiff and PET is noted for its very slow crystallization. If cooled rapidly from the melt to a temperature below T_g, PET solidifies in amorphous form. If PET is reheated

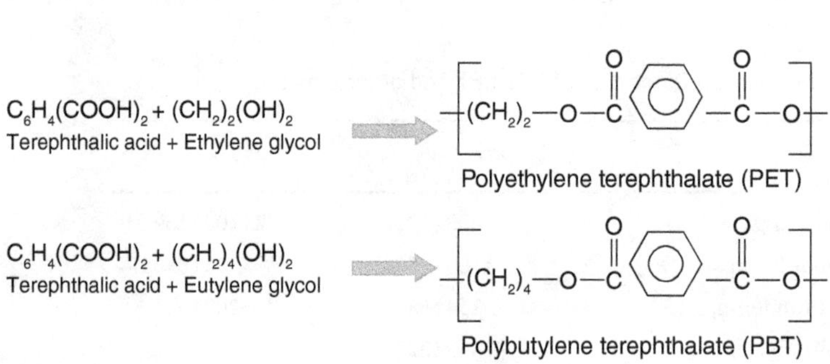

Figure 10.9 Chemical structure of PET and PBT.

above T_g, crystallization takes place to up to 30%. In many applications, PET is first preshaped in the amorphous state and then given a uniaxial (fibers or tapes) or biaxial (film or containers) crystalline orientation. During injection molding, PET can yield amorphous transparent objects (cold mold) or crystalline opaque objects (hot mold).

PET can be used for blow-molded bottles in applications. In fact, 100% of 2-l beverage containers and liquid products are made with PET. The recycling number or resin identification code set by the Society of Plastics Industry (SPI) in 1988 is number "1." PET can be used in fiber applications. Approximately, 25% of the market is in tire cords, rope, thread, cord, belts, and filter cloths. PET can be sued in monofilaments — brushes, sports equipment, clothing, carpet, bristles. In tape form, PET can be uniaxially oriented tape form for strapping.

10.7 Nylon (Polyamide)

Nylon, or polyamide, is considered the first engineering thermoplastic. Nylon is one of many heterochain thermoplastics, which has atoms other than C in the chain. Nylon was invented in 1928 by Wallace Carothers, of DuPont, in search of a "super polyester" fiber with molecular weights greater than 10000. Commercial nylon was brought to market in 1938. Nylon is created when a condensation reaction occurred between amino acids, dibasic acids, and diamines. Nylons are described by a numbering system that indicates the number of carbon atoms in the monomer chains. For example, nylon 6 has 6 carbon atoms. Nylon 6,6 has 12 carbon atoms_ 6 from the diacid and 6 from the amine. Amino acid polymers are designated by a single number, as nylon 6. Diamines and dibasic acids are designated with 2 numbers, the first representing the diamine and the second indicating the adipic acid, as in nylon 6,6 or nylon 6,10 with sebacic acid. Thermoplastic nylons have amide (CONH) repeating link Nylon 6,6 — poly-hexamethylene-diamine is a linear polymer with the following chemical reaction of hexamethylene diamine and adipic acid in a condensation polymerization. It produces a nylon salt, water, and heat as shown in Figure 10.10.

Nylon can have the following chemical structures:

- Nylon 6 — polycaprolactam (linear)
  $[\mathbf{NH}(CH_2)_5\mathbf{CO}]_n$

$$NH_2(CH_2)_6NH_2 + COOH(CH_2)_4COOH \rightarrow$$
$$[NH_2(CH_2)_6\mathbf{NH{\cdot}CO}\ (CH_2)_4CO]_n + nH_2O + \text{(heat)}$$

Figure 10.10 Chemistry of nylon 6,6 formation.

- Nylon 6, 10 — polyhexamethylenesebacamide (linear)
  $[NH_2(CH_2)_6\mathbf{NH.CO}\ (CH_2)_8CO]_n$
- Nylon 11 — poly(11-amino-undecanoic-amide (linear)
  $[\mathbf{NH}(CH_2)_{10}\mathbf{CO}]_n$
- Nylon 12 — poly(11-amino-undecanoic-amide (linear)
  $[\mathbf{NH}(CH_2)_{11}\mathbf{CO}]_n$

Note: Nylon 12 has a significant drop in tensile strength with exposure to moisture and wet environments. (Nylon 12-strength drop 2020).

Nomax PMPI was the first commercial aramid fiber for electrical insulation. Aromatic nylons are used to produce Kevlar fabric. LCP fibers feature straight-chain crystals and are known as Kevlar 24 and Kevlar 49.

- Kevlar 29 PPPT — textile fiber for tire cord, ropes, cables, etc.
- Kevlar 49 PPPT — reinforcing fiber for thermosetting resins

Transparent or amorphous polyamides are commercially available. The reduced crystallization is due to the introduction of side groups. Figure 10.11 shows the chemical structure of amorphous nylons.

Amorphous polyamides have a low shrink rate, excellent dimensional stability, good chemical resistance relative to other amorphous resins, are transparent, and have high cost relative to other amorphous resins. Figure 10.12 shows the moisture absorption by nylon 6, nylon 6,6, nylon 6,10, nylon 11, and nylon 12 (Nylon 2020). The moisture absorption is highest for Nylon 6, Nylon 6,6, Nylon 11, and then Nylon 12. Nylon must be dried for at least four hours before molding. The moisture absorption can cause the mechanical properties of nylon to drop.

The tensile strength of nylon 12 can be significantly reduced as shown in Figure 10.13. By having a 9% absorption of moisture, the tensile strength of Nylon 12 can be reduced 85%. (Nylon-properties drop 2020).

The fiber applications include use into tire cords with nylon 6 and nylon 6,6. Approximately, 50% of the nylon production is used for passenger tires. Nylon van also is used for rope, thread, cord belts, filter cloths, monofilaments- brushes, sports equipment, and bristles with nylon 6,10. The plastic applications can also include bearings, gears, cams, rollers, slides, door latches, thread guides, clothing, light tents, shower curtains, umbrellas, and electrical wire jackets with nylon 11.

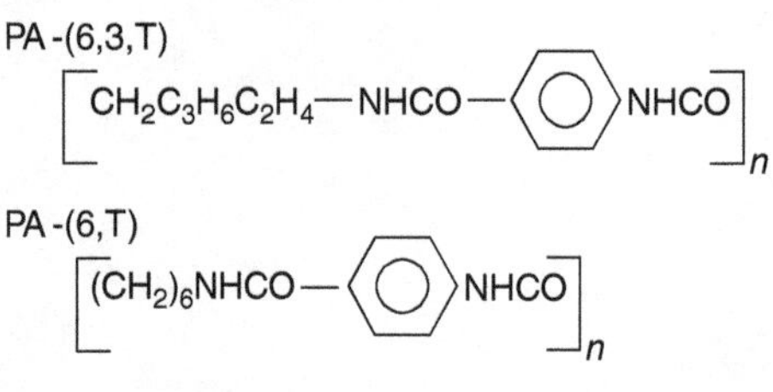

Figure 10.11 Chemical structure of amorphous nylon.

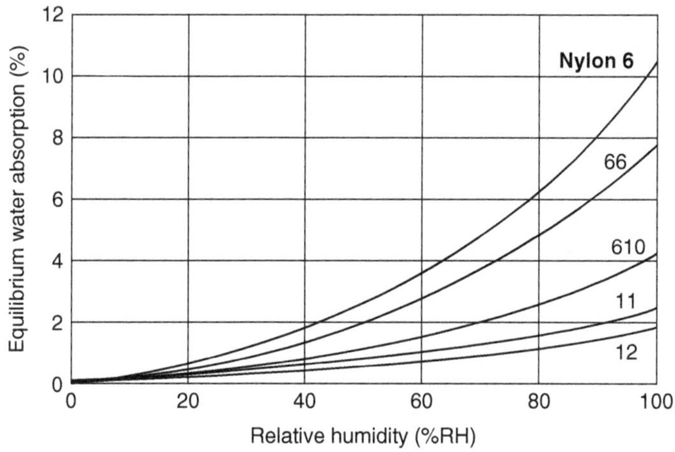

Figure 10.12 Moisture pick-up for several nylon plastics. Source: Nylon moisture 2020.

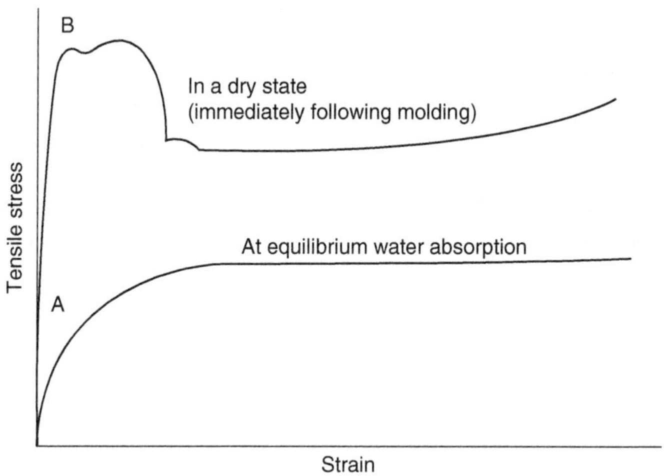

Figure 10.13 Tensile strength of Nylon 12 after exposure to water. Source: Nylon and moisture 2020.

Adhesive applications can include hot melt or solution type. They are thermosets reacting with epoxy or phenolic resins. Some flexible adhesives can be used for bread wrappers, dried soup packets, book bindings. Mechanical properties of nylon are provided in Table 10.7.

The advantages of nylon are that they are tough, strong, impact-resistant plastic with low coefficient of friction and low abrasion resistance. Nylons have

Table 10.7 Mechanical properties of Nylon.

	Nylon 6	Nylon 6,6	Nylon 6,10	Nylon 6,12
Density, g/cc	1.13–1.15	1.13–1.15	1.09	1.05–1.10
Tensile strength, psi	6000–12 000	14 000	8500	6500–8800
Tensile modulus, psi	300k	230k—550k	250k	220k—290k
Tensile elongation, %	30–100%	15–80%	70%	150%
Impact strength, ft-lb/in	0.6–2.2	0.55–1.0	1.2	1.0–1.9
CLTE, 10^{-6} mm/mm/C	50	50	80	80
HDT, @264 psi	160F	160F	160F	160F

Source: Data from Nylon-properties 2020.

high-temperature resistance that can be processed by traditional thermoplastic methods. Nylons have good solvent resistance and are resistant to bases.

The disadvantages of nylon are high moisture absorption with dimensional instability. The tensile strength of wet nylon can be reduced up to 30% and the tensile modulus can be reduced up to 50%. Nylons are subject to attack by strong acids and oxidizing agents. Like all plastics, nylon requires UV stabilization. Nylon can have high shrinkage in molded sections. Electrical and mechanical properties of nylon are influenced by moisture content, Nylon can be dissolved by phenols. Additives for nylon include antioxidants, UV stabilizers, colorants, and lubricants.

The fillers that are blended with nylon include talc and calcium carbonate. Nylon can have glass fibers as reinforcements. The glass fiber is typically short E-glass fiber of 1/8″ in length or long fiber at 1/4″ in length. Fibers, like wollastonite or mica, can be added. Other fibers can include carbon fibers, graphite fibers, metallic flakes, and steel fibers. The properties of glass-filled nylon are shown in Table 10.8. Glass-filled nylon with 33% short glass fiber can be used for camshafts in cars and trucks. Table 10.8 shows that glass fibers double the tensile strength of nylon 6,6 and cause the tensile modulus to increase by 300%. Carbon fiber increases the tensile strength by 250% and increases the tensile modulus by 600%. Glass-filled nylon can replace aluminum in applications for the automobile, like valve covers, as shown in Figure 10.14.

10.8 Polyimide

Polyimides were developed by Du Pont in 1962. Polyimides are obtained from a condensation polymerization of aromatic diamine and an aromatic dianhydride. They are characterized as linear thermoplastics that are difficult to process. Many

Table 10.8 Properties of glass-filled nylon.

	Nylon 6,6	Nylon 6,6 (30% short glass fiber)	Nylon 6,6 (30% long glass fiber)	Nylon 6,6 (30% carbon fiber)
Density, g/cc	1.13–1.15	1.4	1.4	1.05–1.10
Tensile strength, psi	14 000	28 000	28 000	32 000
Tensile modulus, psi	230k—550k	1300k	1400k	3300k
Tensile elongation, %	15–80%	3%	3%	4%
Impact strength, ft-lb/in	0.55–1.0	1.6–4.5	4	1.0–1.9
CLTE, 10^{-6} mm/mm/C	50	12	12	10
HDT, @264 psi	160F	491F	491F	491F

Source: Glass-filled Nylon-Properties 2020.

Figure 10.14 Glass filled nylon valve covers for Ford Mustang. Source: Nylon valve cover 2020.

polyimides do not melt but are fabricated by machining. Molding can occur if enough time for flow is allowed for $T > T_g$. The heat deflection temperature (HDT) of polyetherimide is 210 °C. Heat deflection of several plastics is listed in a recent article (HDT-plastics 2020).

The chemical formula of polyimide is shown in Figure 10.15.

The advantages of polyimides are high-temperature service up to 700 °C. They have excellent barrier, electrical properties, solvent, and wear resistance. Polyimide has good adhesion and is especially suited for composite fabrication. The

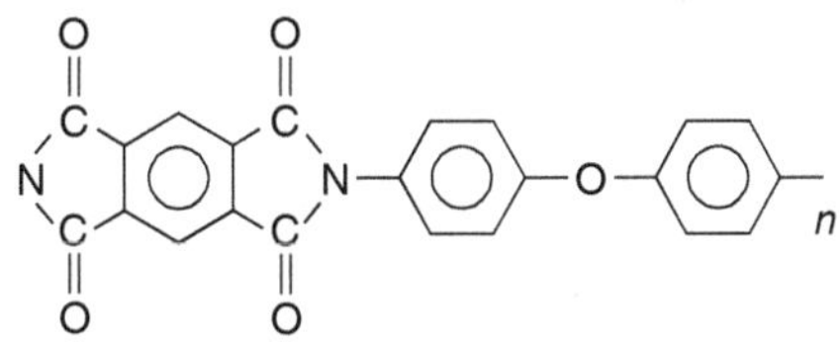

Figure 10.15 Chemical formula of polyimide.

Table 10.9 Mechanical properties of polyimide, nylon 6, and polyamide-imide.

	Nylon 6	Acetal	Polyimide	Poly amide-imide
Density, g/cc	1.13–1.15	1.42	1.43	1.41
Tensile strength, psi	6000–12 000	10 000	10 000	26 000
Tensile modulus, psi	300k	520k	500k	500k
Tensile elongation, %	30–100%	40–80%	90%	80%
Impact strength, ft-lb/in	0.6–2.2	0.07	0.9	2.5
CLTE, 10^{-6} mm/mm/C	50	100	40	45
HDT, @264 psi	160F	320F	410	400F

Source: Data from Polyimide-Properties 2020.

properties of polyimide are compared to nylon 6 and polyamide–imide in Table 10.9. The properties of polyimides are very similar to acetal and higher than Nylon 6.

10.9 Polyarylate

Polyarylesters repeat units feature only aromatic-type groups (phenyl or aryl groups) between ester linkages. Polyarylates cane be called wholly aromatic polyesters and can be based on a combination of suitable chemicals, like,

- *p*-Hydroxybenzoic acid
- Terephthalic acid
- Isophthalic acid
- Bisphenol-A

The mechanical properties correspond to a very stiff and regular chain with high crystallinity and high-temperature stability. Applications can include bearings, high-temperature sensors, and aerospace applications. Polyarylates can be processed with injection molding and compressing molding.

10.10 Polycarbonate

PC was invented in 1898 by F. Bayer in Germany. Commercial production began in the United States in 1959. PC is an amorphous, engineering thermoplastic that is known for toughness, clarity, and high-heat deflection temperatures. PCs are linear, amorphous polyesters because they contain esters of carbonic acid and an

Figure 10.16 Chemical structure of PC.

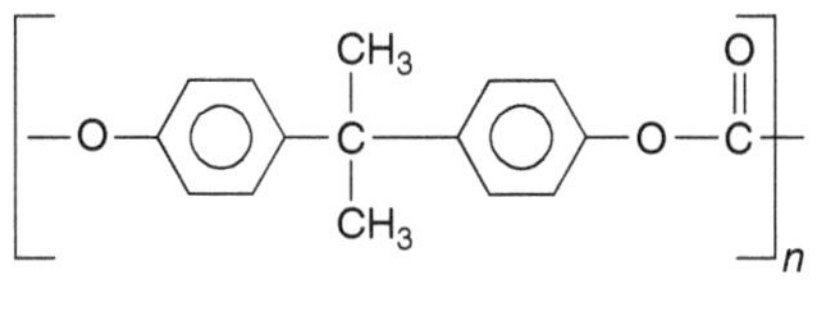

aromatic bisphenol. PC is produced with the chemical reaction of bis-phenol A (BPA) and phosgene. The chemical structure of PC is shown in Figure 10.16.

The mechanical properties of PC are shown in Table 10.10. The table shows that PC has equivalent tensile strength and modulus as acrylics and ABS. PC has the highest elongation and impact strength. PC is a durable material with high impact resistance and low scratch resistance. PC is used very frequently with eye-wear lenses. PC has equivalent mechanical properties as Mechanical properties as Poly methyl methacrylate (PMMA) but significantly higher impact strength and thermal properties as measured with HDT.

PC can be blended with ABS should be used for door panels and fenders are cars. Saturn in the 1990s used PC/ABS for the door panels. Chevy Camaro and Pontiac used the PC/ABS in the door panel and fender in the 1990s as well. PC can be used for appearance housings and structural parts which need stiffness, gloss, impact, and heat resistance which is higher than ABS.

PC housings requiring UL V0, but not 5VA-that is, portable electronic devices, not permanently fixed electric devices are also good applications for PC/ABS. Others include appearance parts requiring color fastness under interior UV light, interior parts where temperature environment is between 180 and 230 F — “above the beltline” applications, and where nonshattering ductility is required

Table 10.10 Mechanical properties of ABS, acrylic, and PC.

	Acrylic	PC	ABS	PC/ABS
Density, g/cc	1.16–1.19	1.2	0.97–1.04	1.18–1.20
Tensile Strength, psi	5000–9000	9500	3300–8000	5800–9300
Tensile Modulus, psi	200k–500k	350k	320k—400k	350k–450k
Tensile elongation, %	20–70%	110%	1.5–25%	50–60%
Impact Strength ft-lb/in	0.65–2.5	16	1.4–12	6.4–11
Hardness	M38-M68	M70	R100-120	R95 -R120
CLTE 10^{-6} mm/mm/C	48–80	68	65–95	67
HDT 264 psi	165–209F	270	190F–225F	225F

Source: Data from LCP 2020.

like instrument panel retainers. Applications can be painted exterior parts that require low-temperature impact strength greater than 10 ft.-lb./in. is another place PC/ABS should be used. Any outdoor application should be painted.

10.11 Thermoplastic Polyurethane

Polyurethane can be either thermoset or thermoplastic. Both are made with the chemical reaction of an isocyanate and alcohol, or polyol. The thermoset polyurethane has low viscosity liquid resin that chemically reacts in the mold. The TPU is made at the production plant and made into plastic pellets or powder that is sent to the molders. The TPU does not have crosslinks in it and can be melted over and over again. The thermoset polyurethane has crosslinks of isocyanate and alcohol, or polyol. This will be further discussed in the thermoset chapter. The practically unlimited amount of possible combinations produced by varying the structure and/or molecular weight of the three reaction compounds allows for an enormous variety of different TPU. This allows urethane chemists to fine-tune the polymer's structure to the desired final properties of the material. The final polyurethane resin consists of linear polymeric chains in block structures. This is shown in Figure 10.17.

- Hard segment (isocyanate)
- Soft segment (polyol) or (alcohol)

The chemistry of polyurethanes involves the chemical reaction of isocyanates and alcohols as shown in Figure 10.18.

The final TPU plastic consists of linear polymeric chains in block structures. The crystalline or pseudo crystalline areas act as physical cross-links, which account for the high elasticity level of TPU, whereas the flexible chains will impart the elongation characteristics to the polymer. These "pseudo crosslinks," however,

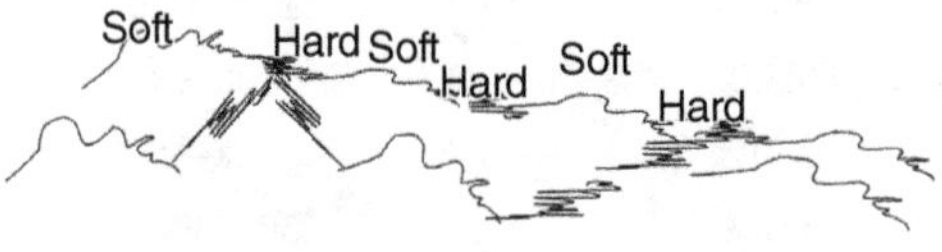

Figure 10.17 Hard and soft segments in polyurethane.

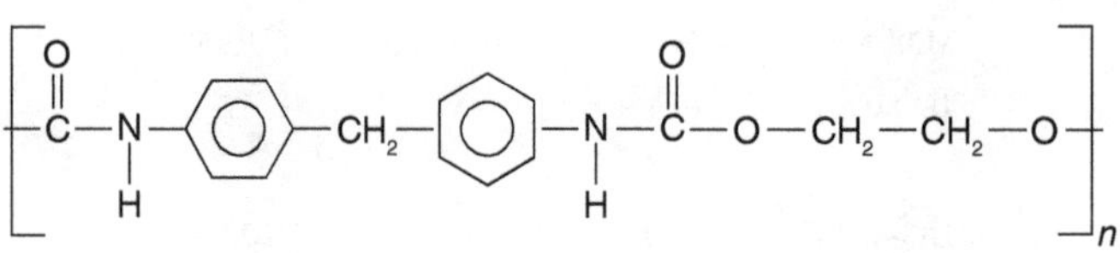

Figure 10.18 Chemistry of polyurethanes.

Table 10.11 Mechanical properties of TPU.

	TPU	LDPE	PP
Density, g/cc	1.16	0.92	0.9
Tensile strength, MPa	54	16	34
Tensile modulus, MPa	145	240	1378
Tensile elongation, %	300%	40–200%	80–100%
Impact strength, kJ/m^2	3	30	20
CLTE, μm/m/C	58	100	80
HDT, @1.8 MPa	85 °C	50 °C	75 °C

Source: Data from TPU properties 2020.

disappear under the effect of heat, and thus the classical extrusion, injection molding processing methods are applicable to these materials. TPU can be recycled through remelting of the recycled plastic.

The mechanical properties of polyurethanes can range from very soft and elastomeric to rigid depending on the ratio of isocyanate to polyol. The higher the amount of isocyanate yields higher mechanical properties. The mechanical properties of TPU are shown in Table 10.11.

TPU has many applications today. TPU can be used for caster wheels, power tools, sporting goods, medical devices, drive belts, footwear, inflatable rafts, and a variety of extruded film, sheet, and profile applications. TPU is also a popular material found in outer cases of mobile electronic devices, such as mobile phones. It is also used to make keyboard protectors for laptops.

10.12 Polyether-Ether-Ketone

PEEK and PEK are thermoplastics with high thermal properties and strength. PEEK was invented by ICI in 1982. PEK was introduced in 1987. PEEK and PEK are aromatic polyketones. PEEK is a linear aromatic polymer that is semicrystalline and is widely regarded as the highest performance thermoplastic material. PEEK has repeating monomers of two ether and ketone groups, as shown in Figures 10.19 and 10.20.

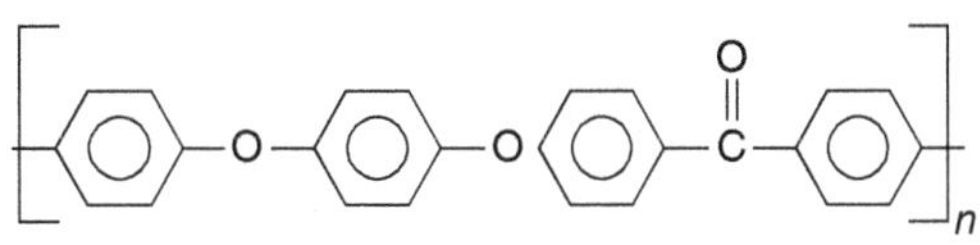

Figure 10.19 Chemistry of PEEK.

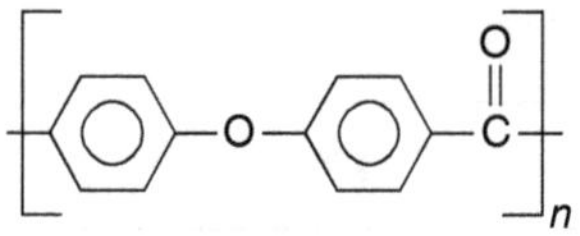

Figure 10.20 Chemistry of PEK.

Table 10.12 Mechanical properties of PEEK.

	PEEK	LCP	Nylon 6
Density, g/cc	1.30–1.32	1.35–1.40	1.13–1.15
Tensile strength, MPa	103	152	62
Tensile modulus, MPa	3447	17 237	2000
Tensile elongation, %	30–150%	1.3–4.5%	30–100%
Impact strength, kJ/m^2	35	140	30
CLTE, μm/m/C	40–47	25–30	50
HDT, @1.8 Mpa	165C	350C	160F

Source: Data from PEEK properties 2020.

The mechanical properties of PEEK are listed in Table 10.12 and are compared with LCP and nylon 6.

PEEK is naturally tan in color and can be pigmented with a wide range of colors, allowing for easy part identification. What sets PEEK apart from fluoropolymers is the fact that PEEK polymer retains its mechanical properties at extremely high temperatures (continuous service temperature of 500 °F/260 °C). The cost is $48 per pound in 2020. The product names are

- ICI: Vivtrex
- BASF: Ultrapak
- Hoechst Celanese: Hostatec
- DuPont: PEKK
- Amoco: Kadel

PEEK replaces metal tubing and is an ideal replacement for stainless steel, other types of metal tubing, and even glass, for weight reduction, comparable strength/mass, chemical resistance, hardness, and low extractables. PEEK most of all is comparable in strength, yet lighter and more cost effective than stainless steel. PEEK is polymer tubing, so the risk of corrosion, outgassing, or leaching (which can cause contamination) is minimal. PEEK is chemically resistant and inert with most acids and bases. PEEK with thin walls can also be made more flexible than stainless steel or titanium tubing, and can easily be cut to length with a razor blade. PEEK is weldable, machinable, and can be used with your existing stainless steel or polymer fittings. PEEK can be bonded with epoxies, cyanoacrylates, polyurethanes. Aerospace industry replaces aluminum with PEEK. It can be used as a fuel line brakes to replacement of primary structure. The electrical applications include wire coating for nuclear applications, oil wells, flammability-critical mass transit. Semiconductor wafer carriers can use PEEK, which can show better rigidity, minimum weight, and chemical resistance to fluoropolymers. Other applications include:

- Chemical and hydrolysis resistant valves (replaced glass).
- Internal combustion engines (replaced thermosets).
- Cooker components (replaced enamel).
- Components (replaced metal).
- High temperature and chemical resistant filters from fiber.
- Low friction bearings.

The advantages of PEEK are that:

- Very high continuous use temperature (240 °C).
- Outstanding chemical resistance.
- Outstanding wear resistance.
- Excellent hydrolysis resistance.
- Excellent mechanical properties.
- Very low flammability and smoke generation.
- Resistant to high levels of gamma radiation.

The biggest disadvantages are the high material costs and high processing temperatures.

10.13 PPO, PPS and PPE

Several plastics have been developed with the benzene ring in the backbone, including, PPO and polyphenylene sulfide. Both are used in aerospace applications. Polyphenylene is shown in Figure 10.21. Polyphenylene oxide is shown in Figure 10.22. Polyphenylene sulfide is shown in Figure 10.23.

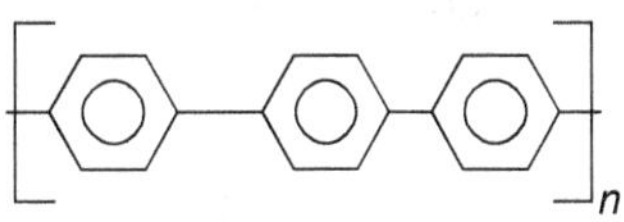

Figure 10.21 Polyphenylene chemical structure. Source: Polyphenylene (2022).

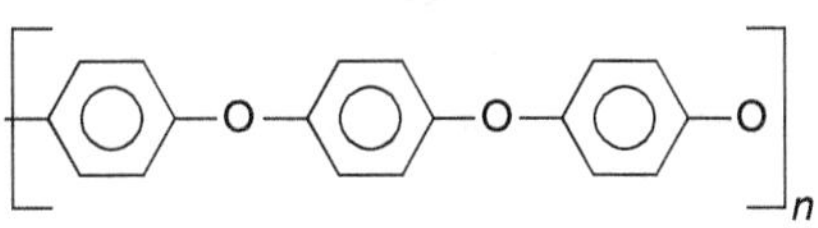

Figure 10.22 Polyphenylene oxide (PPO) chemical structure. Source: PPS-Chemical structure (2022).

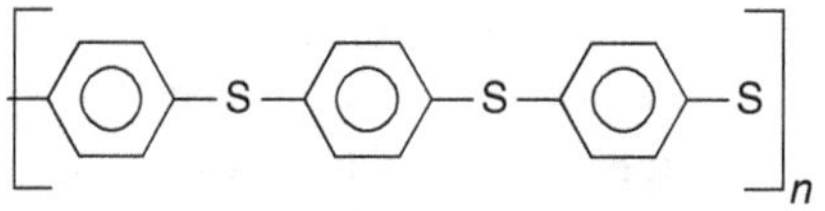

Figure 10.23 Polyphenylene sulfide (PPS) chemical structure. Source: PPS-Chemical structure (2022).

Table 10.13 Mechanical properties of PPS, PPO, and Nylon 6.

	PPS	PPO	Nylon 6
Density, g/cc	1.3	1.04–1010	1.13–1.15
Tensile strength, MPa	65	54	62
Tensile modulus, MPa	3300	2400	2000
Tensile elongation, %	1–2%	60–400%	30–100%
Impact strength, kJ/m^2	5	5	30
CLTE, μm/m/C	49	60	50
HDT, @1.8 Mpa	135C	95C	160F

Source: Data from PPS Properties 2020.

The mechanical properties are listed in Table 10.13 for PS, PPO, and nylon 6. The table shows that PPS has a higher density, tensile strength, and tensile modulus than PPO and nylon 6. PPS has lower tensile elongation, impact strength, and HDT than nylon 6. PPO, also known as polyphenylene ether (PPE), is a thermoplastic, linear, noncrystalline polyether that has high strength, high heat distortion temperature, and high chemical resistance (PPO-Properties 2022). It is also known as Noryl from Sabic Plastics Company. Because of its unique combination of high mechanical property, low moisture absorption, excellent electrical insulation property, excellent dimension stability, and inherent flame resistance, PPO has been widely used for a broad range of applications. For applications, PPO can be used in heater and air conditioner control system components, functional parts in the heating and ventilation sectors, instrument panels, interior mirror housings, seatbacks, grilles, spoilers, exterior mirror housings, wheel trims, and body panels.

The applications for PPS include computer components, range components, hairdryers, submersible pump enclosures, and small appliance housings. The applications for PPO include computer screens, pump impellers, small appliance housings, instrument housing, and some parts.

The advantages of PPS are that it has high usage temperature at 205 °C, good radiation resistance, excellent dimensional stability, low moisture absorption, good solvent, good chemical resistance, and excellent abrasion resistance. The advantages of PPO are that it has good fatigue and impact strength, good radiation resistance, excellent dimensional stability, and low oxidation.

The disadvantages of PPS are that it has high cost, high processing temperatures, and poor resistance to chlorinated hydrocarbons. The disadvantages of PPO are that it has high cost and poor resistance to certain chemicals.

10.14 Polytetrafluoroethylene

PTFE is a synthetic fluoropolymer of tetrafluoroethylene that has numerous applications. The best-known brand name of PTFE-based formulas is Teflon by Chemours. Chemours is a spin-off of DuPont Co., which discovered the compound in 1938. PTFE is a fluorocarbon solid, as it is a high-molecular-weight compound consisting wholly of carbon and fluorine. PTFE is hydrophobic: neither water nor water-containing substances wet PTFE, as fluorocarbons demonstrate mitigated London dispersion forces due to the high electronegativity of fluorine. PTFE has one of the lowest coefficients of friction of any solid.

PTFE is used as a nonstick coating for pans and other cookware. It is very nonreactive, partly because of the strength of carbon–fluorine bonds, and so it is often used in containers and pipework for reactive and corrosive chemicals. Where used as a lubricant, PTFE reduces friction, wear, and energy consumption of machinery. It is commonly used as a graft material in surgical interventions. Also, it is frequently employed as coating on catheters; this interferes with the ability of bacteria and other infectious agents to adhere to catheters and cause hospital-acquired infections. Figure 10.24 shows applications of Teflon.

Figure 10.24 Applications of Teflon. Source: Teflon 2020.

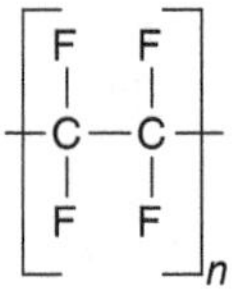

Figure 10.25 Chemistry of PTFE.

PTFE was accidentally discovered in 1938 by Roy Plunkett while he was working in New Jersey for DuPont. (Teflon 2020). As Plunkett attempted to make a new chlorofluorocarbon refrigerant, the tetrafluoroethylene gas in its pressure bottle stopped flowing before the bottle's weight had dropped to the point signaling "empty." Since Plunkett was measuring the amount of gas used by weighing the bottle, he became curious as to the source of the weight, and finally resorted to sawing the bottle apart. He found the bottle's interior coated with a waxy white material that was oddly slippery. Analysis showed that it was polymerized perfluoroethylene, with the iron from the inside of the container having acted as a catalyst at high pressure. Kinetic Chemicals patented the new fluorinated plastic (analogous to the already known polyethylene) in 1941, and registered the Teflon trademark in 1945.

By 1948, DuPont, which founded Kinetic Chemicals in partnership with General Motors, was producing over two million pounds (900 tons) of Teflon brand PTFE per year in Parkersburg, West Virginia. An early use was in the Manhattan Project as a material to coat valves and seals in the pipes holding highly reactive uranium hexafluoride at the vast K-25 uranium enrichment plant in Oak Ridge, Tennessee. The chemistry of PTFE is shown in Figure 10.25.

PTFE is produced by free-radical polymerization of tetrafluoroethylene as shown in Eq. 10.1.

$$nF_2C=CF_2 \rightarrow -\left(F_2C-CF_2\right)_n - \tag{10.1}$$

Because tetrafluoroethylene can explosively decompose to tetrafluoromethane and carbon, special apparatus is required for the polymerization to prevent hot spots that might initiate this dangerous side reaction. The process is typically initiated with persulfate, which homolyzes to generate sulfate radicals as shown in Eq. 10.2:

$$\left[O_3SO-OSO_3\right]^{2-} \rightleftharpoons 2SO_4\bullet^- \tag{10.2}$$

The resulting polymer is terminated with sulfate ester groups, which can be hydrolyzed to give OH end groups. Because PTFE is poorly soluble in almost all solvents, the polymerization is conducted as an emulsion in water. This process gives a suspension of polymer particles. Alternatively, the polymerization is conducted using a surfactant such as Perfluorooctanesulfonic acid (PFOS).

PTFE is a thermoplastic polymer, which is a white solid at room temperature, with a density of about 2.200 g/cm^3. According to DuPont, its melting point is 600 K (327 °C; 620 °F). It maintains high strength, toughness, and self-lubrication

Table 10.14 Properties of Teflon.

	PTFE
Density, g/cc	2.2
Yield strength, MPa	23
Tensile modulus, Mpa	500
Tensile elongation, %	300%
Impact strength, kJ/m^2	50
CLTE, µm/m/C	7.5
Coefficient of friction	0.05–0.10

Source: Data from PTFE 2020.

at low temperatures down to 5 K (−268.15 °C; −450.67 °F), and good flexibility at temperatures above 194 K (−79 °C; −110 °F). PTFE gains its properties from the aggregate effect of carbon–fluorine bonds, as do all fluorocarbons. The only chemicals known to affect these carbon–fluorine bonds are highly reactive metals like the alkali metals, and at higher temperatures also such metals as aluminum and magnesium, and fluorinating agents such as xenon difluoride and cobalt III fluoride. The properties of Teflon are shown in Table 10.14.

References

ABS (2020). https://en.wikipedia.org/wiki/Acrylonitrile_butadiene_styrene (accessed September 2020).

Camaro (2020). https://classiccars.com/listings/view/1229151/1996-chevrolet-camaro-ss-z28-for-sale-in-millburn-new-jersey-07836 (accessed September 2020).

HDT-plastics (2020). https://omnexus.specialchem.com/polymer-properties/properties/hdt-0-46-mpa-67-psi (accessed May 2020).

ICI (2020). http://www.gilgoint.com/pbt-unfilled-glass-fibre/ (accessed September 2020).

LCP (2020). https://en.wikipedia.org/wiki/Liquid-crystal_polymer (accessed September 2020).

LCP-Celanese (2020). https://www.celanese.com/engineered-materials/products/Vectra-LCPZenite-LCP/What-is-LCP.aspx (accessed September 2020).

LCP-Properties (2020). www.matlab.com (accessed September 2020).

Nylon (2020). https://www.toray.jp/plastics/en/amilan/technical/tec_007.html (accessed September 2020).

Nylon 12-strength drop (2020). https://intechpower.wordpress.com/tag/cast-nylon-12/ (accessed September 2020).

Nylon and moisture (2020). https://www.toray.jp/plastics/en/amilan/technical/tec_001.html (accessed September 2020).

Nylon moisture (2020). https://www.toray.jp/plastics/en/amilan/technical/tec_007.html (accessed September 2020).

Nylon valve cover (2020). https://www.moderngenauto.com/2011-2017-Ford-Mustang-50L-V8-Coil-Cover-Kit-w-Roush-Logo--Color-Options_p_26828.html (accessed May 2020).

Nylon-properties drop (2020). https://www.perkinelmer.com/CMSResources/Images/46-74182APP_InfluenceofWaterOnMechPropsandTG.pdf (accessed September 2020).

Nylon-properties (2020). www.matlab.com (accessed September 2020).

PBT (2020). https://omnexus.specialchem.com/selection-guide/polybutylene-terephthalate-pbt-plastic (accessed September 2020).

PBT-properties-PBT-PET properties 2020). https://www.compositesworld.com/articles/pbt-pet-polyester-part-2-the-performance-factor (accessed September 2020).

PEEK-Properties (2020). www.matlab.com (accessed September 2020).

Plastic costs (2021). https://www.plasticsnews.com/resin/currentPricing/engineering-thermoplastics (accessed April 2021).

Polyimide-Properties (2020). www.matlab.com (accessed September 2020).

Polyphenylene chemical structure (2022). Poly phenylene-chemical structure. https://en.wikipedia.org/wiki/Poly(p-phenylene) (accessed February 2022).

Polyphenylene, 2022, https://www.sciencedirect.com/topics/materials-science/polyphenylene (accessed August 2022).

Polyphenylene oxide, 2022. https://en.wikipedia.org/wiki/Poly(p-phenylene_oxide) (accessed August 2022).

POM (2020). https://en.wikipedia.org/wiki/Polyoxymethylene (accessed September 2020).

PPO-Properties (2022). https://en.wikipedia.org/wiki/Poly(p-phenylene_oxide) (accessed February 2022).

PPS-Chemical structure (2022). https://en.wikipedia.org/wiki/Polyphenylene_sulfide (accessed February 2022).

PPS-Properties (2020). www.matweb.com (accessed September 2020).

PTFE (2020). https://www.sukoptfe.com/about-polytetrafluoroethylene-ptfe-details (accessed September 2020).

Saturn (2020). https://www.edmunds.com/saturn/s-series/1997/review/ (accessed September 2020).

Shibaev, V. and Lam, L. (1994). *Liquid Crystalline and Mesomorphic Polymers*. New York: Springer.

Teflon (2020). https://omnexus.specialchem.com/selection-guide/polytetrafluoroethylene-ptfe-fluoropolymer (accessed September 2020).

TG-ABS (2020). www.matweb.com (accessed September 2020).

TPU-Properties (2020). www.matlab.com (accessed September 2020).

11

Thermoset Polymers

Polymer materials can be made using a thermosetting process. These thermoset plastics are the resin materials that can be used to make a polymer composite by chemically reacting the monomer polymers into a crosslinked and long-chain polymer. The thermoset process builds molecular weight in the polymer during the polymer reaction and it can also crosslink the polymer to form a new polymer.

This chapter will describe the thermoset polymer materials in two categories, namely, automotive and aerospace polymers.

11.1 Automotive Thermoset Polymers

The thermoset polymers for automotive applications are thermosets and reacting polymers that are produced with the following characteristics:

1) Cycle times less than two minutes.
2) Service use temperatures of 60–70 °C (140–158 °F).
3) Low-cost materials.
4) Mild chemical resistance.
5) Painted for exterior body use.
6) Good mechanical properties.
7) Good thermal properties.

The following sections discuss the chemistry, polymerization methods, properties, products, and characteristics of automotive thermoset polymers.

Sustainable Plastics: Environmental Assessments of Biobased, Biodegradable, and Recycled Plastics, Second Edition. Joseph P. Greene.

11.1.1 Polyester Resin

Polyester polymer resins are unsaturated synthetic resins formed by the reactions of dibasic organic acids and polyhydric alcohols. The polyesters are formed in a condensation or step-wise polymerization process where the reactants are chemically combined to form intermediates that react to form the polyester resin. Common dibasic acids are as follows:

- Isophthalic acid
- Orthophthalic acid
- Terephthalic acid

Common alcohols for polyester resins are as follows:

- Bisphenol A
- Diethylene glycol
- Ethylene glycol
- Neopentyl glycol
- Propylene glycol

The glycols can all be used to produce epoxies. The ethylene glycol provides low-cost alternative. The propylene glycol provides excellent styrene compatibility. The diethylene glycol provides superior toughness. The neopentyl glycol provides good weathering and chemical resistance. The bisphenol A provides excellent strength, toughness, and weathering, chemical, and water resistance (Figure 11.1).

The reaction for polyester polymerization occurs in two steps when the unsaturated (double) bonds are transformed chemically to saturated (single) bonds as the polymer builds molecular weight in the first step and the polyester resin is crosslinked in the second step. The first step is shown in Figure 11.2 below.

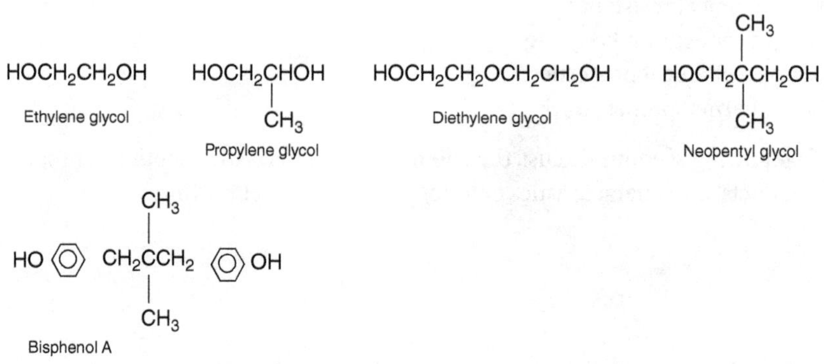

Figure 11.1 Glycol materials for reaction to form polyester polymer.

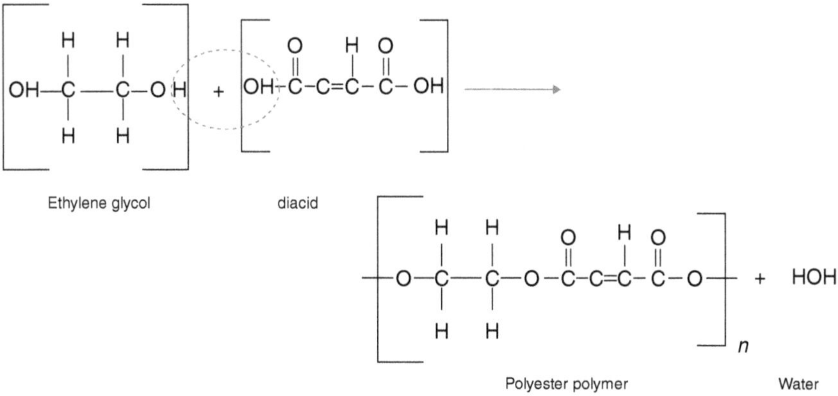

Figure 11.2 Step 1: Chemical reaction of glycol and diacid to form a polyester polymer.

The polymer builds molecular weight based on the reaction of the initiator, glycol, and the diacid. The diacid gives up a hydrogen, H, and bonds with a glycol or hydroxyl group to form an ester linkage and water. The polymerization is a condensation reaction that creates a water by-product. The polymerization continues to build the length of the polymer or molecular weight. The polymerization usually takes several weeks to build molecular weight. The polymer is typically placed in a cold environment like a freezer during the cure process. This prevents the polymer from forming crosslinks. The crosslinking reaction is shown in Figure 11.3.

Figure 11.3 Step 2: Chemical reaction of polyester and styrene to form a crosslinked polymer.

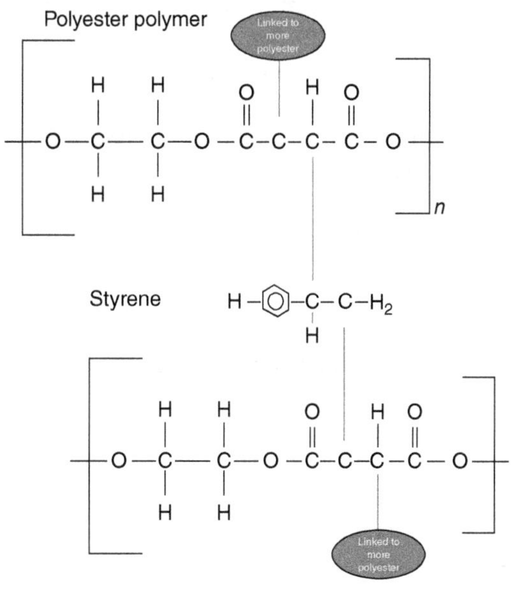

The second step in the polymerization of polyester is the crosslinking operation. In this step, heat or radiation can trigger the crosslinking reaction. The unsaturated bonds in the polyester open up with a peroxide and react with the styrene monomer to form a crosslinked polyester polymer. The crosslinking reaction is an addition polymerization that does not create by-products. The crosslinking step occurs with styrene and polyester resin usually in a heated mold. The chilled polyester resin is usually cut into strips about two inches wide and six inches long and stacked in a heated mold. The polyester resin utilizes a catalyst to initiate the reaction. Also, an accelerator can be used to speed up the reaction and an inhibitor can be used to terminate the crosslinking reaction. Styrene resin is the most common crosslinking agent. Styrene can comprise 30–50% of the volume of the chemicals in the crosslinking reaction. Vinyl toluene can be used to create a vinyl ester polymer. Methyl methacrylate (MMA) can be used for MMA polymer. The initiator can be a peroxide as in MEKP for methyl ethyl ketone peroxide. Typically, 1–2% of peroxide initiates the chemical reaction. The polymerization reaction is exothermic and will add heat to the polymer.

Polyester resin is the most common thermoset resin in use today. It can be used to make body panels on a car, shower enclosures, paints, gold carts, tubs for home, counter tops, and in car part repair (e.g. Bondo). It can be combined with glass fibers and calcium carbonate to make the sheet molding compound (SMC) body panel material. Often times, fillers like talc and calcium carbonate are added to the resin to control shrinkage and thermal stability. Also, glass and carbon fibers can be added to the polymer to create a strong and stiff polymer composite with metal-like properties. The polyester resin must be heated again at about four hours to approximately 65 °C (149 °F) after molding to drive off water in a post-curing process.

Vinyl esters can be produced with chemical reaction of an epoxy with an unsaturated monocarboxylic acid. The polymerization occurs to build molecular weight. It is then crosslinked with styrene to form the crosslinked polymer. The chemistry is shown in Figure 11.4.

The crosslinking in the second step is shown in Figure 11.5.

Other crosslinking reactants can be as follows:

- Styrene
- MMA
- Butyl acrylate and butyl methacrylate
- Alpha methyl styrene
- Vinyl toluene
- Para-methyl styrene
- Diallyl phthalate (DAP)
- Diallyl isophthalate (DAIP)
- Octyl acrylamide

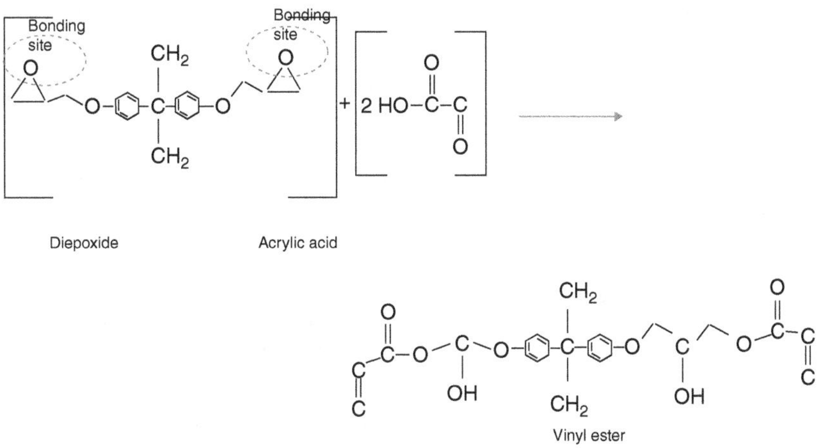

Figure 11.4 Step 1: Polymerization of vinyl ester from epoxy and monocarboxylic acid.

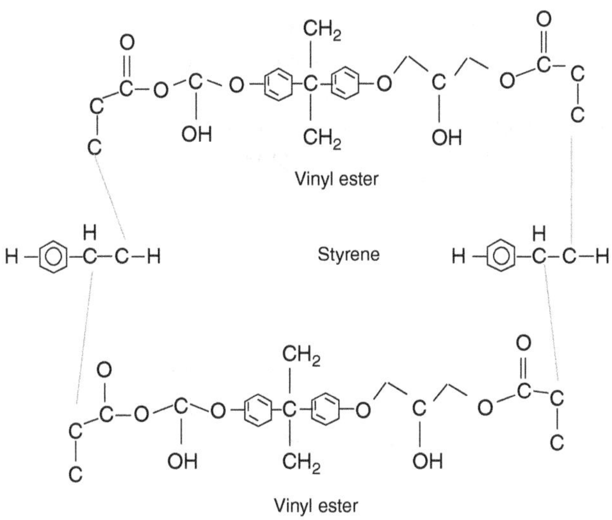

Figure 11.5 Step 2. Crosslinking of vinyl ester with styrene.

- Trimethylol propane triacrylate
- Triallyl isocyanurate
- Diallyl maleate
- Diallyl tetrabromophthlalate

Polyester materials can be cured using an appropriate hardener through a catalyst in order to hasten the process of decomposition. In the case of thermosetting epoxy materials that have ester linkages, hexahydro-4-methylphthalic anhydride (HHMPA) was appropriately adopted as the hardener and later 1-cyanoethyl-2-et hyl-4-methyimidazole was taken, which acted as a catalyst. In the case of materials that have carbamate linkages, the same hardener and catalyst were applied to hasten the decomposition rate.

When polyesters were subjected to temperatures below 350 °C, various insights emerged. From all the materials investigated, degradation was found to begin between 225 and 250 °C. This degradation is linked to thermal breakdown of acetal ester groups within thermosetting material structures. Thermogravimetric analysis (TGA) results indicate these materials have stable structures up to about 150 °C. However, rapid breakdown within their structures happens at temperature of around 200 °C. Additionally, it is apparent that weight loss in thermosetting materials that contain acetal esters, carbonates, carbamates, and esters is 100% at temperatures between 250 and 350 °C.

11.1.1.1 Mechanical Properties

The mechanical properties of polyester are excellent for automotive applications. Polyester use is mostly in automotive body panels, e.g. Corvette door, roof, and hood applications. It is a major component of SMC. The mechanical properties of the Ashland Chemical Arpol polyester are provided in Table 11.1.

Polyester is a liquid with good tensile strength and modulus. It is a brittle resin with low elongation at break. The specific gravity is 1.2, so it will sink even in a cup of water.

11.1.1.2 Processing of Polyesters

Polyesters can be used to produce paints, coverings, and resins for engineering composites. Polyester is a viscous liquid and thus can be pumped in a spray can to produce fiberglass boating. It can be added to glass fibers or carbon fibers using resin transfer molding (RTM). It can be combined with glass fibers and

Table 11.1 Mechanical properties of the Ashland Chemical Arpol polyester.

Property	Arpol NLN 7150	Arpol LP 67400
Tensile strength, psi	30 500	31 000
Tensile modulus, psi	1860	1930
Elongation at break, %	2	2
Flexural strength, psi	41 400	47 300
Flexural modulus, psi	1850	181

Source: Data from Polyester properties (2020).

Table 11.2 Mechanical properties of typical thermoset polymers.

	Polyester, BMC	Vinyl ester	Epoxy	Polyurethane (TPU)	Phenolics
Specific gravity	1.1	1.03	1.18	1.25	1.35
Viscosity, cP	500	575	800	1400	Solid
Ultimate tensile strength, psi	10 900	13 300	14 000	400	8400
Tensile modulus, kpsi	400	590	450	300	1250
Tensile elongation at break, %	2.4	4.6	4	30	12
Izod notched impact strength, ft-lbs/in	7	8.4	2.2	2	0.6
CLTE, μin/(in-F)	55	40	50	65	30
Heat distortion temperature (HDT) (66 psi) °F	284	300	350	220	400

Abbreviations: Bulk Molding Compound (BMC); centipoise (cP); Boeing Material Specification (BMS); Coefficient of Linear Thermal Expansion (CLTE); Heat Distortion Temperature (HDT), Thermoplastic polyurethane (TPU).
Source: Data from Thermoset properties (2020).

calcium carbonate in an SMC and compression molded into a part with a heated mold. The polyester resin will crosslink with styrene in the heated mold to flor a solid part.

11.1.1.3 Mechanical Properties

The mechanical properties of vinyl ester, polyester, polyurethane, and phenolics (Phenolics 2021) can be found at www.matweb.com. The properties listed in Table 11.2 are typical values found on the website.

The values in Table 11.2 show that the thermoset resins have reasonable tensile strength, modulus, and elongation. BMC is a bulk molding compound polyester. Vinyl esters have the highest impact strength than other thermoset polymers. Epoxy has the highest ultimate tensile strength. Vinyl ester has the highest tensile modulus. Phenolics have the highest heat distortion temperature (HDT).

11.1.2 Epoxy

Epoxies are polymerized with a wide range of reactants, which include polyfunctional amines, acids, anhydrides, phenols, alcohols, and thiols.

Epoxy resins are a class of reactive polymers that contain epoxide groups. Epoxy resins may be crosslinked either with themselves or with a wide range of co-reactants, including amines, acids, anhydrides, phenols, alcohols, and thiols. These co-reactants are often referred to as hardeners. The ratio of epoxy resin to

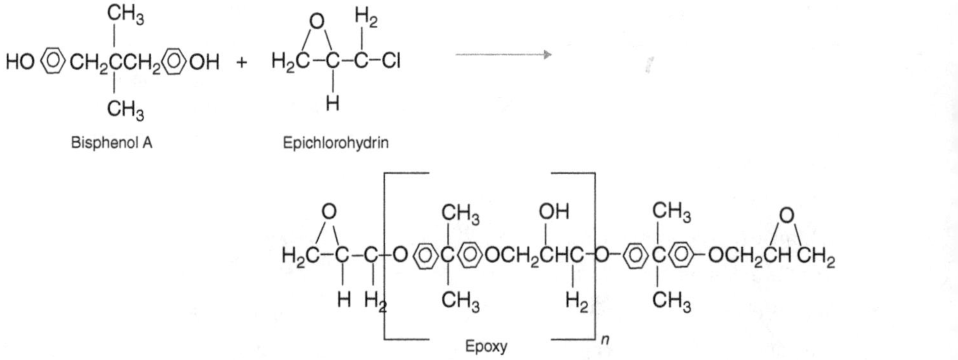

Figure 11.6 Epoxy chemistry.

hardener may be 1:1 or 4:1. The chemistry of epoxies is shown in Figure 11.6. The chemistry of epoxies involves the chemical reaction of bisphenol A and epichlorohydrin as shown below. The degree of polymerization, n, can be between 1 and 15 for epoxy polymers.

11.1.2.1 Epoxy Applications

Epoxy polymers can be used in many applications, including coatings for electronics, paints, coatings, resin for soft tooling, structural adhesives, and as the resin for aerospace high-performance composites when combined with carbon fiber reinforcement. The thermosetting epoxy polymer can achieve high mechanical, thermal, and chemical resistant properties, as shown in Table 11.2. Epoxy can be used in automotive applications for some under the hood parts and for making of composite molds.

In epoxy materials, based on ester linkages, decomposition was notable at temperatures between 250 and 300 °C. This temperature is lower when compared to the same material with primary ester bonds, which shows significant decomposition at 350 °C. Upon degradation test, it was found that both strength and crosslink density showed consistent decrease. Apparent from this is that tertiary esters were responsible for the differences noted. In the case of thermosetting epoxy materials that have carbamate linkages, decomposition began at temperatures between 250 and 300 °C. However, for commercial cycloaliphatic epoxide that lacks the carbamate link, decomposition began at 350 °C. Unlike in the earlier case, complications arising from these materials' possible behavior as catalysts made it hard to account for the emerging differences (Giménez et al. 2005). Similarly, in the thermosetting materials that contain carbonate linkages, decomposition was found to begin at temperatures that are slightly lower than 350 °C. Apparently, this coincides with the decomposition temperature of cycloaliphatic di-epoxide, which is normally found in commercial applications.

11.1.2.2 Processing of Epoxies

Epoxies can be used to produce paints, coverings, and resins for engineering composites. Epoxy is a viscous liquid and thus can be added to glass fibers or carbon fibers using RTM. It can be combined with carbon fibers in RTM process to produce a part with a heated mold. The epoxy resin will crosslink with the hardener in the heated mold to flor a solid part.

11.1.3 Polyurethane

Polyurethanes are a reaction between an isocyanate and an alcohol, which is called a polyol. The polyurethanes are an interesting polymer that has a hard segment from the di-isocyanate and a soft segment from the polyol. The di-isocyanate has a polymer chain in addition to the NCO– functionality. The polyol also has a prepolymer on the polymer chain in addition to the OH– functionality.

The isocyanate can be methyl di-isocyanate (MDI) or toluene di-isocyanate (TDI) based. MDI is used for rigid and flexible polymers. TDI can be used to make flexible polyurethane foams. Flexible polyurethane foam is used in furniture cushions and other common household applications. The di-isocyanate monomer can be MDI and polyol and their chemistry is shown in the chemical reaction of Figure 11.7 below.

Polyurethanes can be used for producing flexible parts, foams, or rigid parts depending on the ratio of di-isocyanate and polyol. For flexible parts the ratio is 1:2 iso:polyol; for foams the ratio is more than 1:4. For rigid parts the ratio is closer to 1:1.

11.1.3.1 Processing of Polyurethane

Polyurethanes can be used to produce paints, coverings, foam, rigid parts, and resins for engineering composites. Polyurethane is a viscous liquid and is usually pumped in a closed molding machine for foam, flexible parts, and firm parts. The

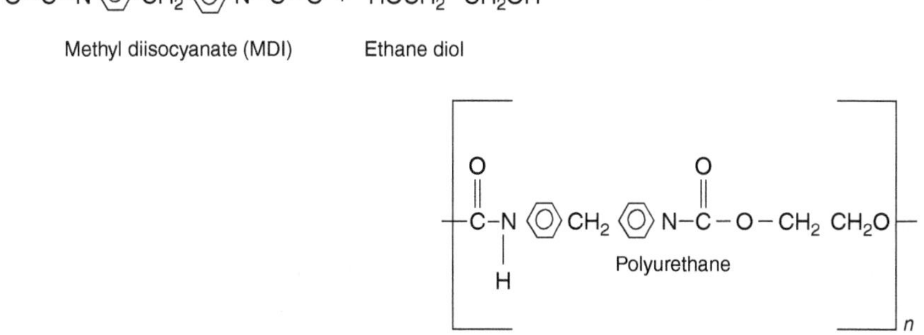

Figure 11.7 Polyurethane chemistry.

Figure 11.8 Polyurethane metering machine for foam. Source: PUR foam (2020).

polyurethane metering machine, typically, has 50 gallon tanks of isocyanate and polyol. The tanks meter the ratio of isocyanate (iso) and polyol to the mixhead that dispenses it into the foam mold. The flow rates are low and the pressures are less than 10 MPa. Figure 11.8 is an example of a typical polyurethane foam metering machine. For more rigid parts, polyurethane is dispensed in a reaction injection molding (RIM) machine. The polyurethane RIM machine, typically, has 50 gallon tanks of isocyanate and polyol. The tanks meter the ratio of isocyanate and polyol to the mixhead that dispenses it into the polyurethane mold for an automotive fascia or fender. The ratios are typically 1:1 of iso and polyol. The flow rates are low and the pressures are less than 20 MPa. The key to proper RIM molding is the use of impingement mixing of the polyol and isocyanate at pressures of 14 MPa (2000 psi) that atomizes the polymers and ensures proper reaction of the two monomers into polyurethane. The cycle times are around two minutes and the parts can be as large as 12 pounds. An example of a typical RIM machine is shown in Figure 11.9. An excellent fundamentals reference is Dr. Chris Macosco's book on RIM (Macosco 2000).

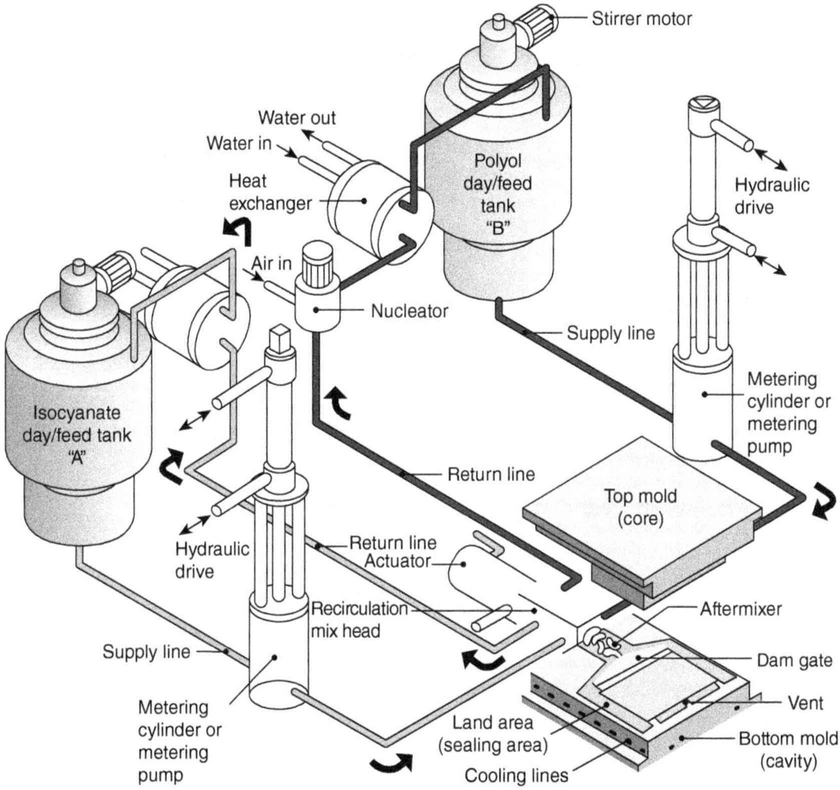

Figure 11.9 Polyurethane RIM machine for molded polyurethane parts. Source: PUR RIM (2020).

11.1.3.2 Polyurethane Automotive Applications

Polyurethanes can be used to make bumper fascia covering, fenders, and quarter panels, e.g. Pontiac Fiero car shown in Figure 11.10.

It also can be made into seats when it is blown into polyurethane foam, as shown in Figure 11.11.

The process to make a foam seat is inexpensive and requires low capital and investments. The process is shown in Figure 11.12.

Molded seats are made by mixing isocyanates, polyols, catalyst, and additives in two large tanks at approximately room temperature. The liquid is pumped through a long hose and pumped to the seat mold shown in Figure 11.12. The mold can be on a carousel to increase efficiency. The liquids are injected to the mold and the top of the mold is placed on the bottom mold. The molds are usually at room temperature. The liquid polyurethane rises in the mold and after the cure time is reached, the polyurethane seat is demolded from the mold. The mold is cleaned and mold release is applied for the next resin shot.

Figure 11.10 Pontiac Fiero car. Source: Pontiac Fiero car (2020).

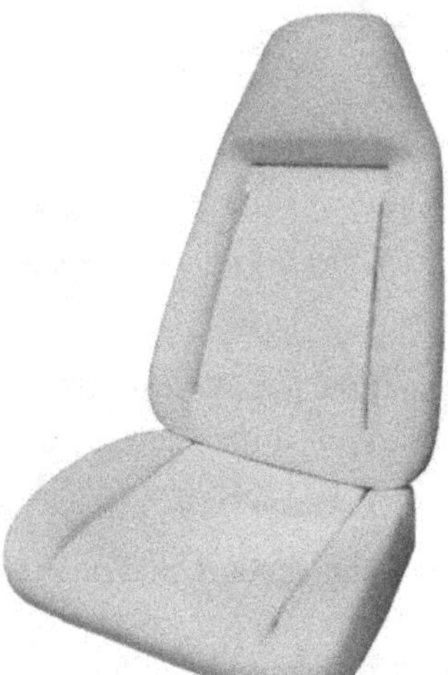

Figure 11.11 Polyurethane foam seat. Source: Foam seat (2020).

11.1.4 Phenolics

Phenolics are thermoset polymers that were first commercialized in the early 1900s and listed as Bakelite. Phenolics are also classified as phenols since they have an aromatic group and a hydroxyl (OH) group in its chemical structure. Phenolics are produced from a chemical condensation reaction of phenol and formaldehyde as shown in Figure 11.13 below. Three active sites on phenol allow for both polymerization and crosslinking reactions to occur. The reactions of phenolics can be carried

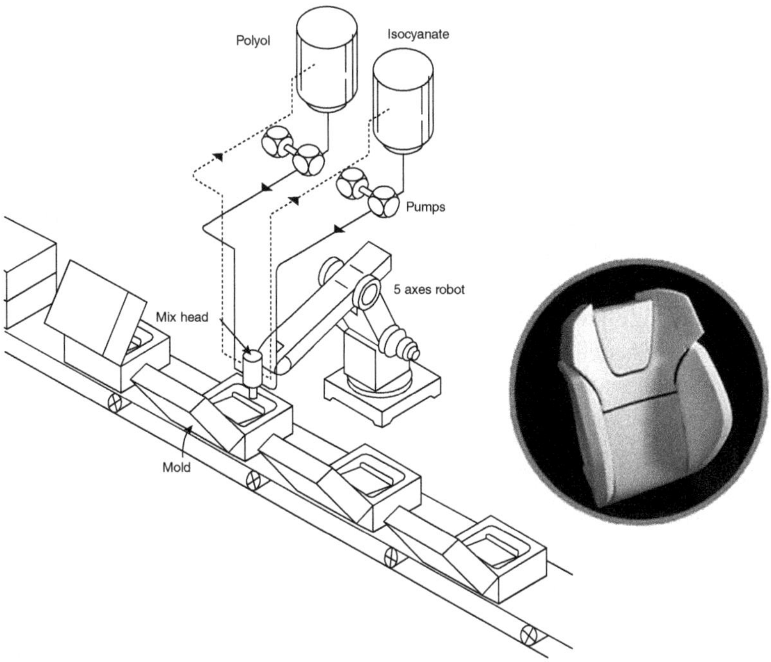

Figure 11.12 Polyurethane foam seat process. Source: Foam seat process (2020).

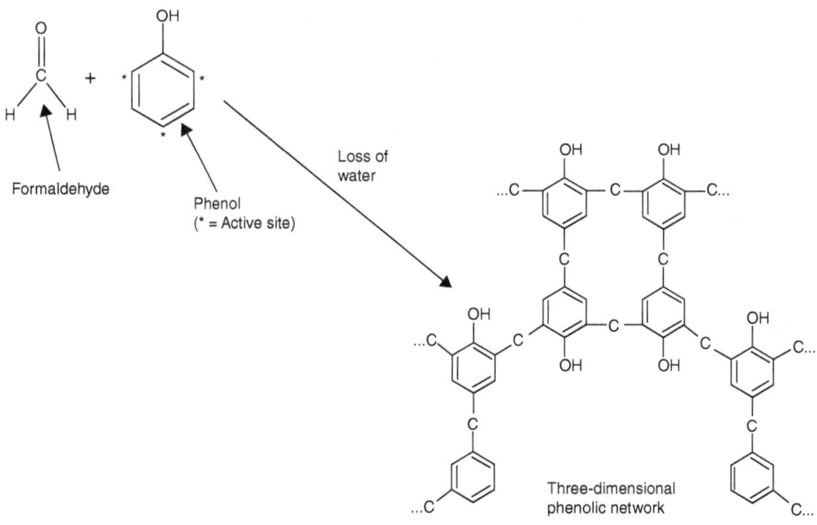

Figure 11.13 Chemistry of phenolics. Source: Phenolics-Wikipedia (2020).

out under different conditions for resole and novolac resins. Resoles are intermediate one-step liquid resins with excess formaldehyde solutions with no crosslinking reactions. Heat can subsequently crosslink the phenolic. Novolac resins are an intermediate with deficient formaldehyde solution that produces a powder without crosslinking. The crosslinking occurs with a curing agent of hexamethylene tetramine in a second step when heat is applied to the phenolic. The chemical structure of phenolics is shown in Figure 11.13. The polymerization reaction is with phenol with an aromatic molecule and formaldehyde. The phenol has three actives sites to react with formaldehyde molecule to form methylene polymer. Multiple reaction sites on phenolic ring allow rings to join together creating a three-dimensional (3-D) polymer structure. Phenolics can be made with prepreg composites with a B-staging of the resin. Phenolics are old thermoset resins. They can be used for general purpose, unreinforced plastics like in the following:

- Electrical switches.
- Junction boxes.
- Automotive molded parts.
- Consumer appliance parts, handles, and billiard balls.

Phenolics require the use of fillers due to their high shrinkage and brittle nature. Filler can include sawdust, nut shells, talc, or carbon black. Fiber-reinforced phenolics have aerospace applications. Phenolics can be used with carbon fiber mats for rocket nozzles and nose cones. Phenolics have an ablative nature, which means they go from solid to gas during burning, and do not have a liquid stage. Thus, there is less residue that can contaminate the rocket engines. It can be used for high-temperature aircraft ducts, wings, fins, and muffler repair kits. The carbon fiber matrices are used in applications requiring excellent heat resistance. Carbon fiber matrices are often made from phenolics.

The carbon-filled phenolics can be made into rocket nozzles or other aerospace parts. The common name for phenolic is Bakelite. Phenolics have excellent flammability properties, are good insulators with low heat transfer, have high thermal stability and electrical resistance, and have excellent resistance to chemicals, especially chlorinated solvents. Phenolics are suitable for transportation including interior parts solutions for commercial aircraft passenger cabins as well as mass transit applications where low flammability and low smoke generation are important.

Phenolics have low density, hence are extensively used with polyaramid honeycomb cores in aircraft interior. Phenolics can be co-molded as phenolic prepreg to dramatically reduce finishing costs of style line, radii, and other geometric variations or on a standalone basis. Phenolics with higher glass or carbon fiber content grades are suitable for aircraft interior structures.

11.1.4.1 Applications for Phenolics

Phenolics can be used for pan handles, appliance knobs, electrical switches, and automotive electrical parts. Phenolics are used in airplanes since it has a low smoke generation. It can be reinforced with carbon fiber and used as an exit nozzle since it has an ablative nature upon heating and transforms from solid to gas without going into a liquid state and has reduced charring. Carbon-fiber-reinforced phenolics can be used for brake linings and shoes for airplanes. Glass-fiber-reinforced phenolics can be used in automotive industry for engine components. Phenolics can be reinforced with glass fiber or carbon fiber and can be used for aircraft interior structures since they generate much lower smoke than polyesters and epoxies. They can be used for high-temperature and high-pressure environments for industrial and oil and gas applications.

11.1.4.2 Processing of Phenolics

Phenolics can be used to produce molded parts or engineering composites. Phenolics are a powder solid that can be combined with carbon fiber or glass fiber. The phenolic can be processed in a compression molding with a heated mold.

11.1.4.3 Properties of Phenolics

Phenolics have good chemical resistance and are resistant to alcohols, esters, ketones, ethers, chlorinated hydrocarbons, benzene, mineral oils, fats, and oils. Phenolics have good resistance to weak acids and bases but poor resistance to strong acids and bases, boiling water, and Ultra violet (UV) light. Phenolics have lower impact toughness compared to epoxies. Phenolics have excellent flammability properties. As shown in Figure 11.14, phenolics emit much less smoke than epoxies or polyesters. Phenolics create much less flame spread (1.0) than typical transportation requirements of less than 35. Phenolics create a smoke density of around 1 compared to 100. Phenolics char is slow to ignite creating a favorable material for the interior of airplanes. The biggest advantage of phenolics is their low smoke generation as shown in Figure 11.14. The smoke generated from phenolics is about 10% of polyester or epoxy composites after 10 minutes.

The Australian government looked at the flammability of composites and rated them accordingly. Table 11.3 shows the performance for flame spread for cabin composite materials. All of the materials were reinforced with glass fibers. The flame spread test was conducted according to American Society for Testing and Materials (ASTM) E162 standard.

Table 11.3 shows that polyether ketone ketone (PEKK) and polyimide have lower flame characteristics than phenolics. Epoxy has twice the flame spread and vinyl ester has five times the flame spread of phenolics. Phenolic item is in bold because of the significance of the material as the choice for aircraft use.

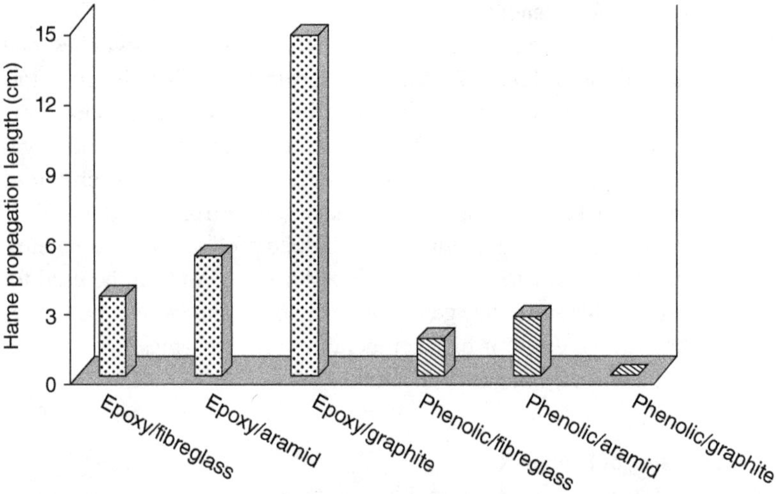

Figure 11.14 Flammability properties of phenolics. Source: Hottocks and Kandola (2005).

Table 11.3 Flame spread for cabin composite materials.

Ranking	Polymer matrix	Flame spread index	Improvement factor	Reference source
1	PMR-15 polyimide	2.0	2.70	Sorathia (1993)
2	Polyimide	2.5	2.16	Sorathia (1993)
3	PEKK	3.0	1.80	Sorathia (1993)
4	**Phenolic**	**5.4 ± 1.4**	1.00	
5	PPS	7.0	0.77	Sorathia (1993)
6	Epoxy	11.0	0.49	Sorathia et al. (1991)
7	Bismaleimide	18.0	0.30	Sorathia et al. (1991)
8	Vinyl ester	27.0	0.20	Sorathia et al. (1991)

Abbreviations: PEKK, polyetherketoneketone; PPS, polyphenylenesulfide.
Note: Entry in bold shows significance of the material as the choice for aircraft use.
Source: Data from Flame spread (2020).

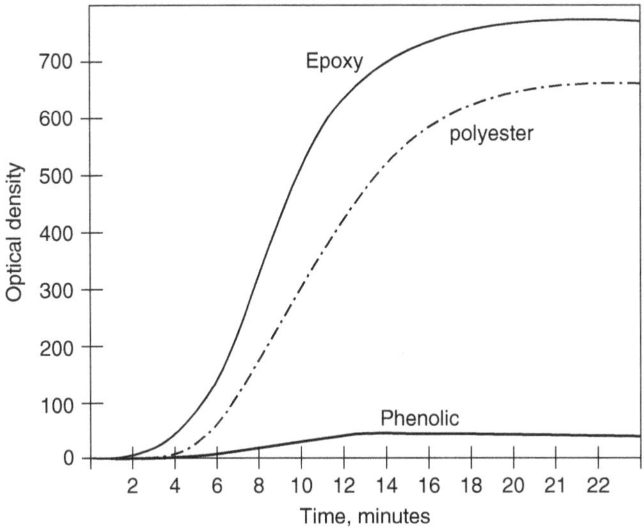

Figure 11.15 Optical density versus time for burning of phenolic, epoxy, and polyester. Source: Johnston et al. (1988).

Phenolics have superior properties compared to epoxy, vinyl ester, and polyester. The properties of phenolic are especially excellent for low flame and smoke. Figures 11.15 and 11.16 show the flame spread and specific optical density for four thermoset resins: phenolic, epoxy, polyester, and vinyl ester.

Phenolics have very low smoke generation compared to epoxy and polyester. Phenolics for char and aromatic rings are slow to ignite.

11.1.5 Silicones

Silicones, also known as polysiloxanes, are polymers that include any inert, synthetic compound made up of repeating units of siloxane, which is a chain of alternating silicon atoms and oxygen atoms, frequently combined with carbon and/or hydrogen. Silicones, a term used for polymers with alternating silicon and oxygen atoms with various organic radicals attached to the silicon, are unique compounds both in terms of their chemistry and in their wide range of useful applications. Some of these properties include low surface tension, high lubricity, enhanced softness, chemical inertness, low toxicity, and nonstick properties.

Silcones are typically colorless, oils, with rubber-like properties. Silicones can be used in sealants, adhesives, lubricants, medicine, cooking utensils, and thermal and electrical insulation. Some common forms include silicone oil, silicone grease, silicone rubber, silicone resin, and silicone caulk (Moretto et al. 2005).

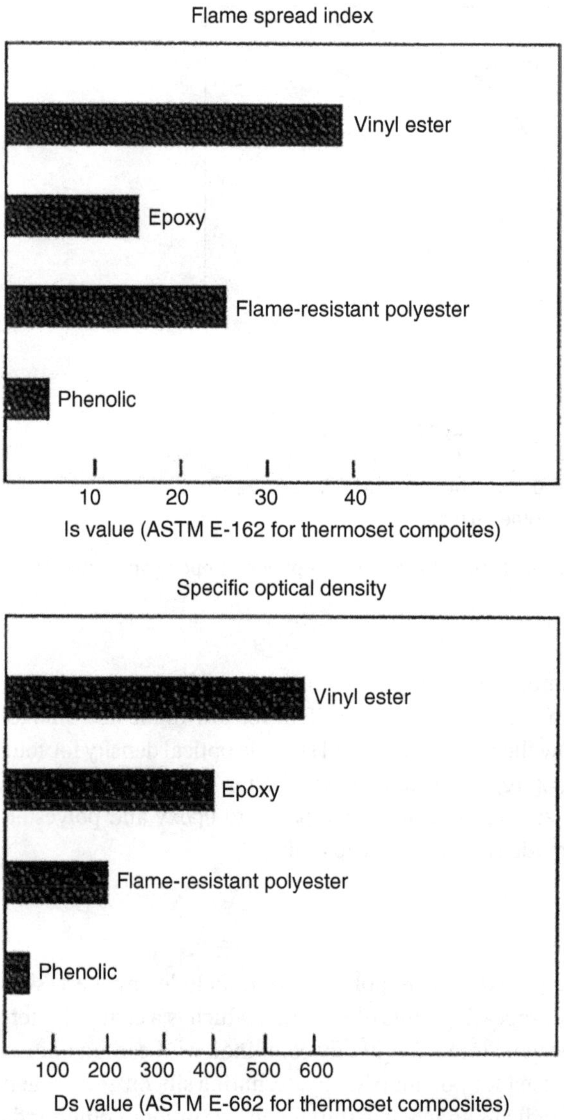

Figure 11.16 Optical density of phenolic, epoxy, vinyl ester, and polyester. Source: Johnston et al. (1988).

Silicones are more precisely called polymerized siloxanes or polysiloxanes; silicones consist of an inorganic silicon–oxygen backbone chain (⋯–Si–O–Si–O–Si–O–⋯) with organic side groups attached to the silicon atoms, as shown in Figure 11.15.

11.1.5.1 Silicone Rubber

Silicone rubber is an elastomer composed of silicone containing silicon together with carbon, hydrogen, and oxygen. Silicone rubbers are widely used in the automotive industry, and there are multiple formulations. Silicone rubbers are often one- or two-part polymers, and may contain fillers to improve properties or reduce cost. Silicone rubber is generally nonreactive, stable, and resistant to extreme environments and temperatures, from −55 to 300 °C (−67 to 572 °F), while still maintaining its useful properties. Due to these properties and its ease of manufacturing and shaping, silicone rubber can be found in a wide variety of products, including: automotive applications, cooking, baking, electronics, and medical devices and implants. Silicone can be used as an automotive silicone sealant (Silicone-rubber 2020).

11.1.5.2 Silicone Resin

Silicone resins are a type of silicone material that are formed by branched, cage-like oligosiloxanes with the general formula of $R_nSiX_mO_y$, where R is a nonreactive substituent, usually methyl or phenyl, and X is a functional group, e.g. hydrogen (H), hydroxyl group (OH), chlorine (Cl), or alkoxy (OR) (Silicone-resin 2020).

These groups are further condensed in many applications, to give highly crosslinked, insoluble polysiloxane networks.

When R is methyl, the four possible functional siloxane monomeric units are described as follows:

- "M" stands for Me_3SiO
- "D" for Me_2SiO_2
- "T" for $MeSiO_3$
- "Q" for SiO_4

Note that a network of only Q groups becomes fused quartz.

Silicone resins represent a broad range of products. Materials of molecular weight in the range of 1000–10 000 are very useful in pressure-sensitive adhesives, silicone rubbers, coatings, and additives. Polysiloxane polymers with reactive side group functionality such as vinyl, acrylate, epoxy, mercaptan, or amine are used to create thermoset polymer composites, coatings, and adhesives.

Silicone resins are prepared by hydrolytic condensation of various silicone precursors. In early processes of preparation of silicone resins, sodium silicate and various chlorosilanes were used as starting materials. Although the starting materials were the least expensive, structural control of the product was very difficult. Silicones can be used in automotive, marine, and vehicle applications. The silicone rubber adhesive sealants provide a durable, tight seal that stands up to rough conditions and high temperatures.

11.1.5.3 Chemistry

The chemical structure of silicones is shown in Figure 11.17. In some cases, organic side groups can be used to link two or more of these –Si–O– backbones together. By varying the –Si–O– chain lengths, side groups, and crosslinking, silicones can be synthesized with a wide variety of properties and compositions. They can vary in consistency from liquid to gel to rubber to hard plastic. The most common siloxane is linear polydimethylsiloxane (PDMS), a silicone oil. The second largest group of silicone materials is based on silicone resins, which are formed by branched and cage-like oligosiloxanes.

11.1.6 Dicyclopentadiene

Dicyclopentadiene, abbreviated DCPD, is a chemical compound with formula $C_{10}H_{12}$ (DCPD 2020). At room temperature, it is a clear, yellow color liquid with an acrid odor. It can be molded into fenders and vertical body panels. DCPD is produced with a steam cracking of naphtha and gas oils to ethylene. The major use is in resins, particularly unsaturated polyester resins. It is also used in inks, adhesives, and paints. DCPD has a density of 0.978, which is lower than epoxy, polyester, and polyurethane. DOW Chemical produces DCPD in large volumes. It is highly reactive and can react with other thermoset polymers like epoxy, polyester, and polyurethane. It can use RIM, like polyurethanes, to make production parts.

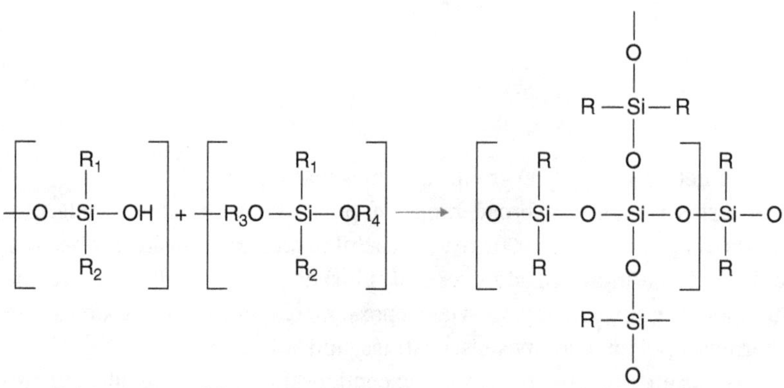

Figure 11.17 Chemistry of silicones.

11.2 Aerospace Thermosets

Aerospace thermoset polymers are those that are used in aircraft. The thermoset polymers have additional property characteristics, which include the following:

- High tensile modulus and stiffness.
- High tensile and impact strength.
- High thermal properties.
- Low coefficient of linear thermal expansion (CLTE).
- High thermal properties as measured with HDT.

These structural, mechanical, and thermal properties can be achieved with advanced polymers, e.g. epoxy, polyamines, polyimides, amino plastics, bismaleimide (BMI), polybenzimidazoles, phenolics, cyanurate esters, polyether ether ketone (PEEK), polyether ketone (PEK), liquid crystal polymer (LCP), and others. They are typically reinforced with carbon fiber at high volume concentrations of fiber, e.g. 60 volume percentage.

Structural thermosets are distinguished from standard thermosets by their use of more specialized resins and higher levels of reinforcement of carbon fibers and aramid fibers. The added reinforcement increases strength and stiffness, while resins protect the fibers and improve the compound's overall physical properties.

The epoxy thermoset resin is molded with vacuum-assisted processes like SCRIMP, VARTM, and others. SCRIMP refers to the Seeman Composites resin infusion molding process where liquid epoxy flows through tubes of plastic to impregnate the carbon fiber mat (SCRIMP 2021). VARTM refers to vacuum-assisted RTM where a vacuum is used to facilitate resin flow into a fiber layup contained within a mold tool covered by a vacuum bag. These will be discussed in more detail in the chapter on polymer composites. After the impregnation occurs, the composite part is allowed to cure at room temperature or elevated temperature in an autoclave (Wikipedia VARTM 2021).

Key to the molding of composites with epoxy resin is the proper viscosity of the epoxy resin. The epoxy resin viscosity is temperature dependent. The resin needs to flow into the carbon fiber mat and completely wet-out the fiber before curing. When the polymer is heated the liquid polymer material forms 3-D covalent bonds between polymer molecules during the crosslinking process. The crosslinked materials cannot be melted and reshaped or recycled.

Three of the most common thermoset resins are polyester, vinyl ester, and epoxy. Polyester and vinyl ester are used in automotive components but not usually in aerospace components. Epoxy is best used in aerospace applications that require high strength, high stiffness, and high thermal properties.

Epoxy was better described in the automotive section of this chapter and will not be repeated here.

11.2.1 Polyimides

Polyimides can also be used in aerospace applications. Polyimide is a polymer of imide monomers (Polyimides 2020). Polyimides can be thermoset or thermoplastic resins. They have high thermal properties and can be used as fuel cells, displays, and in other electrical devices. They are more expensive than epoxies and polyesters. They can be used with glass fiber to make a polyimide composite. They offer excellent mechanical properties and can be used in automotive plastics, films, laminating resins, insulation coatings, and high-temperature structural adhesives. The chemical structure is provided in Figure 11.18.

The most common form of polyimides is PMR-15. The polymerization of monomer reaction creates a 1500 formulated molecular weight. The polymer is made into prepreg and compression molded at 600 °F (316 °C) for several minutes. It can be molded with sintering from powder. Polyimides have high thermal and mechanical properties and low friction with excellent chemical resistance. It has continuous service temperature of 550–600 °F (288–316 °C), which is higher than epoxy-based composites.

BMI is a common polyimide with excellent thermal properties and easier processing conditions. Composite parts can be made with RTM or resin infiltration, vacuum bagging, or other hand-layup methods. The chemical structure is shown in Figure 11.19.

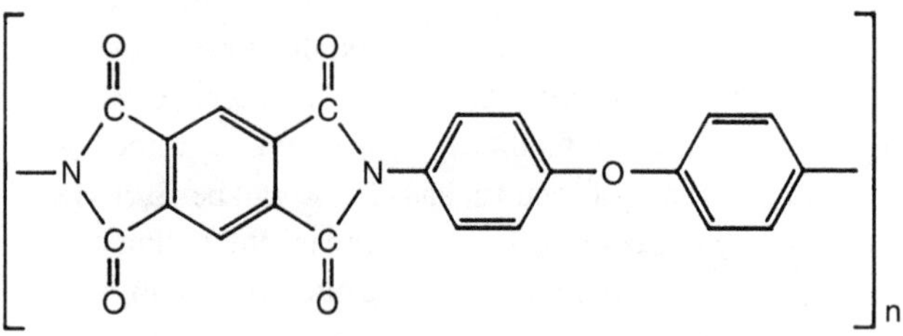

Figure 11.18 Chemical structure of polyimide. Source: Polyimides chemical (2020).

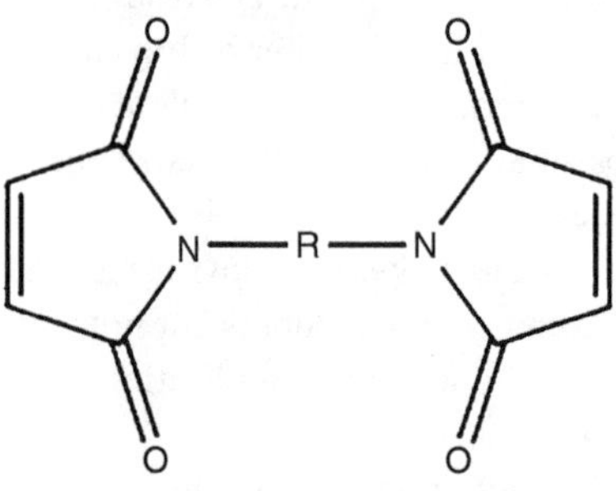

Figure 11.19 Chemical structure of BMI polyimide. Source: Iredale et al. (2017).

The chemical structure is a combination of two (bis)maleic groups with four carbons with C=C bonds.

Crosslinking occurs at C=C bonds. Crosslinking is achieved through exposure to heat. The low-molecular-weight polymer has high crosslinking density. The polymer composite is brittle but has high thermal properties with a glass transition temperature, T_g, greater than 300 °C. It can be used in wings, fuselage, and understructure on jets. It can also be used in flap hinge fairing, tail boom, and engine exhaust parts for airplanes and helicopters.

Polyimides are also used in aerospace applications. Polyimides can be either thermoplastic or thermoset polymers that are produced with carbon fiber reinforcements. Properties of polyimides show excellent thermal values (Iredale et al. 2017).

The thermoset polyimides are end-capped polyimides and BMI. The molecular structure is shown in Chapter 11 – Thermosets. The chemical structure includes four carbon plus nitrogen ring structure group in polymer backbone. The heterocycle group contains more than one type of atoms that are linked with aromatic group to form aromatic heterocycle molecule. It creates a very stiff molecule and is thermally stable due to structure.

BMI can be reinforced with glass fiber to create an excellent polymer composite. The density is 1.27 g/cc. The T_g is greater than 316 °C. The moisture absorption is 4.3%. The CLTE is 4.9×10^{-6} cm/cm/°C. The mechanical properties of the composite are shown in Figure 11.20.

The mechanical properties are listed in Table 11.4.

Polyimides have high thermal properties, high density, and high moisture absorption. The density and moisture absorption are higher than epoxy, polyester, and polyethylene. The thermal properties are higher than epoxy, polyester, and polyethylene.

	RT	350 °F (177 °C)	450 °F(232 °C)	350 °F(W)* (177 °C)	450 °F(W)* (232 °C)
Tension [warp]					
Strength, ksi (MPa)	60.70 (418.5)	49.60 (342.0)	51.60 (355.8)	–	–
Modulus, msi (GPa)	4.28 (29.5)	3.70 (25.5)	3.35 (23.1)	–	–
Compression [warp]					
Strength, ksi (MPa)	69.70 (480.6)	56.90 (392.3)	56.90 (392.3)	32.50 (224.1)	25.40 (175.1)
Modulus, msi (MPa)	4.56 (31.4)	4.06 (28.0)	4.07 (28.1)	3.61 (24.9)	3.35 (23.1)
Short Beam Shear					
Strength, ksi (MPa)	7.70 (53.1)	7.75 (53.4)	5.71 (39.4)	2.31 (15.9)	2.06 (14.2)
Flexure [warp]					
Strength, ksi (MPa)	92.50 (637.8)	78.60 (541.9)	78.50 (541.2)	46.30 (319.2)	31.90 (219.9)
Modulus, msi (GPa)	4.51 (31.1)	4.09 (28.2)	3.95 (27.8)	3.60 (24.8)	3.26 (22.5)

(w)–Wet condition–96 hours in boiling H_2O*

Figure 11.20 Mechanical properties of glass-reinforced HexPly F650 BMI. Source: HexPLY (2020).

Table 11.4 Properties of thermoset polyimide.

Property	Value
CLTE, $\times 10^{-6}$/°C	55
Shrinkage, %	0.2–1.2
Water absorption after 24 h, %	1.36
Density, g/cc	1.31–1.43
Tg, °C	250–240
HDT (1.8 MPa), °C	240–360
Maximum service temperature, °C	260–360

Abbreviations: CLTE; HDT, heat distortion temperature.
Source: Data from Iredale et al. (2017).

11.2.2 Amino Plastics

Amino resins can be made from urea and melamine formaldehyde resins. They are usually made by a batch process. The resins have low toxicity. Melamine resin or melamine formaldehyde (also shortened to melamine) is a resin with melamine rings terminated with multiple hydroxyl groups derived from formaldehyde. This thermosetting plastic material is made from melamine and formaldehyde. It can be used to crosslink with alkyd, epoxy, acrylic, and polyester resins, and is used in surface coatings (Melamine 2020). It can be used with polyester as a thermosetting resin for a composite material that uses glass fibers. Melamine formaldehyde can be used as counter tops for kitchens under the name of Formica®.

The chemical structure is shown in Figure 11.21. The properties are similar to polyester resin.

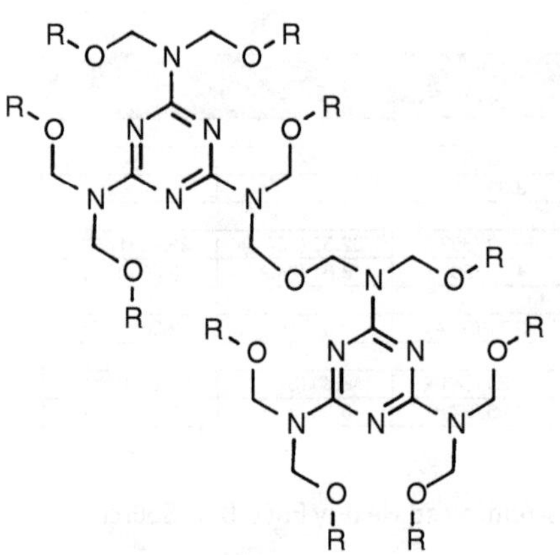

Figure 11.21 Chemical structure of melamine. Source: Melamine (2020).

In the case of thermosetting maleimides, they were first polymerized using 1,1-di(tert-amyl peroxy)cyclohexane (Khosravi, Ezat, and Osama M. Musa) with a process that led to the formation of thermosetting material with acetal ester linkages. Analysis was later done with the help of a hot-stage Fourier transform infrared spectroscopy (FTIR), gas chromatography-mass spectroscopy (GC-MS), thermogravimetric analysis (TGA), and differential scanning chromatography (DSC). On the other hand, preparation of the thermosetting ring opening metathesis polymerization (ROMP) material involved synthesis from two monomers, i.e. mono- and di-functional norbornene dicarboximide monomers. Apparently, these two contain the acetal ester groups, as shown in Figure 11.22.

The reaction resulted in the formation of *N*-caproic acid-exo-norbornene dicarboximide. This was later reacted with both 1,4-butanediol divinyl ether and butyl vinyl ether to form both mono- and di-functional monomers shown in Figure 11.23. The presence of acetal C–O bonds was confirmed using elemental analysis and nuclear magnetic resonance (NMR) spectroscopy.

Maleimide-based thermosetting materials that are extensively used in packaging of electronics exhibited almost similar behaviors once subjected to the test. Upon analysis by GC-MS, TGA, and hot-stage FTIR, degradation was found as

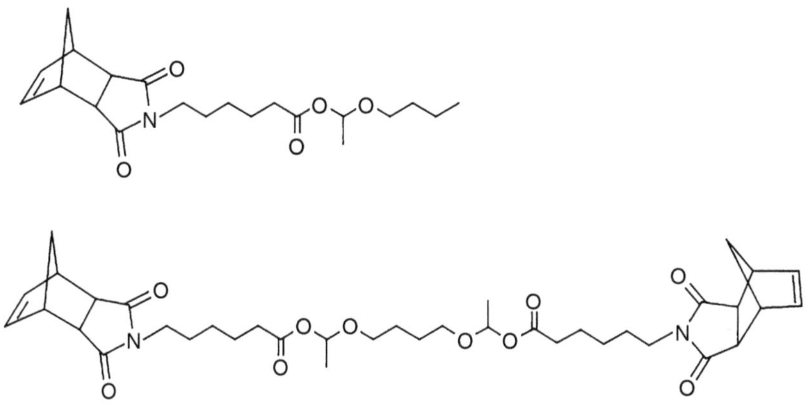

Figure 11.22 Primary thermosetting materials containing acetal ester group. Source: Khosravi, Ezat and Musa, Osama, (2011).

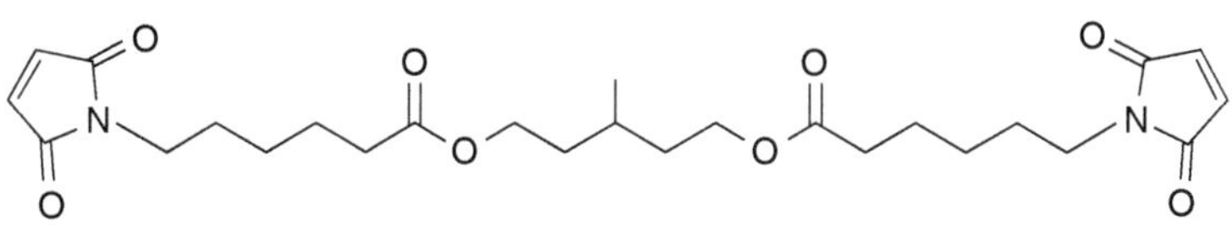

Figure 11.23 Secondary thermosetting material containing acetal ester group. Source: F. Sldivar-Gonzalez, et. al. (2019).

rapid at over 225 °C. When this thermosetting material was introduced into an adhesive, adhesion was improved. This is because of the adhesively merged substrates (González et al. 2007).

Additionally, the same material is said to have undergone a 10% loss in weight at a temperature of 242 °C. Figure 11.24 illustrates this.

Figure 11.24 shows that the thermosetting compound weight loss is 5% at 300 °C. Data captured from the hot-stage FTIR indicated that the intensity of the C–O peak reduced significantly. The article associates this to network breakdown due to degradation of linkages of acetal ester at temperature of about 250 °C. In the case of thermosetting ROMP materials, the structure of the thermosetting material investigated is as shown in Figure 11.25.

The results indicate that the stability of the material was good until the temperatures of about 150 °C. However, increase in temperature by three steps of 50 °C each led to weight losses of 14, 30, and 45%, respectively.

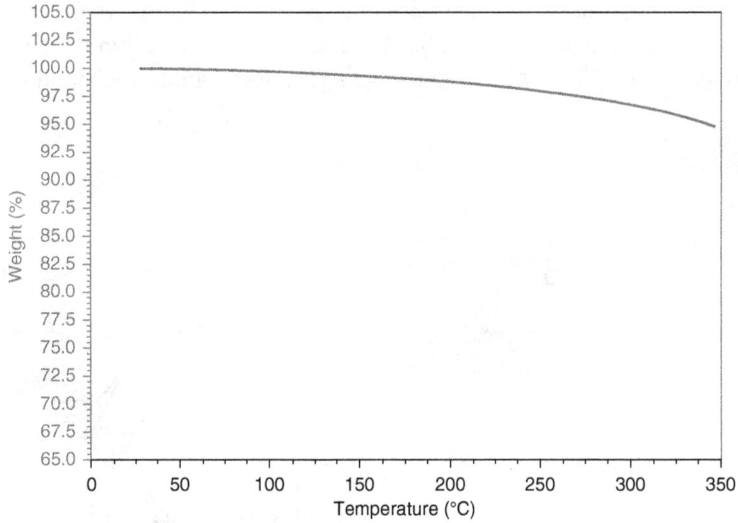

Figure 11.24 Thermogram for thermosetting acetal material in Figure 11.19. Source: Khosravi and Musa (2011).

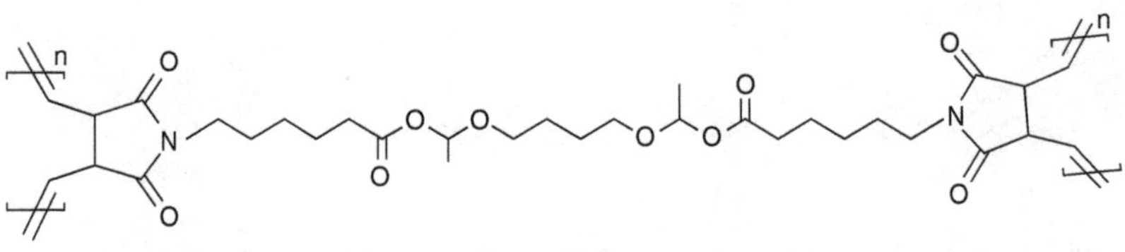

Figure 11.25 ROMP thermosetting material. Source: Sheng et al. (2009).

ROMP thermosetting materials have only partial thermal decomposition occurring within acetal ester groups. This leads to conversion of these thermosetting materials to thermoplastic material. This property of the ROMP thermosetting materials makes them recyclable.

11.3 Bio-Based Thermoset Polymers

Many polymers can be produced from bio-based sources. Corn products can be used for the alcohol portion of the reacting chemicals to form bio-based thermoset polymers. These polymers include the following:

- Bio-based polyesters
  - Alcohol (bio-based) + acids
- Bio-based epoxies
  - Bisphenol A + amines
- Bio-based polyurethanes
  - Alcohol (bio-based) + di-isocyanates
- Bio-based nylons
  - Amine

11.3.1 Bio-Based Polyesters

Bio-based polyesters can be produced with 100% organic sources. Polyesters are a chemical reaction between an alcohol and an acid. Both can be made from corn or from other natural products. Alcohol can be bio-based and made from corn. The bio-succinic acids can be made from bio-based sources. These bio-based polyesters can be made with renewable content up to 100% based on choice of diol. The performance is comparable to adipic polyols. The price is competitive with adipic polyols. They can be made with a smaller carbon footprint. GC Innovation America Company can produce bio-based polyesters (Bio-based polyester 2021). An excellent book on bio-based polymers discusses thermoplastics and thermoset polymers that can be produced from corn, sugar cane, and other organic sources (Voet, Vicent, Jager, Jan, and Folkersma, Rudy, *Plastics in the Circular Economy*, Bio-based polyester. https://www.ieabioenergy.com/wp-content/uploads/2020/02/Bio-based-chemicals-a-2020-update-final-200213.pdf April 2021).

Polyesters are mostly used in laminating resins, molding compositions, films, surface coatings, rubbers, fibers, plasticizers, composites, etc. Polyesters can be prepared by different methods such as through a polycondensation reaction of diacid, hydroxyacid, or anhydride and a secondary alcohol, or via a lactone ring opening polymerization. Bio-based polyesters can be made for drug release and tissue engineering (Naiwen et al. 2013).

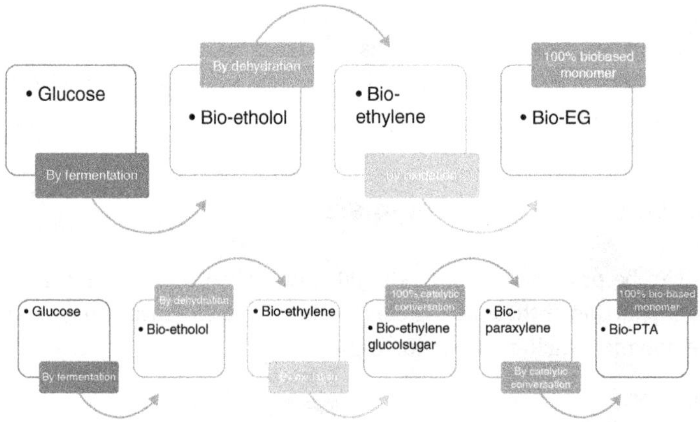

Figure 11.26 Chemical formula to produce bio-based ethylene glycol (bio-EG) and bio-based purified terephthalic acid (bio-PTA). Source: Siracusa and Blanco (2020).

In addition, several other thermoplastic polyesters are prepared by bio-based diols and carboxylic acids, such as poly(ethylene terephthalate [PET]) (Siracusa and Blanco 2020). Bio-based PET can be produced from bio-based feedstocks. The bio-based PET can be produced from bio-based diol, while the diacid or di-ester can be produced from bio-based succinic or adipic acid. The chemical formulation is provided in Figure 11.26.

The aliphatic monomer bio-based ethylene glycol can be synthesized from the hydrolysis of ethylene oxide, obtained via the oxidization of bio-ethylene obtained from the fermentation of glucose, followed by dehydration. The percentage of bio-based monomers can be varied, according to the chemistry, to obtain a partial bio-based polymer. In 2015, the Coca-Cola Company introduced a 100% bio-based beverage bottle, named "PlantBottle™," made of 100% bio-based ethylene glycol (EG) and terephthalic acid (Coca-cola 2021). PlantBottle product maintains the high-quality bottle with the added benefit of being made from renewable materials. The PlantBottle can be used for a variety of packaging sizes and across water, sparkling, juice, and tea beverage brands. Cocoa-Cola utilizes sugarcane to create PlantBottle products.

The bio-based polyesters can be used in shoe soles, belts, hoses, paints, adhesives, coatings, wheels, electronics, and foams.

11.3.2 Bio-Based Epoxies

Bio-based epoxies can be produced from unsaturated vegetable oils, saccharides, tannins, cardanols, terpenes, rosins, and lignin. Bio-based epoxies can be cured with a hardener. One company has produced bio-based epoxies with lower environmental impacts, such as lower carbon footprint. (Bio-based epoxies. April 2021).

The bio-based epoxy is combined with a hardener to create an excellent thermoset polymer. The mechanical properties of the bio-based epoxy should be very similar to the petroleum-based epoxy. A bio-based epoxy adhesive was made that produced excellent lap-shear strengths (Patel et al. 2019). The lap-shear strengths exceeded 20 MPa. The bio-based adhesive used a bio-based (bis-(furfurylamine) curing agent. The epoxy exhibited stable crack growth and excellent fracture behavior. It also demonstrated excellent adhesive strength. They used a bio-based epoxy formulation and a bio-based curing agent as described in Figure 11.27.

Bio-based epoxies can be produced by Entropy ECO-resin. It consists of poly(propylene glycol) bis(2-aminopropyl ether) (D230) and methyl hexahydrophthalic anhydride (MHHPA). Figure 11.28 shows the production process for

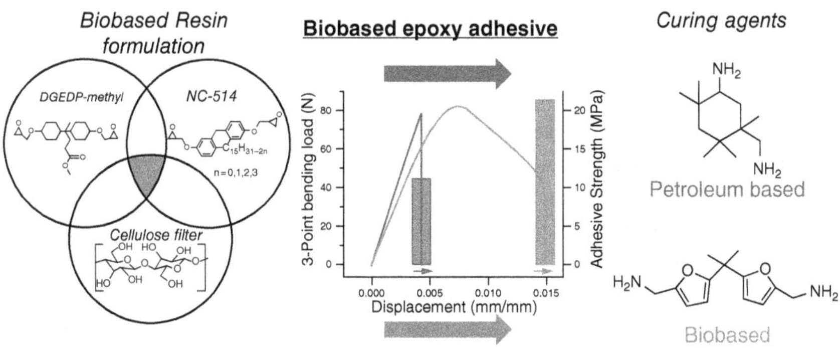

Figure 11.27 Bio-based epoxy formulation. Source: Patel et al. (2019).

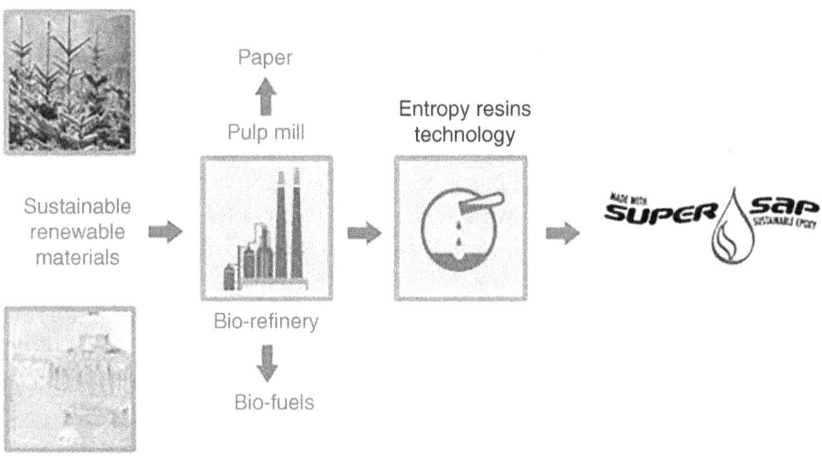

Figure 11.28 Production process of bio-based Entropy ECO-epoxy resin. Source: Mitxelena et al. (2015).

the Entropy ECO-resin. The eco-friendly resin is the Super-Sap Clear Laminating Resin/Clear Laminating System epoxy system produced by Entropy resins Inc. The eco-epoxy/flax system provided excellent mechanical properties with equivalent properties as the conventional polyester/glass composite material. The new material uses sustainable polymer and natural fibers with lower carbon footprint than conventional petroleum-based epoxy polymers.

The Entropy resins can be made with 50–100% bio-based materials. The bio-based materials can obtain up to 5% reduction in greenhouse gas emissions. The green chemistry eliminates harmful by-products and no bisphenol A. They also can be made with reduced power requirements and water consumption.

Figure 11.29 provides the typical working properties of the SuperSap^{TM} hardener system.

11.3.3 Bio-Based Polyurethanes

Bio-based polyurethanes can be produced for paints and surface coatings. Bio-based polyurethanes can be prepared by two different processes. The first is a prepolymer process and the second is a one-shot process. In both cases, polyurethanes are formed by a rearrangement reaction, which is not with the usual condensation polymerization. In the one-shot process, all of the components including di/polyol and di-isocyanate are reacted together at the desired ratio at 70–110 °C. The reactivity depends on the catalyst. Urethane oils can be made with vegetable oils, that react first with alcohol or glycerol, and then by reaction with the di-isocyanates. In the two-step method, polyols are reacted with the di-isocyanate at a ratio of 1:1 of iso:polyol. In the second step, this polyurethane prepolymer is reacted with a chain extender or di-isocyanate to obtain the desired ratio of reactants in the final polyurethane.

A large number of vegetable oil derivatives based on castor, soybean, sunflower, olive, peanut, canola, corn, and safflower oils, among others, are used as naturally obtained bioresources to produce a wide variety of bio-based polyurethanes with different structures and chemical compositions (Karak 2016). Low-molecular-weight polyurethane is obtained by the reaction of soybean oil polyol, a product of methanolysis of epoxidized oil with TDI at 50 °C. Both poly(urethane ester) and poly(urethane amide) resins are produced from seed oil derivatives, such as monoglyceride and fatty amide diol, respectively, by the reaction with TDI in the presence of polyethylene glycol as a chain extender and dibutyltin dilaurate as the catalyst using a one-shot process (Deka et al. 2010).

Bio-based polyurethane is also prepared by reaction of castor and linseed oils using lipase as the catalyst and TDI (Athawale and Joshi 2004). Polyurethane resins were produced from castor oil and different types of aromatic and aliphatic isocyanates. The reaction of castor oil and linseed oil was carried out at near

Typical Working Properties using compatible Super Sap® Hardener Systems		
Property	**1000 Hardener**	**HVA Hardener**
Hardener Product Rev	ONE01	HVA01
Mix Ratio by Weight	100:48	100:70–100:50
Mix Ratio by Volume	2:1	4:3
System Bio-Content by Mass[3]	39% – 48%	51% – 61%
Mix Viscosity (cPs @ 77 °F)	2000-3000	4000-7000
Gel Time (min, 150g @ 77 °F)	22	150
Thin Film Set (hrs @ 77 °F)	2	–
Tack Free Time (hrs @ 77 °F)	3	–
Cure Cycle (see notes below)	7 days @ 77 °F	30 min @ 180 °F
Typical Performance Properties[4]		
Tensile Molulus[5] (psi)	3.6×10^5	2.9×10^5
Tensile Strength[5] (psi)	9,000	6,200
Flexural Modulus[6] (psi)	3.0×10^5	2.5×10^5
Flexural Strength[6] (psi)	12,000	7,500
Elongation at Break[5] (%)	7	15
Tg (°F)	180	180

Figure 11.29 SuperSap™ hardener system. Source: SuperSap. https://entropyresins.com/ (2021).

ambient temperature using lipase as a catalyst. The materials were reacted with isocyanates to form polyurethane resins. The different chemicals resulted in significant differences in the hardness, adhesion, flexibility, impact resistance, and chemical resistance.

Several US companies are active in the development of soybean oil-based polyurethanes, including Cargill, Bio-based Technologies, Urethane Soy Systems, and DOW Chemical. DOW Chemical produces bio-based polyurethanes (DOW-polyurethane 2021). DOW refers to the new technology as RENUVA™ Renewable Resource Technology. RENUVA technology is used to produce bio-based polyols with high renewable content in the finished product with excellent performance and providing lower environmental impacts than petroleum-based polyurethanes. According to car manufacturer Ford Motors, all of its vehicles built in North America now use soy oil-based foams in the seat cushions, backs, and head restraints.

Ford first used soy oil-derived seating foam in its 2008 Mustang, in collaboration with automotive management systems supplier Lear Corporation. With the soy oil-based foams, Ford said it has helped reduce its annual petroleum oil usage by more than three million lbs (Ford Soy. http://www.scribd.com/doc/106613407/Bio-based-polyurethane#scribd April 2021).

According to industry analysts, soybean oil-based polyols can offer between 40 and 100% bio-based content, resulting in soy oil-based polyurethane (PU) that can have bio-based content of 30–100%, allowing for the PU to have a bio-based content between 22 and 70%, according to industry analysts. Last year, German chemical firm BASF teamed up with Ford to develop the automotive industry's first castor oil-based foam for the 2012 Ford Focus instrument panel. The PU foam is said to contain more than 10% bio-based content. Over time, Ford says it plans to incorporate castor oil-based foam solutions across more products in its global portfolio.

BASF Polyurethane Systems has been producing bio-based PUs for about a year now and bio-based thermoplastic polyurethane (TPU) elastomers for six months; TPUs have bio-based content ranging between 4 and 40% depending on the grade.

Bio-based polyurethanes are sustainable with lower carbon footprint than petroleum-based polyurethane, as shown in Figure 11.30.

11.3.4 Bio-Based Nylon-6

Nylon can also be made with natural bio-based materials. Every year, the world makes 5 million tons of nylon-6, which results in an estimated 60 million tons of greenhouse gas emissions. Producing nylon creates nitrous oxide, a greenhouse gas that is 300 times more potent than carbon dioxide. Manufacturing nylon also requires large amounts of water and energy, further contributing to

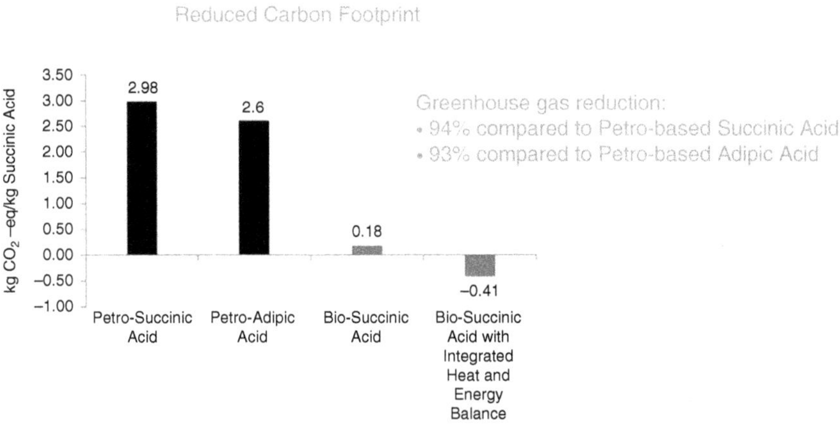

Figure 11.30 Carbon footprint of bio-based polyurethane. Source: Bio-based polyester (2021).

environmental degradation and global warming (Bio-based nylon 2021). Caprolactam can be produced from plant sugars to create nylon. Genomatica Company engineered a microorganism and production process that ferments the sugars found in plants to make the chemical intermediate for nylon-6. The chemical is then converted into nylon-6 polymer chips and yarn by Aquafil in Slovenia (Genomatica 2021). The properties of bio-based nylon are similar to petroleum-based nylon-6. Recently, researchers found that 2% loading of carbon fiber increased the tensile modulus, flexural modulus, and creep recovery of bio-based nylon-11 (Magniez et al. 2014). The nylon-11 was made with Rilsan BESNO TL, an injection molding grade of nylon-11 that was produced by Arkema. It has a melting point of 189 °C, a glass transition of approximately 55 °C, and a density of 1.02 g/cm^3. It is made from greater than 98% renewable materials. The data sheet is provided in Figure 11.31 and the tensile properties are provided in Figure 11.32.

11.4 Conclusions

Thermoset polymers can be very sustainable if made from bio-based sources, like agricultural waste. Thermoset epoxies, polyurethanes, nylons, and polyesters can be made with organic sources. Thermosets have high mechanical properties and lower density than aluminum and steel. When combined with glass or carbon fiber, they can replace aluminum and steel in automotive, industrial, home, and aerospace applications.

CAMPUS® Datasheet

Rilsan® BESNO TL - PA11
ARKEMA

ARKEMA

Product Texts
PA11,EHL,22-010
Rilsan® BESNO TL resin is a polyamide 11 produced from a renewable source. This natural grade is designed for extrusion.

The percentage of renewable carbon according to ASTM D 6866 (calculated) is **>98%.**

Main applications:

- Fluid transportation

Packaging:
This grade is delivered dried in sealed packaging (25 kg bags, 44 lb bags, 1000 lb rigid containers) ready to be processed.

Shelf Life:
Two years from the delivery. For any use above this limit, please refer to our technical services.

Rheological properties	**dry / cond**	**Unit**	**Test Standard**
Melt volume-flow rate, MVR	**1 / ***	cm³/10min	ISO 1133
Temperature	**235 / ***	°C	ISO 1133
Load	**2.16 / ***	kg	ISO 1133
Mechanical properties	**dry / cond**	**Unit**	**Test Standard**
Tensile modulus	**- / 1280**	MPa	ISO 527-1/-2
Yield stress	**- / 38**	MPa	ISO 527-1/-2
Yield strain	**- / 5**	%	ISO 527-1/-2
Nominal strain at break	**- / >50**	%	ISO 527-1/-2
Charpy impact strength, +23°C	**- / N**	kJ/m²	ISO 179/1eU
Charpy impact strength, -30°C	**- / N**	kJ/m²	ISO 179/1eU
Charpy notched impact strength, +23°C	**- / 11**	kJ/m²	ISO 179/1eA
Charpy notched impact strength, -30°C	**- / 12**	kJ/m²	ISO 179/1eA
Puncture - maximum force, -30°C	**- / 6000**	N	ISO 6603-2
Puncture energy, -30°C	**- / 70**	J	ISO 6603-2
Shore D hardness, 15s	**72 / ***	-	ISO 7619-1
Thermal properties	**dry / cond**	**Unit**	**Test Standard**
Melting temperature, 10°C/min	**186 / ***	°C	ISO 11357-1/-3
Temp. of deflection under load, 1.80 MPa	**50 / ***	°C	ISO 75-1/-2
Temp. of deflection under load, 0.45 MPa	**145 / ***	°C	ISO 75-1/-2
Vicat softening temperature, 50°C/h 50N	**160 / ***	°C	ISO 306
Coeff. of linear therm. expansion, parallel	**85 / ***	E-6/K	ISO 11359-1/-2
Burning Behav. at 1.5 mm nom. thickn.	**HB / ***	class	IEC 60695-11-10
Thickness tested (1.5)	**1.6 / ***	mm	IEC 60695-11-10
Yellow Card available	**Yes / ***	-	-
Electrical properties	**dry / cond**	**Unit**	**Test Standard**
Relative permittivity, 100Hz	**3 / -**	-	IEC 62631-2-1
Relative permittivity, 1MHz	**3 / -**	-	IEC 62631-2-1
Dissipation factor, 100Hz	**308 / -**	E-4	IEC 62631-2-1

Figure 11.31 Data sheet for bio-based nylon 11. Source: Properties of bio-based nylon (2021).

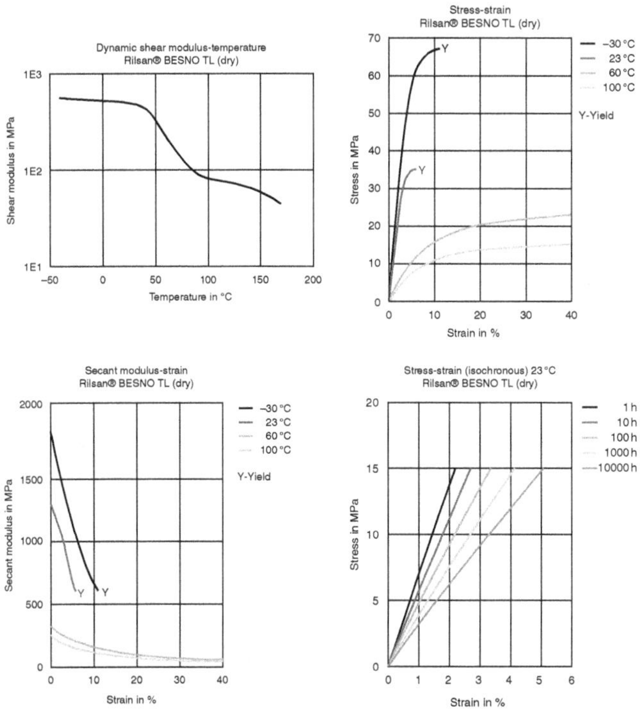

Figure 11.32 Tensile properties for bio-based nylon 11. Source: Properties of bio-based nylon (2021).

References

Athawale and Joshi (2004). Structure-property relationship in polyurethane coatings synthesized from chemoenzymatically modified caster oil. *J. Polym. Mater.* 21 (2): 165–173.

Deka, H. et al. (2010). Bio-based thermostable, biodegradable, and biocompatible hyperbranched polyurethane/Ag nanocomposites with antimicrobial activity. *Polym. Degrad. Stab.* 95: 1509–1517.

F. Sldivar-Gonzalez, et. al. (2019). Bicyclic acetals: biological relevance, scaffold analysis, and applications in diversity-oriented synthesis, *Organic & Biomolecular Chemistry*, Issue 5.

Giménez, R., Ferandex-Francos, Xavier, Salla, Josep, Serra, Angela, MAntecon, Ana, and Ramis, Xavier (2005). New degradable thermosets obtained by cationic copolymerization of DGEBA with an s (γ-butyrolactone). *Polymer* 46 (24): 10637–10647.

González, L. et al. (2007). The degradation of new thermally degradable thermosets obtained by cationic curing of mixtures of DGEBA and 6, 6-dimethyl (4, 8-dioxaspiro [2.5] octane-5, 7-dione). *Polym. Degrad. Stab.* 92 (4): 596–604. https://www.researchgate.net/publication/229103724_The_degradation_of_new_thermally_degradable_thermosets_obtained_by_cationic_curing_of_mixtures_of_DGEBA_and_66-dimethyl_48-dioxaspiro25octane-57-dione (accessed February 2022).

Horrocks, A.R. and Baljinder, K. (2005). Flamability of Fire Resistant Composites. https://www.researchgate.net/publication/282758409_Flammability_and_fire_resistance_of_composites. (accessed September 2022).

Iredale, R. et al. (2017). Modern advances in bismalemide resin technology: A 21st century perspective on the chemistry of addition polyimides. *Prog. Polym. Sci.* 69: 1–21.

Johnston, P.K. et al. (1988). Phenolics: a literature review of thermal decomposition products and toxicity. *J. Am. Coll. Toxicol.* 7 (2): 202.

Karak, N. (2016). Biopolymers for paints and surface coatings. In: *Biopolymers and Biotech Admixtures for Eco-Efficient Construction Materials*, (ed. Karak), Woodhead Publishing. 2018. Kindle Edition. 333–368, https://www.sciencedirect.com/science/article/pii/B9780081002148000154 (Accessed May 2022).

Khosravi, E. and Musa, O. (2011). Thermally degradable thermosetting materials. *Polym. J.* 47 (4): 465–473.

Macosco, Chris. (2000). *RIM Fundamentals of Reaction Injection Molding*. Hanser Gardner Publications **ISBN-10:** 1569900558.

Magniez, Kevin, et al., Properties of bio-based polymer nylon 11 reinforced with short carbon fiber composites, *Polym. Compos.*, Volume 36, Issue 4. pg 668–674. (2014).

Mitxelena, A. et al. (2015). Eco-friendly epoxy resin/flax composite system as a material for surf boards production. *April 2015 Conference: 2nd International Conference on Natural Fibers – From Nature to Market (ICNF2015) April 2015*. Sao Miguel, Azores, Portugal. https://www.researchgate.net/publication/314102334_ECO-FRIENDLY_EPOXY_RESINFLAX_FIBRE_COMPOSITE_SYSTEM_AS_A_MATERIAL_FOR_SURF_BOARDS_PRODUCTION (accessed February 2022).

Moretto, H. et al. (2005). Silicones. In: *Ulmann's Encyclopedia of Industrial Chemistry*. (ed. Wiley VCH). Weinheim: Wiley-VCH. 2002 Wiley-VCH Verlag GmbH & Co. KGaA. (accessed February 2022).

Naiwen, Z. et al. (2013). Preparation and properties of biodegradable poly(lactic acid)/poly(butylene adipate-*co*-terephthalate) blend with epoxy-functional styrene acrylic copolymer as reactive agent. *J. Polym. Environ.* **21**: 286–292.

Patel, A. et al. (2019). Design and formulation of a completely bio-based epoxy structural adhesive. *ACS Sustain. Chem. Eng.* 7 (19): 16382–16391.

Sheng, X. et al. (2009). Influence of cross-link density on the properties of ROMP thermosets. *Polymer* 50 (5): 1264–1269.

Siracusa, V. and Blanco, I. (2020). Bio-Polyethylene (BioPE), Bio Polypropylene (Bio-PP) and Bio-poly(Ethylene Terephthalate (Bio-PET)): recent developments in bio-based polymers analogous to petroleum-derived ones for packaging and engineering applications. *Polymers* 12 (8): 1641.

Sorathia, U. (1993). Flammability and fire safety of composite materials. *Proceedings of the 1st International Workshop on Composite Materials for Offshore Operations*, Houston, Texas (26–28 October), pp. 309–317.

Sorathia, U., Dapp, T, and Kerr, J. (1991). Flammability characteristics of composites for shipboard and submarine internal applications. *Proceedings of the 36th International SAMPE Symposium*, San Diego, CA (15–18 April), pp. 1868–1878. https://www.sciencedirect.com/science/article/pii/S1359835X0900267X (accessed February 2022).

Websites

Bio-based epoxies (2021). https://entropyresins.com/sustainability (accessed April 2021).

Bio-based nylon (2021). https://www.forbes.com/sites/johncumbers/2020/02/11/bio-nylon-is-the-new-green-how-one-company-is-fermenting-your-future-materials/?sh=3bd716044030 (accessed April 2021).

Bio-based polyester (2021). https://www.ieabioenergy.com/wp-content/uploads/2020/02/Bio-based-chemicals-a-2020-update-final-200213.pdf (accessed April 2021).

Coca-cola (2021). https://www.coca-colacompany.com/press-releases/coca-cola-produces-first-pet-bottle-made-from-plants (accessed April 2021).

DCPD (2020). https://en.wikipedia.org/wiki/Dicyclopentadiene (accessed September 2020).

DOW-polyurethane (2021). https://news.bio-based.eu/media/news-images/20070926-04/Breakthrough_Technology_from_Dow_Polyurethanes_Promotes_Susta. . .pdf (accessed April 2021).

Flame spread (2020). https://www.atsb.gov.au/media/32739/grant_20040046.pdf (accessed September 2020).

Foam seat (2020). https://www.indiamart.com/proddetail/car-seat-polyurethane-foam-15513147697.html (accessed September 2020).

Ford Soy (2021). http://www.scribd.com/doc/106613407/Bio-based-polyurethane#scribd (accessed April 2021).

Genomatica (2021). https://www.genomatica.com/worlds-first-ton-of-renewable-nylon-intermediate/ (accessed April 2021).

HexPLY (2020). https://www.hexcel.com/user_area/content_media/raw/HexPly_F650_us_DataSheet.pdf (accessed September 2020).

Melamine (2020). https://en.wikipedia.org/wiki/Melamine_resin (accessed September 2020).

Phenolics (2021). https://en.wikipedia.org/wiki/Phenol_formaldehyde_resin (accessed April 2021).
Phenolics-Wikipedia (2020). https://en.wikipedia.org/wiki/Phenol_formaldehyde_resin (accessed September 2020).
Polyester properties (2020). https://www.compositesworld.com/products/ineos-composites-aropol-resin-emphasizes-cosmetics-low-shrinkage (accessed September 2020).
Polyimides (2020). https://en.wikipedia.org/wiki/Polyimide (accessed September 2020).
Polyimides chemical (2020). https://omnexus.specialchem.com/selection-guide/polyimide-pi-plastic (accessed September 2020).
Pontiac Fiero car (2020). https://www.hagerty.com/media/valuation/pontiac-fiero-affordable-forever/ (accessed September 2020).
Properties of bio-based nylon (2021). https://www.campusplastics.com/material/pdf/138789/RilsanBESNOTL?sLg=en (accessed May 2021).
PUR foam (2020). https://hunterpure.com/companies/dut-korea/dut-high-pressure-foaming-machine/ (accessed September 2020).
PUR RIM (2020). https://www.sciencedirect.com/topics/engineering/mixing-head (accessed September 2020).
SCRIMP (2021). http://seemanncomposites.com/scrimp.htm. (accessed April 2021).
Silicone-resin (2020). https://en.wikipedia.org/wiki/Silicone_resin (accessed September 2020).
Silicone-rubber (2020). https://en.wikipedia.org/wiki/Silicone_rubber (accessed September 2020).
SuperSap (2021). https://entropyresins.com/ (accessed April 2021).
Thermoset properties (2020). www.matweb.com. (accessed September 2020).
Wikipedia VARTM (2021). https://en.wikipedia.org/wiki/Vacuum_assisted_resin_transfer_molding#:~:text=VARTM%20is%20a%20variation%20of,to%20assist%20in%20resin%20flow (accessed April 2021).

12

Polymer Composites

Polymer composites are polymer materials with a reinforcement, wherein the polymer acts a matrix resin that penetrates the reinforcement bundles and bonds to the reinforcement. The polymer composite can have thermoplastic or heat-forming polymers or thermoset polymers binding the glass, carbon, nanocomposite, Kevlar, or other reinforcement. The polymer composite materials can be categorized into two areas: one for automotive and one for aerospace. The two different areas typically have different polymer matrix resins and different reinforcements. The automotive composites are described in greater detail in the following section.

12.1 Automotive Polymer Composites

The polymer composite for automotive are thermoset-reacting polymers or thermoplastics that combine with typically glass fibers and are made with the following characteristics:

1) Cycle times less than two minutes.
2) Service use temperatures of 60–70 °C. (140–158 °F).
3) Low-cost glass fiber and resin materials around $1 per pound.
4) Glass fiber loadings usually between 10 and 50% by weight.
5) Mild chemical resistance.
6) Usually painted if used as a body panel.
7) Usually a large molded part of a door, fender, hood, door, or bumper beam.
8) Good mechanical properties.
9) Good thermal properties.

Sustainable Plastics: Environmental Assessments of Biobased, Biodegradable, and Recycled Plastics, Second Edition. Joseph P. Greene.

The following sections discuss the chemistry, polymerization methods, properties, products, and characteristics of automotive polymer composites. The polymer composites for automotive can be split into two areas based on the type of polymer resin that is used, either thermoset or thermoplastic polymers. The fibers can be also broken into two areas: continuous and discontinuous fibers. Automotive composites usually use discontinuous glass fibers. Aerospace composites usually use continuous carbon fiber.

12.2 Thermoset Polymer Composites

Polyester polymer resin is the most common polymer composite in the automotive industry. Polyester is the resin used in the most common composite material in the automotive industry, namely, sheet molding compounds (SMC). SMC is a polyester paste material made up of equal amounts of polyester resin, glass fibers, and calcium carbonate. The glass fibers are typically greater than 1 inch in length and are about 33% by weight of the polymer composite. The thick liquid paste material is blended on an SMC machine shown in Figure 12.1.

With the SMC machine, the polyester resin is combined with glass fibers and calcium carbonate, and then placed on a polyethylene sheet that is sent through rollers for compaction. The compacted sheet is wound up onto a roll. During the process, styrene is a significant part of the polyester resin. Also, curing agents, inhibitors, and catalysts are added to the polyester resin. During the compaction, the polyester resin reacts with itself to form a larger molecular weight of polyester resin. It can take up to 30 days for the molecular weight to grow large enough to achieve high tensile

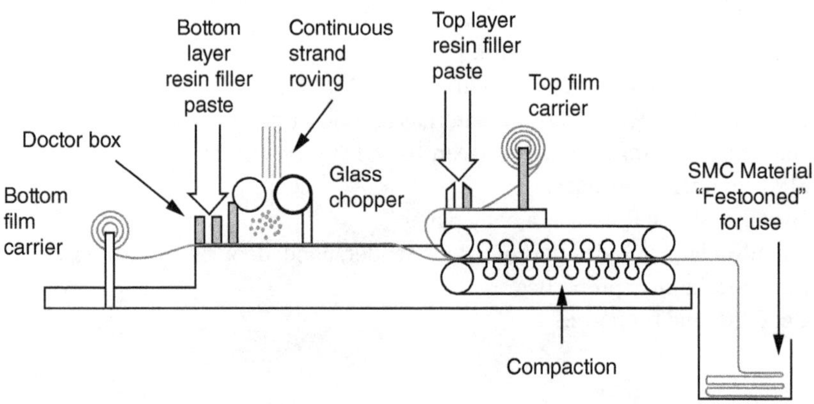

Figure 12.1 SMC machine components. Source: SMC (2020).

strength and tensile modulus. The rolls are usually placed in a cold environment to prevent the crosslinking of the polyester with the styrene. After about a month, the SMC is removed from the freezer and the SMC is cut into charges and then placed in a hot mold for curing. The SMC material can be compression molded in a hot mold to make hoods, doors, fenders, and roofs for automotive cars and trucks. SMC was formulated with a recipe that has minimal fiber read-through and can be used for exterior panels for the automotive industry when painted. SMC can be used with 33% glass fibers for the outer panel of a hood or door. SMC can be used with 50% glass loadings for the inner door panel. The outer door or hood panel has lower glass fibers so that it can be painted. The SMC crosslinks with the styrene resin in the hot mold. The cycle time for SMC is usually less than two minutes for a large molded part like a hood, bumper, roof, or door panel.

Polyester resin can also be combined with glass fibers and mineral fillers to create a bulk molding compound or BMC. The BMC is suitable for either compression or injection molding. Injection molding of BMC is used to produce complex components such as electrical equipment, car components, housings for electrical appliances and tools, in large industrial volumes.

Unlike SMC, the BMC does not have a month-long maturation step. Consequently, the BMC can contain higher filler contents of calcium carbonate or other fillers. The chopped glass strands can vary in length of ½ inch to 1 inch and at a reinforcement level between 15 and 25%. BMC uses a lower reinforcement content than SMC and has lower costs.

Other polyester resins can be combined with glass fiber preforms to make a polyester composite for hoods in automotive applications. The glass preform is first formed to the shape of the large part like a hood with a die and then the resin is added to it to make the polymer composite. The glass fibers can be heated and then thermoformed to the shape of the hood and then placed in a two-sided molded. The polyester resin is pumped with a meter into the glass preform in a process called resin transfer molding (RTM). RTM will be discussed in more detail in the later sections of this chapter.

- Other thermoset polymer composites can be made with polyurethane, vinyl ester, or epoxy resins. The glass preform is usually used rather than the discontinuous glass fibers as in SMC and BMC. The glass mat is usually preformed into a shape with a preforming metal tool that creates a glass fiber preform with near net shape of the molded part. The glass preform is shaped and placed into the molding tool where the thermoset polymer is added to the glass in a reaction injection molding (RIM) process to form the composite part. The RIM process was described in Chapter 13. Or, an RTM process can be used to meter the polyester, vinyl ester, polyurethane, or epoxy into a mold that has the shape of the large part. The RTM process is shown in Figure 12.2. The cycle times for RTM parts are usually five to

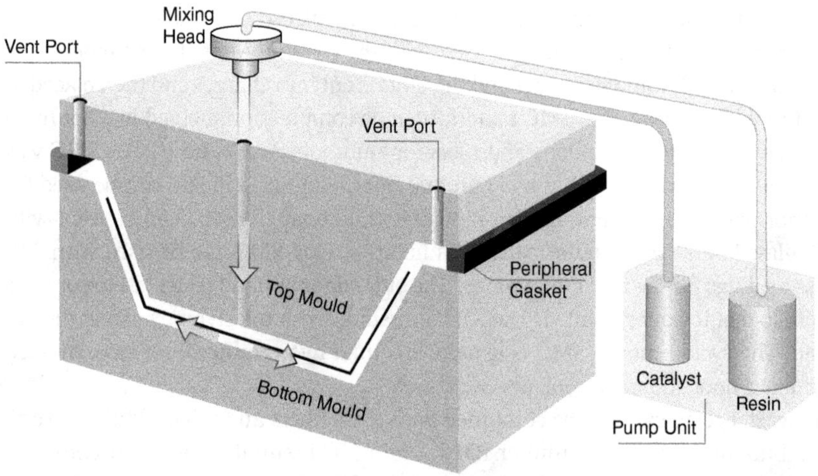

Figure 12.2 Resin transfer molding process. Source: RTM (2020).

ten minutes due to the pumping of the resin through the glass preform. The resin is usually added to a catalyst to cure the polyester composite part. An excellent reference for the RIM process is provided. (Macosko 2000).

12.2.1 Thermoplastic Polymer Composites

Thermoplastic polymer composites can be made from glass fibers and thermoplastic resins. The most common plastics are polypropylene (PP) and nylon. The glass fibers can be short length glass fibers of 1/8 inch for the injection molding process or large length glass fibers of one inch for compression molding of PP. The injection molding process is described in later chapters of this book but involves a screw heater that melts the plastic and an injection unit and press that injects the liquid into the mold. The injection molded parts are usually small parts that can be made. The nylon plastic can have glass fibers in it (up to 33%) and molded into automotive headers or engine parts. Glass-filled nylons need larger diameter runners and thicker part walls by as much as 50% thicker than unfilled nylon parts. The design book published several years ago by Monsanto (now Dow Solutia) contained a flow ratio chart. It indicated that when you divide the wall thickness into the flow length, you will come up with a flow ratio number that can be used to determine the number of gates you will need on a given part and where the gate or gates might be located.

The flow ratio of unfilled nylon is 220 vs. 120 for filled nylon. You can see that working with filled nylon is like working with many of the stiffer flowing materials we commonly mold today.

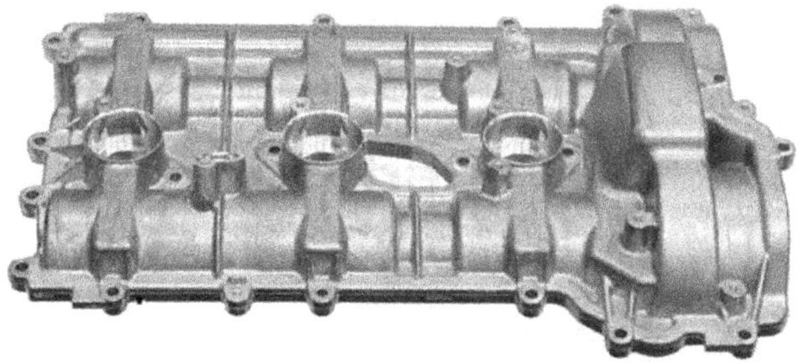

Figure 12.3 Injection molded valve cover with 33% glass and 66% nylon plastic.

This part is short enough and the wall thickness is generous enough to allow the part to be filled and packed from one gate location. But we still have thick to thin transitions to be concerned about, and this part is gated into a thin section, which is causing some, but not all, of the trouble. To compensate for the design shortcomings, the molding technicians have raised the barrel melt temperatures, resulting in longer cycle times.

Glass-filled nylon can be used as valve covers on many cars and trucks as shown in Figure 12.3.

The glass-filled valve cover offers several performance characteristics that create greater rigidity, along with tighter tolerances, than parts made from metal. That allows engineers to reduce a number of metal parts in the valve cover to a single structural part molded out of glass-filled nylon. Snap-fits, bosses, ribs, and other features, such as brackets and attachment points, can be molded in to eliminate assembly steps, reduce weight, and reduce part count.

Polypropylene can also be injection molded with short length fibers of 1/8 inch or long glass fivers of ¼ inch. The PP parts can be used for many automotive parts and have glass loadings up to 40%. The injection molding process is fast with low cycle times. A typical injection molded PP part with glass fibers is shown in Figure 12.4.

Figure 12.4 Injection molded valve cover with 40% glass and 60% PP plastic. Source: PP valve cover (2020).

Table 12.1 Properties of glass filled plastics.

Material	Nylon neat	Nylon 33% short glass fibers	PP neat	PP 32% short glass fibers	PP 50% long glass fibers	PEEK neat	PEEK 33% carbon fibers
Specific gravity	1.3	1.4	0.9	1.14	1.35	1.35	1.42
Tensile strength, MPa	2	8 (4X)	1.2	4 (3X)	11 (8X)	4	10 (2.5X)
Tensile modulus, GPa	50	150 (3X)	35	80 (2X)	130 (4X)	120	150 (1.25X

The properties of injection-molded glass-reinforced parts are listed in Table 12.1. Glass fibers increase the tensile strength of nylon four times and the tensile modulus three times versus neat nylon. Glass fibers increase the tensile strength of PP three times and the tensile modulus two times versus neat PP. Glass fibers increase the tensile strength of PEEK 2.5 times and the tensile modulus 1.25 times versus neat PEEK.

Glass-filled polypropylene can be used for a valve cover as shown in Figure 12.4. The PP valve cover will not have the thermal stability of the glass-filled nylon valve cover. Figure 12.4 shows a glass mat thermoplastic (GMT) part with 40% glass fiber and 60% PP.

Polypropylene can also have longer 1 inch glass fibers at loadings up to 50% with glass mat thermoplastic (GMT) parts. The PP glass-filled material can be compression molded into large parts like door inners and bumper beams. The process can have up to two-minute cycle times. The PP is usually heated with an infrared oven and then placed into a compression mold. The press closes the mold and the compression molded part is usually made with a two-minute cycle time. The compression molder can be as low as 200 tons and as high as 2000 tons of pressure.

GMT PP can also be used for bumper beams for automotive applications. The GMT process can include a hot PP plastic that is placed on a cold mold and molded into a part. The glass-filled PP is usually provided as a sheet with a thickness of approximately 3 mm. The plastic is cut into strips and then placed in a heater, like an infrared heater, on a conveyer belt. When the cut strips swell, they can be placed in the mold and then compressed in a compression press. The parts are molded with a compression molder usually. An example of a compression molding press is shown in Figure 12.5.

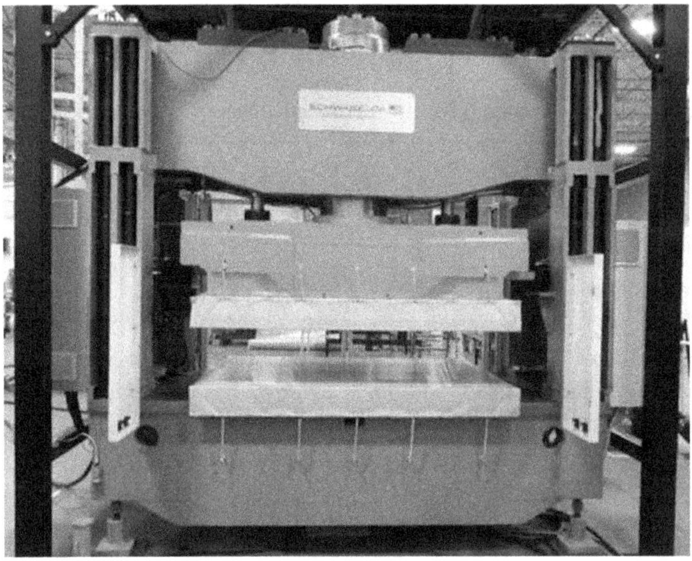

Figure 12.5 Compression molding machine. Source: Compression press (2020).

12.2.2 Kevlar Composites

Polymer composites can use Kevlar as a reinforcement instead of glass or carbon fiber. The aromatic nylon fiber, Kevlar, can be combined with polyamides (nylon), polyesters, epoxies, polypropylene, or PEEK. The plastics can be combined with the Kevlar and then compression molded into a plastic part. The chemistry to make Kevlar is shown in Figure 12.6. Kevlar is produced with a reaction of 1,4-phenylene-diamine (*para*-phenylenediamine) and terephthaloyl chloride in a condensation reaction yielding hydrochloric acid as a byproduct and 1,4-phenylene-diamine (*para*-phenylenediamine) (Kevlar).

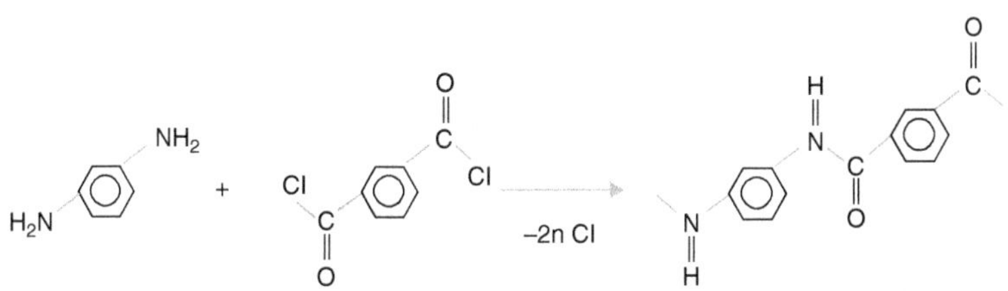

para-phenylenediamine   terephthaloyl chloride   1,4-phenylene-diamine (*para*-phenylenediamine) (Kevlar)

Figure 12.6 Kevlar chemistry.

Table 12.2 Kevlar properties.

Grade	Density (g/cm^3)	Tensile modulus (GPa)	Tensile strength (GPa)	Tensile elongation (%)
29	1.44	83	2.92	3.5
49	1.44	124	3.45	2.9
149	1.47	179	3.60	1.5

Source: Hallad and Banapurmath (2018).

The properties of Kevlar depend upon the Kevlar type, as shown in Table 12.2. Kevlar 149 has the highest tensile modulus. Kevlar 49 has the highest tensile strength. Kevlar 29 has the largest tensile elongation.

12.3 Nanocomposite

Nanocomposite is a multiphase solid material where the reinforcement from clay has one, two, or three dimensions of less than 100 nanometers (nm). For plastics, the nanocomposites are made from Montmorillonite clay. The size of clay is on pico (10^{-9}) meter scale, which causes swelling due to exfoliation in the plastic and results in increased strength and stiffness at concentration less than with glass fibers. Nanocomposites are added in concentrations of 2–10%, typically. Glass and mineral fibers are usually added in concentrations of 10–40%. An example of a nanocomposite is shown in Figure 12.7.

Nanocomposites provide high strength and stiffness at lower concentrations than glass fiber. They can be used for many automotive parts including engine covers, timing belt covers, and body trim pieces. The nanocomposites typically are 0.5 micron in size. The nanoparticle is very small in diameter and can be 50 000 times smaller than the plastic component. The properties of a nanocomposite particle with nylon is shown in Table 12.3.

12.4 Fiber Materials for Composites

For automotive parts and aerospace parts, many composites use a continuous fiber reinforcement as a mat structure rather than discontinuous short fibers for injection molding. The following section describes the glass mats that are used for automotive applications.

Polymer composites can have a fiber preform of glass or carbon to mold with the polymer resin to form a polymer composite part in a die. The fiber preform is

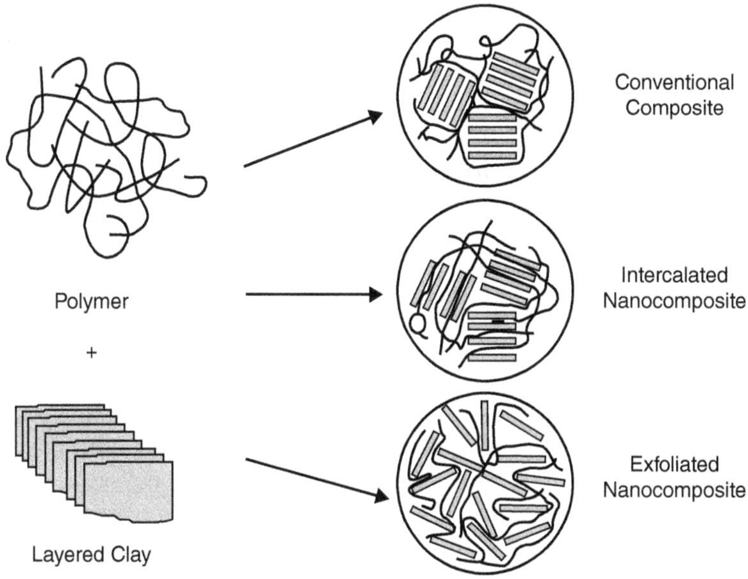

Figure 12.7 Nanocomposite example. Source: Gacitua et al. (2005).

Table 12.3 Mechanical properties in-reactor polymerization.

Material	Nanomer I.24T% (wt/wt)	FlexStr (Mpa)	FlexMod (Mpa)	TenStr (Mpa)	TenMod (Mpa)	HDT@ 264psi
Nylon 6(control)	0	107	2885	72	1821	55
Nanocomposite	2.5	144	3644	91	2372	112

Source: Han et al. (2017).

a continuous type usually in the shape of a mat form. The fiber mat can be made from a chopped glass that has isotropic properties.

The properties of common fibers are shown in Table 12.4.

The fiber composite must have a resin component and a fiber component. The fibers must be preformed into a shape that resembles the final product. For bumper beams using glass mat and a urethane from the RIM process, the e-glass fiber mat, like a twill, must be heated and then formed into a bumper shape. The preform is subsequently placed in the mold and then the urethane resin is injected onto the glass preform. The process is called structural RIM or SRIM.

The continuous mat can have several constructions that are used in the apparel industry. The glass fiber, carbon fiber, or Kevlar fiber can be woven from the spool of fiber into a fabric of many configurations, including twill, woven, satin, knit,

Table 12.4 Tensile properties of fibers.

Property	Glass, E	Carbon	Aramid (Kevlar 49)
Tensile strength, ksi (MPa)	450 (3100)	500 (3450)	535 (3600)
Tensile modulus, Mpsi (GPa)	11.0 (76)	33 (228)	19 (131)
Elongation at break (%)	4.5	1.6	2.8
Density, g/cc	2.54	1.8	1.44

Source: Fibers (2020).

biaxial weave, and so on. The type of fiber pattern depends on the amount of fibers made in the weft (perpendicular) or fill (parallel) directions of the fabric. Common amounts are 50–50%, 60–40%, and 80–20% in the fill and warp directions. More fiber in the fill direction can lead to a unidirectional fiber. The orientation of the fibers can affect the mechanical properties of the polymer composite in the parallel and perpendicular directions.

The fiber yarn manufacturing process is the first step in the preform construction. The glass fiber is initially made by melting glass spheres in a furnace at elevated temperatures. Figure 12.8 illustrates the process of glass fiber manufacturing where the glass marbles are melted down and drawn into a fiber with an extrusion process. The melted glass is drawn out of the bottom of the furnace and pulled through dies and cooled into a glass strand. The most common glass types are in three versions: "E" glass for electrical glass, "S" glass for high-strength glass, and "C" glass for chemical-resistant glass. The most common types of glass fiber used

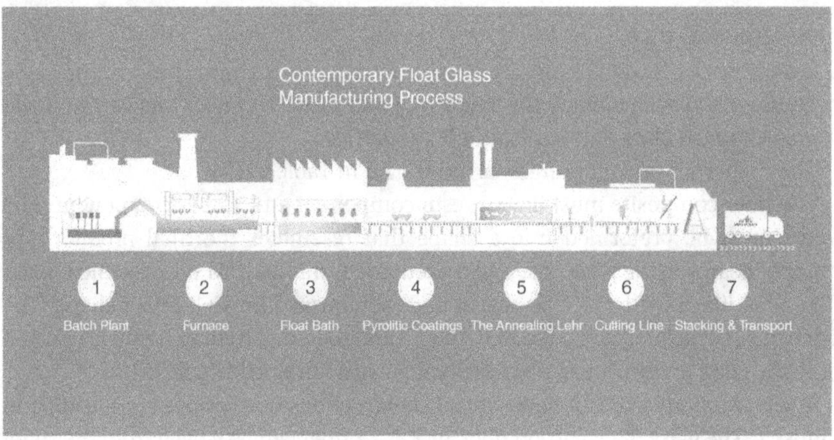

Figure 12.8 Glass manufacturing process. Source: Glass (2020).

in fiberglass is E-glass, which is alumino-borosilicate glass that is used for glass-reinforced plastics. Other types of glass used are A-glass (**A**lkali-lime glass with little or no boron oxide), E-CR-glass (**E**lectrical/**C**hemical **R**esistance; alumino-lime silicate with less than 1% w/w alkali oxides, with high acid resistance), C-glass (alkali-lime glass with high boron oxide content, used for glass staple fibers and insulation), D-glass (borosilicate glass, named for its low **d**ielectric constant), R-glass (alumino silicate glass without MgO and CaO with high mechanical requirements as **r**einforcement), and S-glass (alumino silicate glass without CaO but with high MgO content with high tensile strength). (Glass et al. 2000).

Pure silica (silicon dioxide), when cooled as fused quartz into a glass with no true melting point, can be used as a glass fiber for fiberglass, but has the drawback that it must be worked at very high temperatures. In order to lower the necessary work temperature, other materials are introduced as "fluxing agents" (i.e., components to lower the melting point). Ordinary A-glass ("A" for "alkali-lime") or soda lime glass, crushed and ready to be remelted, as so-called cullet glass, was the first type of glass used for fiberglass. E-glass ("E" because of initial **e**lectrical application), is alkali free, and was the first glass formulation used for continuous filament formation. It now makes up most of the fiberglass production in the world, and also is the single largest consumer of boron minerals globally. It is susceptible to chloride ion attack and is a poor choice for marine applications. S-glass ("S" for "Strength") is used when high tensile strength (modulus) is important, and is thus important in composites for building and aircraft construction. The same substance is known as R-glass ("R" for "reinforcement") in Europe. C-glass ("C" for "chemical resistance") and T-glass ("T" is for "thermal insulator"—a North American variant of C-glass) are resistant to chemical attack; both are often found in insulation grades of blown fiberglass. (Fiberglass 2020).

The protective coating and sizing agents are applied to the fiber glass. The strands are pulled off of the forming tube and into the roving fabricating steps with creels to the winding machine to create the glass roving package that resembles a cotton yarn ball. The fiber rovings are wound onto a machine that creates the different types of weaves and fabrics for woven, twill, knit, satin, and other fabric types.

The fiber yarn manufacturing for carbon fiber is similar to the glass fiber process where the carbon fiber is created and then melted in a furnace. The fiber is drawn from the bottom of the furnace in a die and then wrapped onto a spool for a fiber roving package.

The glass fiber can also be created in a discontinuous process to create single fibers that are not connected with other fibers. The glass roving package can be chopped for short or long fibers for injection molding or chopped for sheet molding compound as explained earlier.

12.5 Carbon Fiber Manufacturing

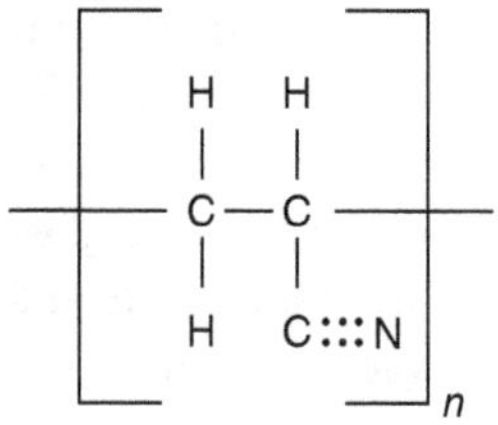

Figure 12.9 Chemistry of acrylonitrile plastic fiber.

The carbon fiber manufacturing process involves starting with a PAN fiber, rayon, or a petroleum pitch fiber. PAN is the abbreviation for polyacrylonitrile plastic fiber. PAN plastic is show chemically in Figure 12.9.

The fiber has carbon, hydrogen, and nitrogen. All of the hydrogen and nitrogen must be thermally removed from the molecule.

The PAN fibers have good properties with a lower cost for standard modulus carbon fiber. PAN or rayon are first spun into filament yarns by using chemical and mechanical processes that initially align the polymer in a single direction. This enhances the final physical properties of the completed carbon fiber. After drawing or spinning, the polymer filament yarns are heated to drive off hydrogen and nitrogen in the polymer fiber in a process called carbonization. This produces the final carbon fiber. The manufacturing process to produce carbon fibers is shown in Figure 12.10.

Polyacrylonitrile (PAN) is a commercially available textile fiber and is a good starting material for PAN-based carbon fibers. The steps to produce PAN carbon fibers are shown in Figure 12.11 and listed below:

1) Stabilize by thermosetting (crosslinking) so that the polymers do not melt in subsequent processing steps. PAN fibers are stretched as well.
2) Carbonize: Fibers are pyrolyzed until they transform into all-carbon
   a) Heating Fibers to 1800 °F yield PAN fibers at 94% carbon and 6% nitrogen
   b) Heating fibers to 2300 °F removes nitrogen and yields 99.7% carbon
3) Graphitize: Carried out at temperatures greater than 3200 °F to improve the tensile modulus by improving the crystalline structure and the three-dimensional nature of the structure.

The fibers are surface treated and a sizing agent, a coupling agent, and finish are applied. Fibers are wound up for shipment.

A common method of carbon fiber manufacture, as shown in Figure 12.11, involves heating the spun PAN filaments to approximately 300 °C in air, which breaks many of the hydrogen bonds and oxidizes the material. The oxidized PAN is then placed into a furnace having an inert atmosphere of a gas such as argon, and heated to approximately 2000 °C, which induces graphitization of the material, which changes the molecular bond structure.

When heated in the correct conditions, these chains bond side-to-side (ladder polymers), forming narrow graphene sheets which eventually merge to form a single, columnar filament. The result is usually 93–95% carbon. Lower-quality

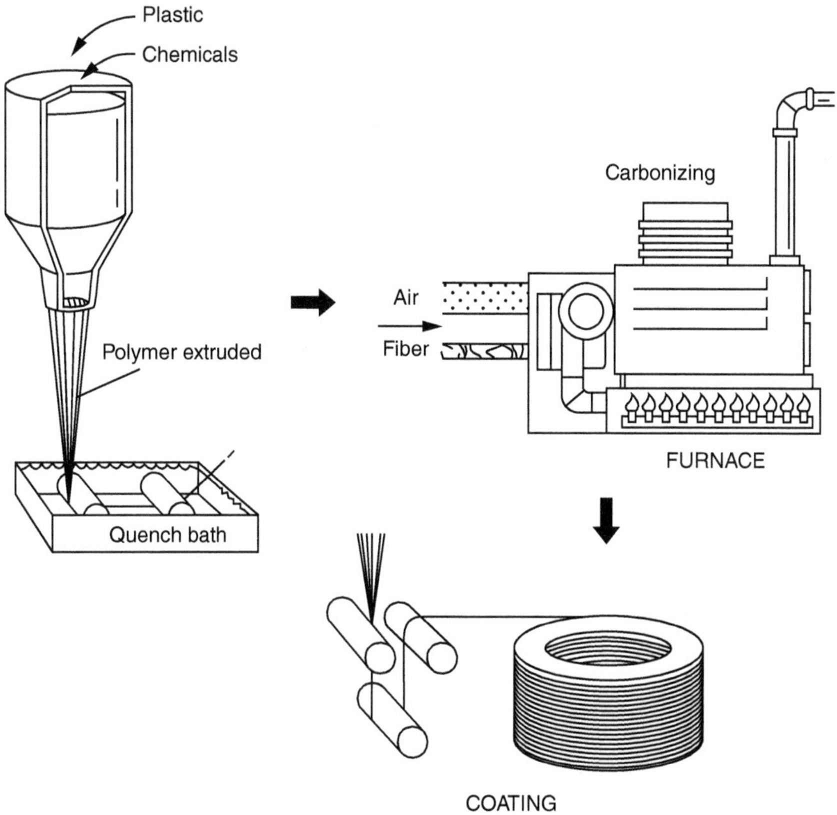

Figure 12.10 Carbon fiber manufacturing process. Source: Carbon fiber (2020).

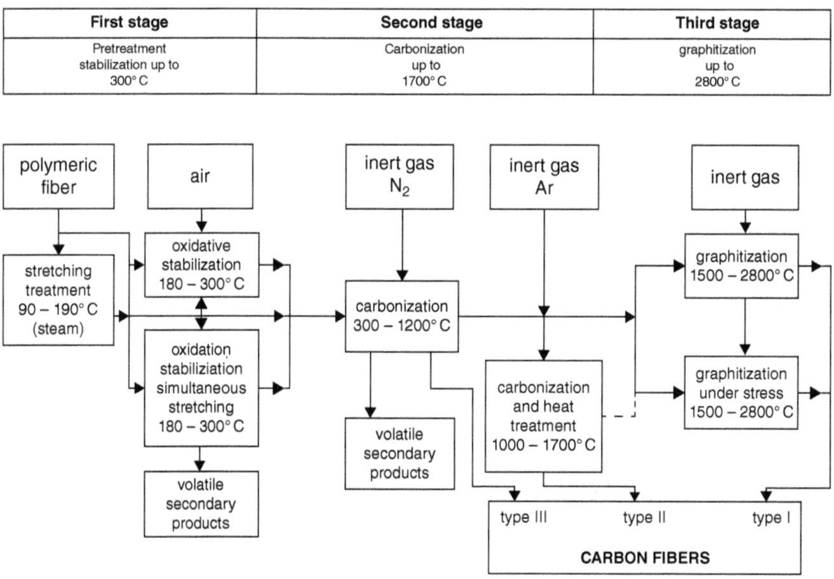

Figure 12.11 Carbon fiber thermal manufacturing process. Source: Carbon fiber process (2020).

fiber can be manufactured using pitch or rayon as the precursor instead of PAN. The carbon can become further enhanced, as high-modulus, or high-strength carbon, by heat treatment processes.

Carbon heated in the range of 1500–2000 °C (carbonization) exhibits the highest tensile strength (820 000 psi, 5650 MPa or N/mm^2), while carbon fiber heated from 2500 to 3000 °C (graphitizing) exhibits a higher modulus of elasticity (77 000 000 psi or 531 GPa or 531 kN/mm^2).

The pitch manufacturing process is similar to the PAN manufacturing process. The pitch is a viscoelastic material that is composed of aromatic hydrocarbons. Pitch is created by distillation of plants, crude oil, and coal. Pitch must be converted into a suitable fiber from petroleum tar. Pitch fibers are converted to a fiber by passing them through a meso-phase where the polymer chains are somewhat oriented and is in a liquid state, called a liquid crystal phase. Orientation is responsible for the ease of consolidation of pitch into carbon. The pitch fibers are first stabilized by crosslinking the polymer so that the polymers do not melt in subsequent processing steps. The processing steps include the following:

1) Carbonize: Fibers are pyrolyzed until transformed into all-carbon
   a) Heated fibers 1800 °F
   b) Heated to 2300 °F
2) Graphitize: Carried out at temperatures greater than 3200 °F to improve the tensile modulus by improving the crystalline structure and the three-dimensional nature of the structure.

The fibers are surface treated and a sizing agent, a coupling agent, and finish are applied. Fibers are wound up for shipment.

To produce a carbon fiber, the carbon atoms are bonded together in crystals that are more or less aligned parallel to the long axis of the fiber as the crystal alignment gives the fiber a high strength-to-volume ratio (making it strong for its size). Several thousand carbon fibers are bundled together to form a tow, which may be used by itself or woven into a fabric.

Carbon fibers are usually combined with other materials to form a composite. When impregnated with a plastic resin and baked, it forms carbon-fiber-reinforced polymer, which has a very high strength-to-weight ratio and is extremely rigid although somewhat brittle. Carbon fibers are also composited with other materials, such as graphite, to form reinforced carbon-carbon composites, which have a very high heat tolerance.

Carbon fiber can also be referred to as graphite fiber if the carbon percentage is greater than 90%. The diameter of the carbon fiber is about 5–10 micrometers. Carbon fibers have several advantages including high stiffness, high tensile strength, low weight, high chemical resistance, high temperature tolerance, and low thermal expansion. These properties have made carbon fiber very popular in aerospace, civil engineering, military, motorsports, and competition sports. However, they are relatively expensive compared with similar fibers, such as glass fibers or plastic fibers.

12.6 Properties of Fibers

For tensile strength, the same calculations show that that carbon fiber has seven times the specific strength as steel and Kevlar has 11 times the specific strength as steel. The tensile strengths of the materials are listed in Table 12.5.

Glass, carbon, and aramid fibers have high strength, high stiffness, and low elongation. The properties of fibers can be compared with metal materials as shown in Table 12.6.

Table 12.6 shows that carbon fiber has the highest tensile modulus and low density. If we divide the tensile modulus by the density, we obtain a specific

Table 12.5 Tensile strength of engineering materials.

Material	Tensile strength, MPa, (ksi)	Density Mg/m^3	Specific strength m^2/s^2
Low alloy steel	1500 (220)	7.85	191.08
Aluminum	500 (75)	2.8	178.57
Carbon fiber	2400 (360)	1.8	1333.33
Glass fiber	2000 (300)	2.56	781.25
Aramid fiber (Kevlar)	3000 (450)	1.4	2142.86

Source: Matweb (2020).

Table 12.6 Tensile modulus of engineering materials.

Material	Tensile modulus, GPa, (Mpsi)	Density Mg/m^3	Specific modulus, m^2/s^2
Low alloy steel	207 (30)	7.85	26.37
Aluminum	72 (10)	2.8	25.71
Carbon fiber	300 (40)	1.8	166.67
Glass fiber	76 (10)	2.56	29.69
Aramid fiber (Kevlar)	125 (20)	1.4	89.29

Source: Matweb (2020).

modulus. The specific modulus is a good way to compare the highest properties with the lowest mass or weight. The results show that carbon fiber has six times the specific modulus of steel. Glass fiber has an equivalent specific modulus. Kevlar has three times the specific modulus.

The relationships between specific modulus and strength for several fiber materials are shown in Figure 12.12. The fibers with high specific tensile modulus and specific tensile strength are Kevlar and high-tensile graphite. Boron has lower specific strength but a higher specific tensile modulus than "S" glass fiber. "E" glass fiber has an equivalent specific modulus as steel and aluminum but higher specific tensile strength. Kevlar, graphite, and boron have higher specific tensile modulus and specific tensile strength than steel and aluminum.

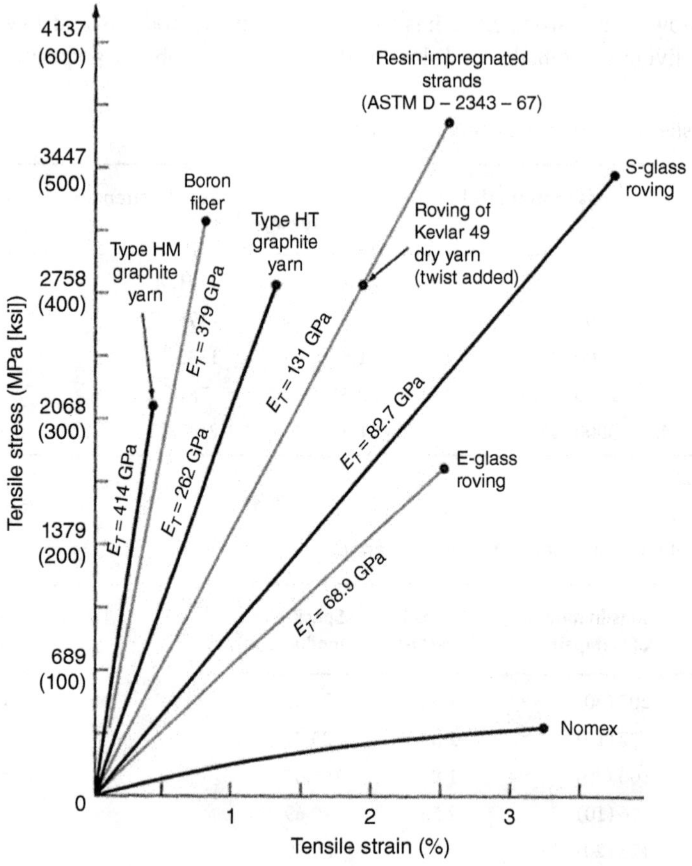

Figure 12.12 Stress strain behavior of reinforcing fibers. Source: Stress strain fibers (2020).

The tensile modulus and tensile strain of many fibers are shown in Figure 12.12. The results show that the high-modulus carbon fiber has a tensile modulus of 414 GPa with a strain of less than 1%. Boron fiber is also compared with carbon, Kevlar, and glass fibers. The stiff fibers are graphite (carbon) and boron with strain less than 1%. Glass fibers and Kevlar fibers have lower tensile modulus but have larger strains of up to 3%.

The fibers must be combined with a resin to create a polymer composite. The polymers can be nylon, polypropylene, polyester, polyurethane, phenolic, PEEK, and others. The tensile moduli range from 1 to 7 GPa. The tensile strengths range from 33 to 145 MPa. The densities of the polymers range from 0.9 to 1.3 g/cc. The properties of the polymer composites are not homogenous and are directional based on the amount of fibers in that direction. If a unidirectional fiber is used with the polymer, the properties are much higher in the direction of the fiber. In fact, the strength can be reduced as much as 90% in the transverse direction as compared to the parallel direction. The stiffness and tensile modulus in the transverse direction can be reduced by 95% compared to the parallel direction. If a quasi-isotropic fiber preform is used with 50% of the fiber in the weft direction and 50% of the fiber in the fill direction, the properties in both directions can be considered the same or iso-tropic.

The properties of polymer composites are dependent upon the type of reinforcement and the percentage of reinforcement are listed in Table 12.7. For carbon fiber and glass fiber composites that have mat or woven fiber layups, the fiber mat is stacked in different orientations in the layup that can result in high strength mechanical properties. Glass and carbon fiber mats have ply orientation

Table 12.7 Mechanical properties of unidirectional and quasi isotropic polymer laminates with carbon fiber and epoxy resin.

	Unidirectional laminate			**Quasi-isotropic laminate**		
Testing angle	**0°**	**90°**	**0°/90° ratio**	**0°**	**90°**	**0°/90° ratio**
Tensile strength (MPa)	793	20	39.65	379	241	1.57
Tensile modulus (GPa)	303	3.3	91.82	104	97	1.07
Tensile ultimate strain, %	0.25	0.5	0.50	0.27	0.23	1.17
Compressive strength (MPa)	400	158	2.53	172	200	0.86
Compressive modulus (GPa)	255	6.7	38.06	76	88	0.86
Compressive ultimate strain, %	—	—		0.55	0.86	0.64

Source: Czel et al. (2017).

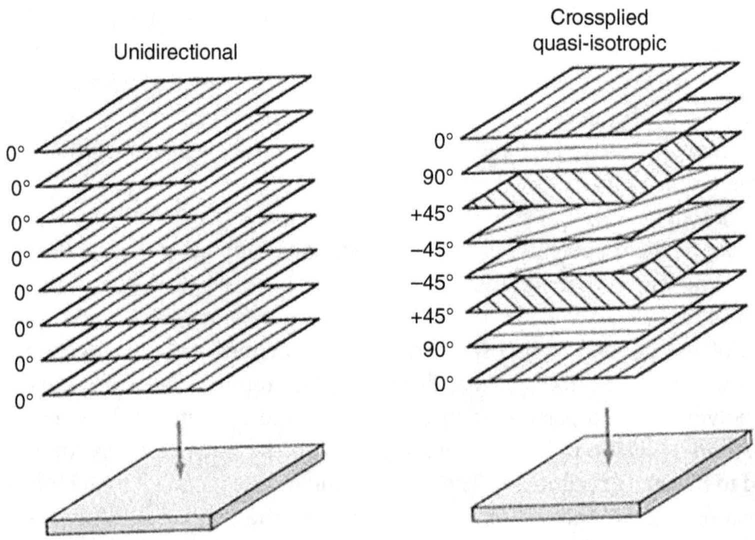

Figure 12.13 Unidirectional and multidirectional laminate plies for a composite laminate. Source: Composite laminate (2020).

stack-ups. The top and bottom have different fiber orientations (0/90°) than the middle fiber mats (=/−45°). Figure 12.13 shows the layup of different fiber mats or woven products. The unidirectional composite has a layup of only 0° fibers and thus the properties of the composite are very stiff and strong in the direction of the fibers. The properties are weak in the transverse direction of the fibers. For the quasi-isotropic laminate, the fiber layup is a 0°/90° orientation in the top and bottom layers and a +/−45° orientation in the middle of the composite. This will yield a polymer composite with homogeneous properties in the parallel direction and transverse directions in the part.

Properties of an unidirectional laminate and quasi-isotropic laminate are listed in Table 12.7.

In the table, the unidirectional laminate is 40 times stronger and 90 times stiffer in the 0° direction versus the transverse 90° direction in tensile strength. The quasi-isotropic laminate is stronger and stiffer in tension in the 0° direction than in the 90° direction. The opposite is true for compression. The mechanical properties of several glass-filled and carbon-filled composites are shown in Table 12.8.

Table 12.8 shows that T300 and T100 carbon fiber and epoxy composites have the highest mechanical properties because they have the highest volume fraction of fibers. And the fibers are carbon fiber and not glass fiber. The SMC composite materials have higher mechanical properties than BMC composites.

Table 12.8 Mechanical properties of several glass filled and carbon filled composites.

Material	Tensile modulus, GPa	Tensile strength, MPa	Density, g/cm^3
T300. Carbon fiber (60Vol%) and epoxy	138	2790	1.31
T1000. Carbon fiber (60 Vol%) and epoxy	234	3475	1.31
SMC. Glass (33wt%) and polyester resin	11	105	1.6
SMC. Glass (50 wt%) and polyester resin	16	220	1.74
SMC lite	8	80	1.2
SMC carbon fiber (20 wt%)	15	120	1.2
GMT PP (40% glass)	6	105	1.19
BMC (30% glass)	10	80	1.8

Source: MatWeb (2020).

Carbon-fiber polymer composites also have excellent fatigue properties due to the excellent fatigue properties of the carbon fiber. Figure 12.14 shows the fatigue properties of several engineering materials.

Figure 12.14 data is also listed in Table 12.9. The table shows that boron fiber with epoxy resin and graphite fiber with epoxy resin retain 80% of their maximum

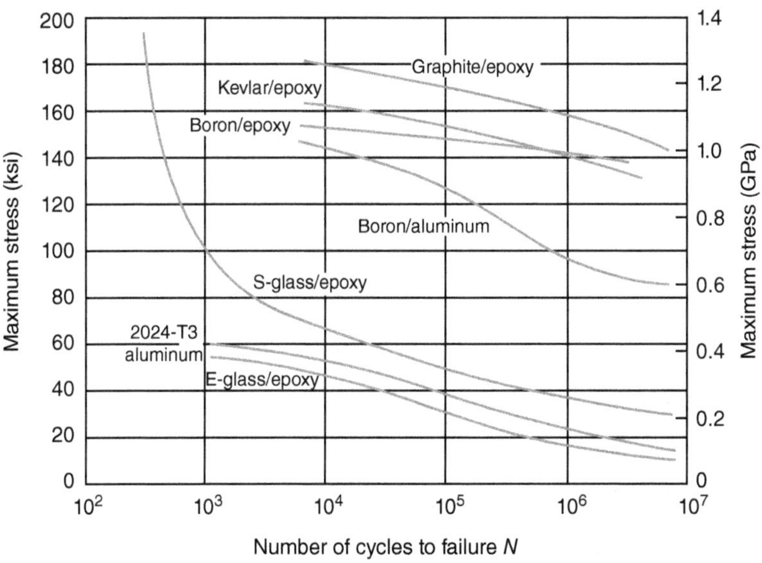

Figure 12.14 Fatigue properties of several engineering materials. Source: Fatigue (2020).

Table 12.9 Fatigue properties of engineering materials.

Material	Maximum stress original (ksi)	Maximum stress after 10 million cycles (ksi)	Ratio
Boron/epoxy	158	135	0.85
Graphite/epoxy	180	145	0.81
Kevlar/epoxy	165	130	0.79
Aluminum	60	15	0.25
E-glass/epoxy	55	12	0.22
S glass/epoxy	190	30	0.16

Source: Hercules (2011).

stress after 10 million cycles. Comparatively, aluminum and glass fiber retain 22–25% of their stress. Boron and carbon fiber composite structures do very well under fatigue conditions.

12.7 Rule of Mixtures

The combination of the polymers with the fibers can create a polymer composite with a range of properties depending upon the percentage of fibers. The properties of the composite can be estimated based on the volume fraction of the fiber and the volume fraction of the polymer resin. The Rules-of-Mixtures equations can be used to determine mechanical properties. Equation 12.1 is an equation of the rule of mixtures for tensile modulus. Equation 12.2 is an equation of the rule of mixtures for tensile strength. Equation 12.3 is an equation of the rule of mixtures for density. Equation 12.4 is an equation of the rule of mixtures for Poisson's ratio.

$$\mathrm{Ec} = \mathrm{Ef} * \mathrm{Vf} + \mathrm{EmVm} = \mathrm{Ef} * \mathrm{Vf} + \mathrm{Em}(1 - \mathrm{Vf}) \tag{12.1}$$

Where, E is tensile modulus, f is fiber, m is matrix, and c is composite.

$$\mathrm{TSc} = \mathrm{TSf} * \mathrm{Vf} + \mathrm{TSmVm} = \mathrm{TSf} * \mathrm{Vf} + \mathrm{TSm}(1 - \mathrm{Vf}) \tag{12.2}$$

Where, TS is tensile strength, f is fiber, m is matrix, and c is composite.

$$\rho\mathrm{c} = \rho\mathrm{f} * \mathrm{Vf} + \rho\mathrm{mVm} = \rho\mathrm{f} * \mathrm{Vf} + \rho\mathrm{m}(1 - \mathrm{Vf}) \tag{12.3}$$

Where, ρ is density, f is fiber, m is matrix, and c is composite.

$$\mathrm{vc} = \mathrm{vf} * \mathrm{Vf} + \mathrm{vmVm} = \mathrm{vf} * \mathrm{Vf} + \rho\mathrm{m}(1 - \mathrm{Vf}) \tag{12.4}$$

Where, ν is the Poisson's ratio, f is fiber, m is matrix, and c is composite.

The Rule-of-Mixtures calculations can be used to predict the tensile properties and densities of polymer composites at different fiber percentages or loadings. Figures 12.15–12.17 show the tensile modulus, tensile strength, and density of polyester composites with glass fibers.

Most polymer composites are created by weighing the glass or fiber amount and then adding the resin amount to the composite part. Thus, we can calculate a weight percentage as the weight of the fiber divided by the weight of the composite part.The amount of thermoset resin or plastic is also weighed during the processing of the polymer composite. The transformation from weight percentage to volume

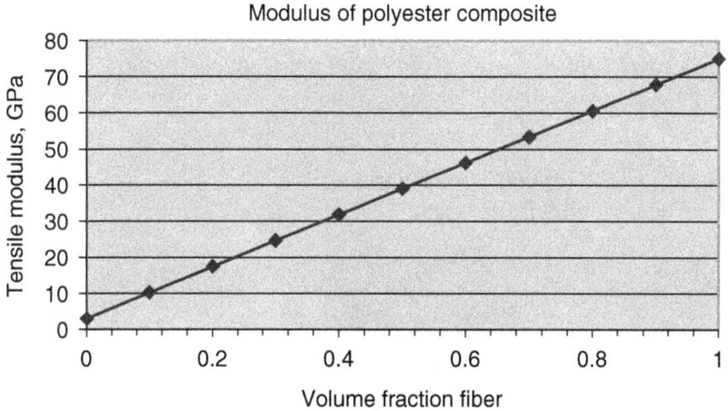

Figure 12.15 Tensile modulus prediction of polyester composite from Rule-of-Mixtures calculations.

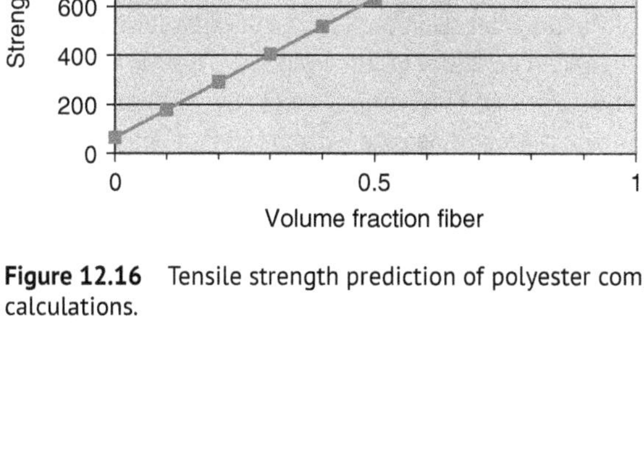

Figure 12.16 Tensile strength prediction of polyester composite from Rule-of-Mixtures calculations.

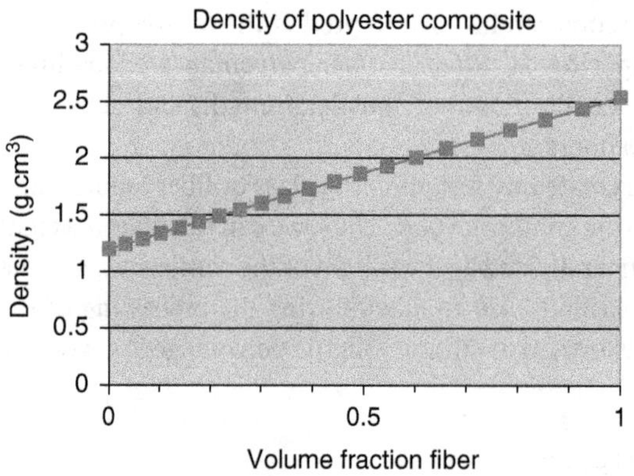

Figure 12.17 Density prediction of polyester composite from Rule-of-Mixtures calculations.

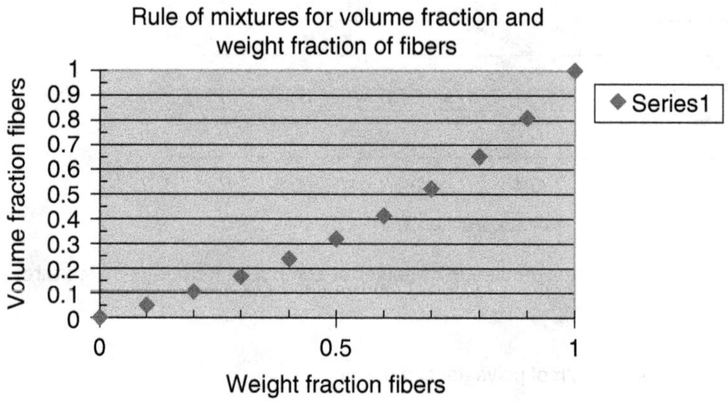

Figure 12.18 Weight fraction and volume fraction for polyester resin and E-glass fiber with the Rule-of-Mixtures calculations.

percentage is dependent of the density of the glass fiber of 2.54 g/cm^3 or carbon fiber at 1.8 g/cm^3 plus the density of the polymer resin or plastic. For polyester resin with a density of 1.2 g/cm^3 and E-glass with a density of 2.54 g/cm^3, the results are shown in Figure 12.18 for glass fiber and polyester. The example below illustrates the weight percentage and volume percentage calculations with carbon fiber an epoxy resin.

The calculations assume:

- 1 cm^3 of composite and
- 60% volume of carbon fiber and 40% volume of epoxy.

Thus, for 1 cm^3 of composite would have:

- 0.60 cm^3 of carbon fiber times 1.8 g/cm^3 yields 1.08 g of carbon fiber.

Similarly, for 1 cm^3 of composite would have:

- 0.40 cm^3 of carbon fiber times 1.2 g/cm^3 yields 0.48 g of epoxy.

Then the total mass would be 1.08 g carbon fiber plus 0.48 g of epoxy to yield 1.56 g of composite for 1 cm^3 of composite.

The mass percentages would be:

- for carbon fiber: 1.08 g/1.56 g total that yields 69% by weight carbon fiber, and
- for epoxy: 0.48 g/1.56 g of composite total would yield 30.77% by weight epoxy.

The density of the composite would be the mass divided by the volume or 1.56 g/1 cm^3.

The Rule-of-Mixtures calculations can be compared with actual examples of carbon fiber composites in the following example. The T300 data sheet shows that the epoxy resin is 60 volume % carbon fiber. This yields a polymer composite material with a high tensile strength and tensile modulus because the carbon fiber has a tensile strength of 3.53 GPa and a tensile modulus of 230 GPa.

For density:

- The actual density of the composite is 1.52 g/cm^3.
- The calculation for the T300 at 60 volume % was 1.54 g/cm^3.
- This represents a 1.29% difference.

For tensile modulus:

- The actual tensile modulus of the composite was 135 GPa.
- The calculation for the T300 at 60 volume % was 168 GPa.
- The represents a 24.44% difference.

For tensile strength:

- The actual tensile strength of the composite was 2.603 GPa.
- The calculation for the T300 at 60 volume % was 1.860 GPa.
- The represents a −39.94% difference.

Thus, the Rule-of-Mixture over-predicts the density and tensile modulus and under-predicts the tensile strength for the carbon fiber and epoxy combination at 60% carbon fiber.

For polyester and glass combinations in an automotive composites, the composite is made with 33% glass fiber by weight and 67% polyester resin by weight.

For tensile modulus:

- The actual tensile modulus of the composite was 17 GPa.
- The calculation for the glass fiber at 33 volume % was 26 GPa.
- The represents a 52.2% difference.

For tensile strength:

- The actual tensile strength of the composite was 360 MPa.
- The calculation for the glass fiber at 33 volume % was 395 MPa.
- The represents a 9.7% difference.

Thus, the Rule-of-Mixture over-predicts the tensile strength and tensile modulus for the glass fiber and polyester combination at 33% glass fiber.

Example 12.1 Calculate the density, tensile modulus, and tensile strength of a composite with 60 volume % carbon fiber and 40 volume % epoxy resin. The density of carbon fiber is 1.8 g/cc and the density of epoxy is 1.2 g/cc. The tensile modulus of carbon fiber is 43 GPa and that of epoxy resin is 0.5 GPa. The tensile strength of carbon fiber is 3530 MPa and that of epoxy resin is 1200 MPa.

Answer
Assume 1 cc of composite.

Then 60 volume % carbon fiber = 0.6 cc of carbon fiber × 1.8 g/cc = 1.08 g carbon fiber and 0.4 cc of epoxy × 1.2 g/cc = 0.48 g epoxy. The total mass of the composite is 1.56 g. The density of the composite is 1.56 g/cc.

The tensile modulus of the composite is Ec = Vf*Ef + Ve*Ee = 0.6(43 GPa) + 0.4(0.5 GPa) = 26 GPa. Here Vf is volume fraction of fiber, Ef is tensile modulus of the fiber, Ve is the volume fraction of epoxy, Ee is the tensile modulus of the epoxy. The tensile strength of the composite is Vf*TSf + Ve*TSe = 0.6(3530 MPa + 0.4*1200 MPa = 2598 MPa.

The above calculations can be done for other fibers and resins, and plastic matrices. The calculations can be done for CLTE, the Poisson's ratio, impact strength, compressive strength, and other mechanical properties.

12.8 Sandwich and Cored Polymer Composite Structures

Core structures are added to composites to provide stiffness with sectional properties. The cored structures provide increased thickness in the part that generates higher stiffness. The stiffness of a composite part with a board-like shape increases as thickness increases. The engineering analysis with the moment of inertia calculations generates a relationship between stiffness and thickness cubed, as shown in Equation 12.5.

$$\text{Stiffness} = \text{E}\,\text{t}^3, \tag{12.5}$$

where, E is the tensile modulus and t is the part thickness.

Thus, a part that has thickness of 3 mm, that uses a cored structure with thickness of 13 mm, increases the stiffness by 81 times. This is shown below.

Example,

Stiffness 1 = E (3 mm)^3 = 27E
Stiffness 2 = E (13)^3 = 2197E
The ratio of stiffness 1 to stiffness 2 = 2197E/27E = 81.37

When thickness is doubled, stiffness increases by a factor of 8. When thickness is tripled, stiffness increases by 64 times.

Cored structures can be used for thick composites that are greater than 10 mm, up to 100 mm. These consist of the following:

- Two face sheets made with fiber and polymer resin (epoxy or polyester)
- Core made with foam or wood
- Adhesives

Cored structures have greater stiffness and bending strength due to the increased section, as shown in Figure 12.19.

The core materials can be made from wood, polyurethane foam, or PP bead foam. The most common core material is balsa wood. Bamboo can also be used. Figure 12.20 lists the common core materials. Foams can be urethane, PP, PVC, PMI, PEI, and SAN. These foams have good properties: water absorption in some and water resistance in others. Ceramic foams have excellent thermal resistance.

The two types of foam cores are either open-cell and closed-cell types depending if the foam is open in the middle. Polyurethane is an example of an open-cell foam, and polypropylene and expanded polystyrene (EPS) are examples of closed-cell foam. The foams are formed by trapping a gas or air in a polymer. Open cell is a type of foam in which the cells are open and not closed off. Household sponge for the kitchen is an example of open-cell foam. Water or liquids can easily flow in and out of open-cell foam. Polyurethane open-cell foam can be used in composites as a core material but can cause the resin to fill in the open-cell foam.

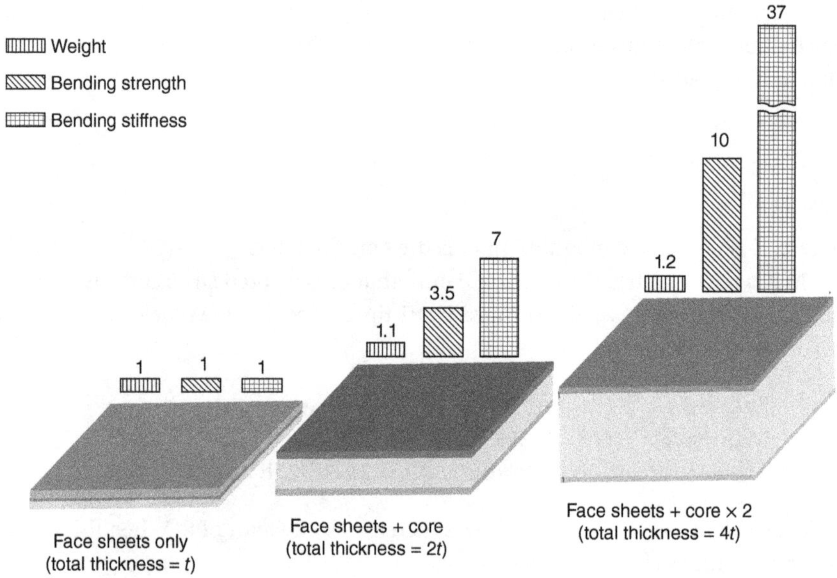

Figure 12.19 Increased bending strength and stiffness for core structures in polymer composites. Source: Sutcliffe (1998).

Closed cell is a type of foam where the gas forms pockets in the polymer and the cell walls are closed off and are not connected. Polypropylene, polyethylene, neoprene, and EPS are examples of closed-cell foam.

Honeycomb-type foam cores can be made from paper, polyimide, polyolefin (PE, PP), engineering plastic, metal, carbon, aramid, and quartz.

The properties of foam core materials are shown in Figures 12.21–12.24. Figure 12.22 shows that polyisocyanurate (PIR) foam core had the highest tensile strength and the lowest density. Figure 12.23 shows that balsa wood core had the highest tensile strength.

Figure 12.23 shows that polyurethane (PUR) foam core had the highest compressive strength and polyisocyanurate (PIR) had the lowest density. Figure 12.24 shows that balsa wood core had the highest compressive strength.

The core materials are usually placed inside the composite part with glass fibers or carbon fibers on the outside. The polymer resin can be either polyester or epoxy. The core is usually surrounded by fabric and resin is cured around it.

Resin penetration into the core is not desired and must be prevented. Adhesives can be added to the core material with an epoxy resin that is coated with adhesive and then a fiber mat is added to the structure. The composite cored structure is

Core Material	Characteristics and Benefits
Balsa wood (end grain)	Good shear strength, high fatigue endurance, low cost, easily bonded, easily finished, good temperature range
Polyvinyl chloride (PVC) foam (crosslinked)	High strength, high stiffness, low cost, easily bonded
PVC foam (linear)	Low cost, easily bonded, good impact resistance
Polymethacrylimide (PMI) foam	High dimensional stability under heat, excellent mechanical properties, solvent resistance, low thermal conductivity, high strength and stiffness
Polyetherimide (PEI) foam	Low water absorption, high thermal stability, high strength, fire resistant, good dielectirc properties
Styrene-acrylonitrile copolymer (SAN) foam	No outgassing, high stiffness, high impact and fatigue strength, no environmental problems with resin or recycling
Ceramic foams	Unsurpassed thermal resistance, excellent thermal insulation, solvent resistance
Paper honeycomb	Low cost, easily bonded, strong for weight
Polyimide paper honeycomb	High strength to weight, corrosion resistant, good thermal insulation, fire resistant, easily shaped, excellent dielectric properties, easily bonded
Polyolefin honeycomb	Rigid and elastic, high toughness, sound and vibration dampening, explosion containment vessels, scrim cloth available, high strength to weight, corrosion resistant, fungi resistant, can be thermoformed, recyclable
Engineering plastic honeycomb	Tough, relatively high-temperature tolerant, excellent dielectric properties, good thermal insulator, fire resistant, fungi resistant, highly variable cell sizes and densities
High-performance honeycomb	Carbon fiber reinforced, carbon-carbon, aramid, quartz, superior strength, superior thermal resistance
Metal honeycomb	Aluminum, titanium, stainless, nickel available, no outgassing, high-temperature tolerant, fire resistant, fungi resistant, high thermal conductivity
Stitched/compressed	Excellent drape, needs to be fully wetted, high impact resistance, reduces cracks, absorbs resin for added strength

Figure 12.20 Common core materials.

placed in autoclave for temperature and pressure. Pressure must not be too low in the molding. If the pressure results in a vacuum, the core might collapse. Milled fibers and other fillers, like calcium carbonate, can be added to the resin to form a part on the outside of the core.

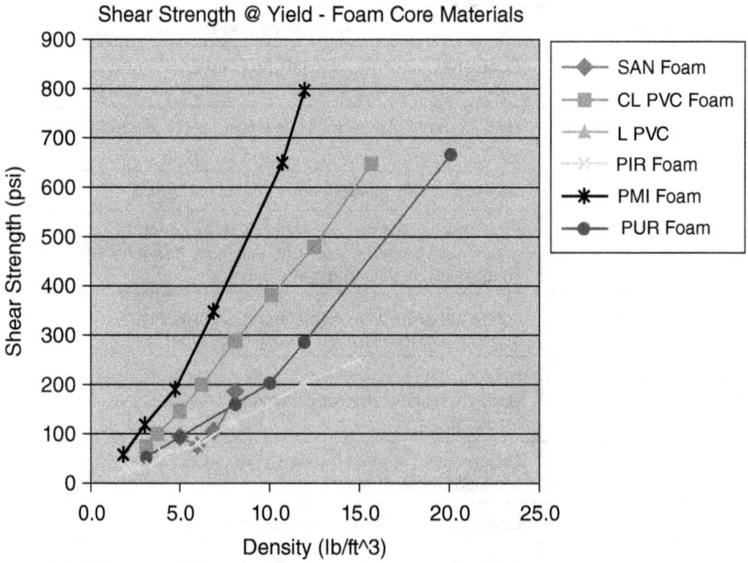

Figure 12.21 Shear strength and density of foam cores. Source: Foam cores strength (2020).

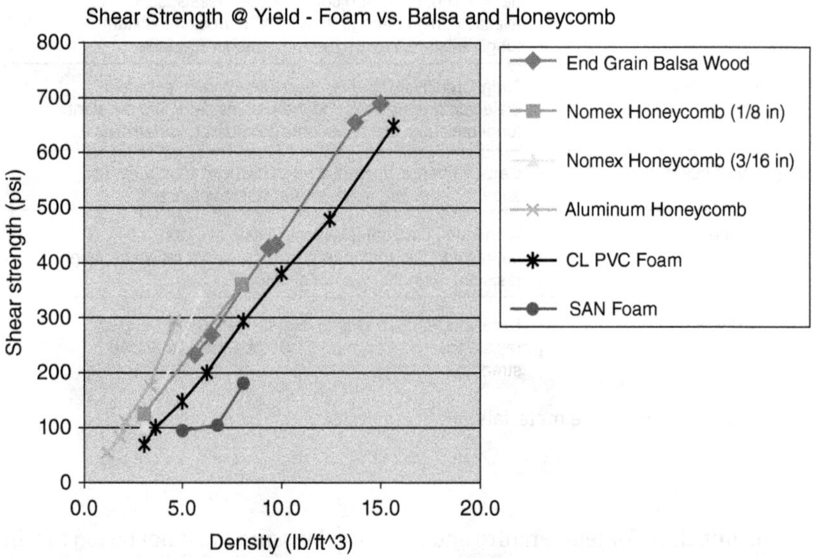

Figure 12.22 Shear strength and density of balsa wood and honeycomb cores. Source: Foam cores strength (2020).

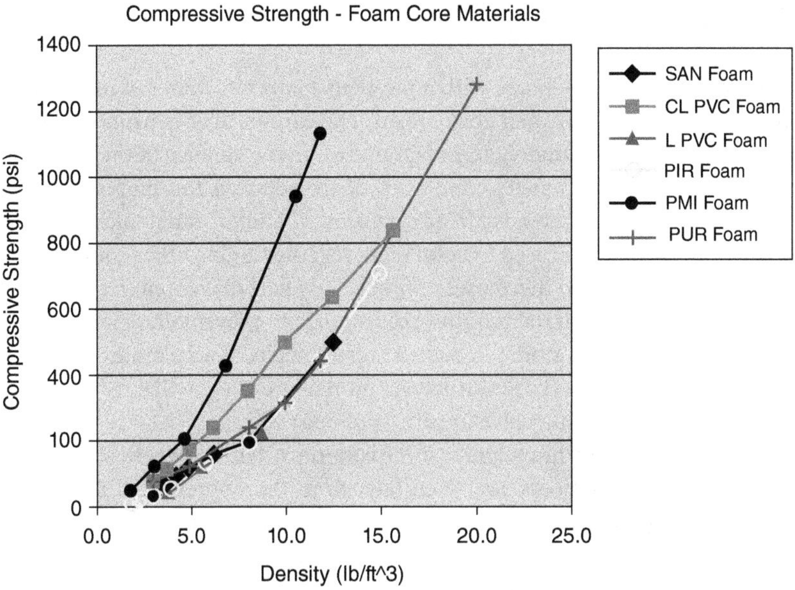

Figure 12.23 Compressive strength and density of foam cores. Source: Foam cores strength (2020).

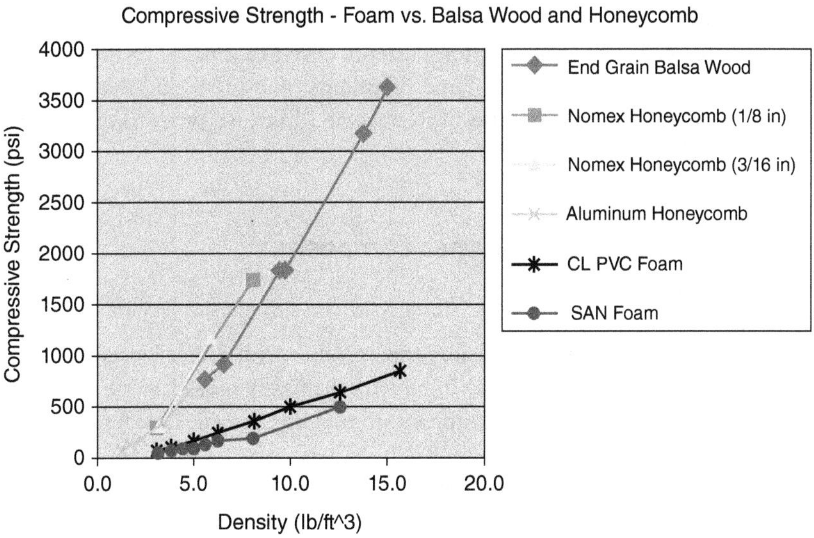

Figure 12.24 Compressive strength and density of balsa wood and honeycomb cores. Source: Foam cores strength (2020).

12.9 Polymer Pre-Preg Composites

Polymer composites can be made with a pre-preg composite material that is usually made from carbon fiber and epoxy resin. The carbon fiber is usually at high loadings of 50–60% by volume. Epoxy resin makes up the balance of the material.

Pre-preg is "pre-impregnated" composite fibers where a thermoset polymer matrix material, such as epoxy, is already present. The fibers often take the form of a weave and the matrix is used to bond them together and to other components during manufacturing. The thermoset matrix is only partially cured to allow easy handling; this B-Stage material requires cold storage to prevent complete curing. B-Stage pre-preg is always stored in cooled areas since heat accelerates complete polymerization. Hence, composite structures built of pre-pregs will mostly require an oven, heated mold, or autoclave to cure. (Prepreg)

Pre-preg allows one to impregnate the fibers on a flat workable surface, or rather in an industrial process, and then later form the impregnated fibers to a shape which could prove to be problematic for the hot injection process. Pre-preg also allows one to impregnate a bulk amount of fiber and then store it in a cooled area (below 20 °C) for an extended period of time to cure later. Unfortunately the process can also be time consuming in comparison to the hot injection process and the added value for pre-preg preparation is at the stage of the material supplier.

Pre-pregs are resin-impregnated cloth, mat, or filaments in flat form that can be stored for later use. The resin is often partially cured to a tack-free state called "B-staging." Catalysts, inhibitors, flame retardants, and other additives may be included to obtain specific end-use properties and improve processing, storage, and handling characteristics.

12.10 Processing of Polymer Composites for Automotive Parts

The processing of polymer composites can be done with either closed-mold or open-mold processes. The automotive industry primarily uses close-mold operations with injection molding, compression molding, resin transfer molding (RTM), and structural RIM molding (SRIM).

Plastic injection molding equipment, typically, comprises an extruder and a compression press. The plastic pellets are placed in the hopper and fed through the extruder in three main heated zones. The temperature profile in the three zones is dependent on the plastic type. The first zone is called the feed zone. The second zone is called the compaction zone. The third zone is called the metering zone.

Short or long fibers are compounded with a twin screw extruder in concentrations, usually between 20 and 40% glass fibers and 60–80% plastic resin, like PP, nylon, phenolic, PVC, PBT, silicone, and others. The glass-filled plastic composite is made with the injection molding process to form a part as shown in Figure 12.25. The part is usually molded with a cycle time of less than one minute for big parts, like bumper fascia, or less than 30 seconds for smaller parts. The injection molding process utilizes steel tooling to produce polymer composites under higher pressure.

The glass-filled polymer composite can also be compression molded in a process described in Figure 12.26. The polymer can either be a thermoplastic material, like polypropylene (PP), or a thermoset material like SMC or BMC. For the thermoplastic material, the plastic is hot and the mold is cold. The glass fibers are impregnated with the polypropylene plastic to create a glass-mat-thermoplastic (GMT) material in a mat form. Plastics can be PP, nylon, polycarbonate (PC), PVC, and polyphenylene sulfide (PPS).

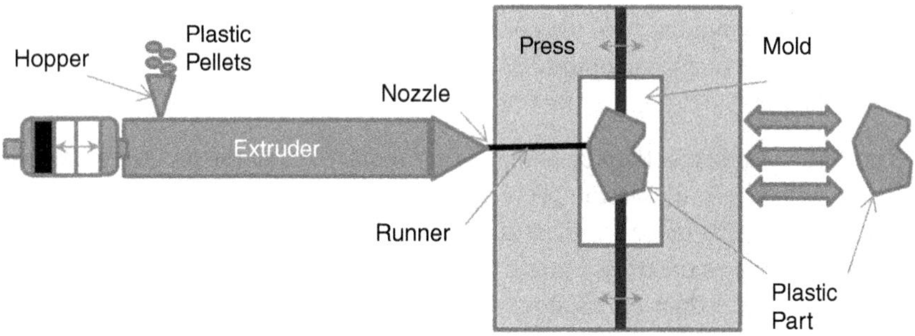

Figure 12.25 The injection molding process.

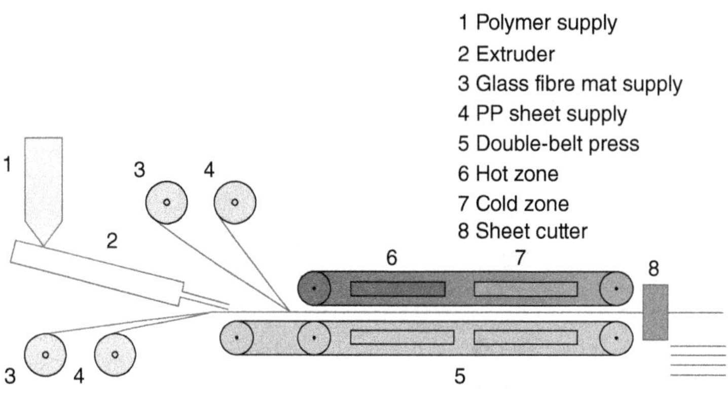

Figure 12.26 GMT melt impregnation process. Source: GMT (2020).

Typically, random glass-fiber mats are impregnated with molten plastic by sandwiching the plastic extrudate between two layers of glass mat, which are themselves retained within outer thin sheets of polyethylene, prior to entry into a continuous double-belt press. as shown schematically in Figure 12.26. The belts are heated to impregnate the fiber with the plastic resin. Temperature, pressure, and residence time are the key factors in controlling the quality of resin impregnation. This process can produce a sheet of up to 4 mm in thickness with fiber loadings between 20 and 50% by weight. (GMT 2020).

For GMT plastic materials, the sheet of plastic is cut into strips and then placed in an infrared oven near the compression molder, as shown in Figure 12.27. The GMT is molded in a compression press with the following process.

- Place the strips of GMT stacked and place them in the IR oven
- Transfer the stacked GMT into the oven for two minutes
- When GMT is heated to about 200F, remove it from the oven
- Place the stack of GMT on mold
- Compress the mold at 1500 tons press
- Hold for 60 seconds to cool the GMT
- Open the press and remove the GMT part

The GMT plastic can then be compression molded using the process shown in Figure 12.28. The compression press has two platens that close at a certain rate. Usually, the platens move at different speeds when it opens with the platen moving at a slow rate as the press initially opens, then moves quickly to near the final open location and then opens slowly the last few millimeters to the final open position. The process is repeated in reverse with at a slow initial close speed, followed by a rapid close speed, and then finally a slow final close speed.

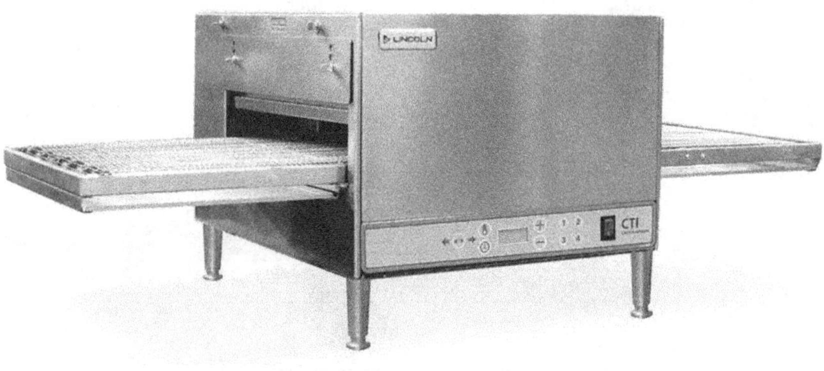

Figure 12.27 Infrared heating machine for GMT. Source: GMT oven (2020).

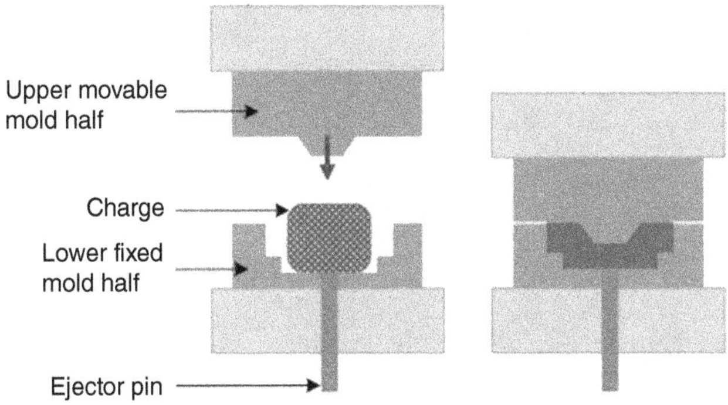

Figure 12.28 The compression molding process. Source: Compression molding (2020).

The thermoset polymers can be molded with other closed molding operations. During these operations, the thermoset polymers are usually cold and kept refrigerated before molding on a hot (200 °C) mold. The most common thermoset polymer for the automotive industry are polyester materials. The SMC is the most common thermoset polymer composite material that is usually compression molded into a part like a hood, floor, roof, or bumper beam. The SMC can be made with two glass percentages depending on the application. Typically, 33% glass fibers (25 mm in length) are used to make body panels that are painted to body color. Typically, 50% glass fibers are used to make inner panels for roofs, hoods, and doors. Also, 50% glass fibers are used for bumper beam parts. The higher glass-loaded parts are typically not painted and surface appearance is not critical. The cycle times for SMC and compression molding are usually less than two minutes. The compression molding process utilizes a steel tool to form the product under moderate pressure.

Other polyester polymer composites can be compression molded like bulk-molded compound or BMC. Also, polyester can be molded using a glass mat or a woven product to make a hood for a car using a resin transfer (RTM) process. The RTM process, Figure 12.29, consists of a small pump that transfers the resin from the holding tank to the fiber mat product. The RTP process utilizes a steel tool to form the low-pressure part. In the RTM process,

- Dry, or unimpregnated, reinforcement is pre-shaped and oriented into the skeleton of the actual part, known as the preform, which is inserted into a matched die mold.
- The heated mold is closed and the liquid resin is injected.
- The part is cured in mold.
- The mold is opened and the part is removed from the mold.

Large parts can be made with this process, though the cycle time can be five to ten minutes per part. Other large parts can be made using a modified RIM

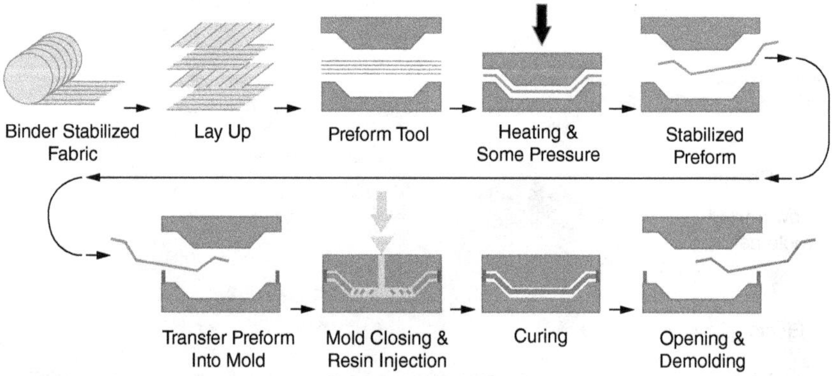

Figure 12.29 Resin transfer molding (RTM) process. Source: RTM process (2020).

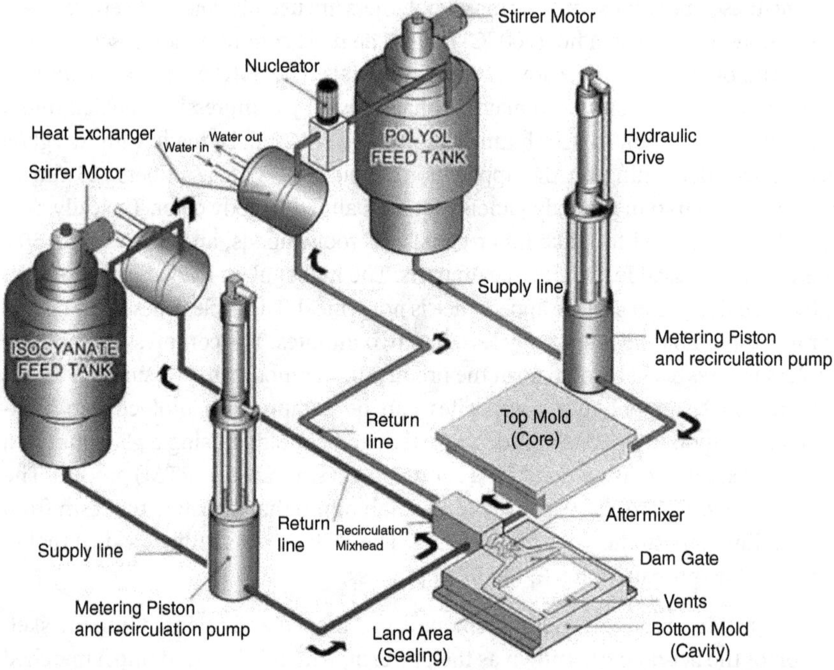

Figure 12.30 Structural Reaction Injection Molding (SRIM) process. Source: SRIM (2020).

process. The structural RIM process is similar to the RTM process in that a liquid alcohol and isocyanates are injected into a preformed mat or woven product to form a glass-filled polyurethane polymer composite. Figure 12.30 provides an illustration of the SRIM process.

The process uses the Reaction Injection Molding (RIM) process to pump, mix, and inject the polyurethane resin into the mold at moderate pressure. The SRIM process first preforms the glass matt or woven product into a shape of the final part, like bumper beam with a steel preform tool. Then the fiber preform is placed in a heated mold and the resin is injected onto it. The SRIM process molds the polyurethane liquid and glass mat in a steel tool to form the polymer composite with cycle times less than two minutes.

12.11 Aerospace Polymer Composites

Thermoset composites are mostly made with carbon fiber and epoxy resin to form shapes that can be placed on an aircraft. Polymer composites for the aerospace industry are thermoset-reacting polymers or thermoplastics that typically combine with carbon fibers and are made with the following characteristics:

1) Cycle times more than 10 minutes
2) Service use temperatures of 200–300 °C (140–158 °F)
3) High-cost carbon fiber and resin materials between \$2 and \$20 per pound
4) Carbon fiber loadings usually between 30 and 60% by volume
5) Excellent chemical resistance
6) Usually not painted
7) Can be a small molded part like a rocket booster or a large molded part like an aircraft wing
8) Excellent mechanical properties
9) Excellent thermal properties

12.12 Processing of Polymer Composites for Aerospace Parts

Aerospace parts are usually made with processes with long cycle times, like RTM, SCRIMP, Vacuum infusion, other hand-layup processes, and filament winding. RTM was described earlier. VARTM (Vacuum Assisted Resin Transfer Molding) is the name of the process used by Lockheed Martin that is similar to SCRIMP, but does not use a flow media. The entire fuselage of the AGM 159 JASSM missile is made using VARTM, as shown in Figure 12.31.

SCRIMP (Seemann Composites Resin Infusion Molding Process) is one of the earliest patented infusion methods. It is used in many marine and wind blade applications, but was also licensed by some aerospace firms. It relies upon the use of a flow or "distribution media" with high permeability between the layup and

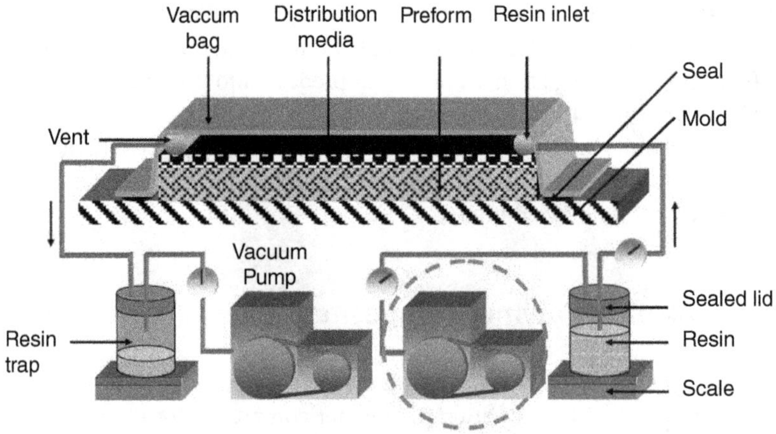

Figure 12.31 VARTM molding system. Source: Dominik et al. (2006).

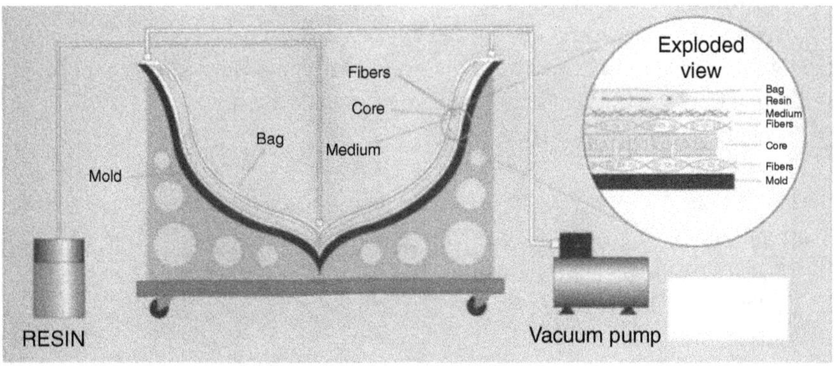

Figure 12.32 SCRIMP molding system. Source: SCRIMP (2020).

vacuum bag to rapidly and evenly distribute resin laterally across the part. The SCRIMP process is shown in Figure 12.32.

CAPRI (Controlled Atmospheric Pressure Resin Infusion) was patented by Boeing and is said to reduce thickness variation and result in fiber volumes and mechanical properties equivalent to pre-preg/autoclave materials. First, it uses vacuum debulking cycles on the dry preform to reduce compressed thickness prior to infusion. During infusion, the resin supply is held at partial vacuum, which assists in degassing the bulk resin but also reduces the pressure differential driving resin into the preform.

RTI (Resin Transfer Infusion) is a Bombardier-patented process used to produce the wing skins of its C-Series aircraft. Infusion of resin into the preform is performed using vacuum pressure only. However, the mold is located in an un-pressurized autoclave during the infusion step. After the preform is fully infused, the autoclave is pressurized and heated to perform cure. This makes it easier to achieve high laminate quality because positive cure pressure (>14 psi) helps prevent void formation from entrapped air and volatiles. It has the drawback that a suitable size autoclave is still requited. All other methods cited above are true out-of-autoclave processes.

The Vacuum Infusion Process or VIP is one of many closed-mold processes. It distinguishes itself by being the only process that utilizes only atmospheric pressure to push the resin into the mold cavity. The mold cavity can be a one-sided mold with bagging film being utilized for the "B" side, a two-sided mold, or even a soft "envelope" bag. The process is highly controllable, because it is governed by the principles of D'Arcy's Law. This means that there are only three variables affecting the flow of the resin: (i) permeability of the laminate, (ii) viscosity of the resin, and (iii) pressure differential in the cavity in relation to atmospheric pressure. If all three of these variables are unchanged, then the infusion process will consistently flow the same way with every injection for a given part. This also equates to a very accurate bill of material for a given part since the resin and fiberglass usage will not change.

Filament winding is a fabrication technique mainly used for manufacturing open- (cylinders) or closed-end structures (pressure vessels or tanks). The process involves winding filaments under tension over a rotating mandrel. The mandrel rotates around the spindle (Axis 1 or X: Spindle), while a delivery eye on a carriage (Axis 2 or Y: Horizontal) traverses horizontally in line with the axis of the rotating mandrel, laying down fibers in the desired pattern or angle. The most common filaments are carbon and are impregnated in a bath with resin as they are wound onto the mandrel. Once the mandrel is completely covered to the desired thickness, the resin is cured. Depending on the resin system and its cure characteristics, often the rotating mandrel is placed in an oven or placed under radiant heaters until the part is cured. Once the resin has cured, the mandrel is removed or extracted, leaving the hollow final product. For some products such as gas bottles, the "mandrel" is a permanent part of the finished product forming a liner to prevent gas leakage or as a barrier to protect the composite from the fluid to be stored.

Filament winding, shown in Figure 12.33, is well suited to automation, and there are many applications, such as pipe and small pressure vessel, that are wound and cured without any human intervention. The controlled variables for winding are fiber type, resin content, wind angle, tow or bandwidth, and thickness of the fiber bundle. The angle at which the fiber is laid down on the composite has an effect on the properties of the final product. The "hoop" will provide circumferential strength, while lower angle patterns (polar or helical) will provide greater longitudinal/axial tensile strength.

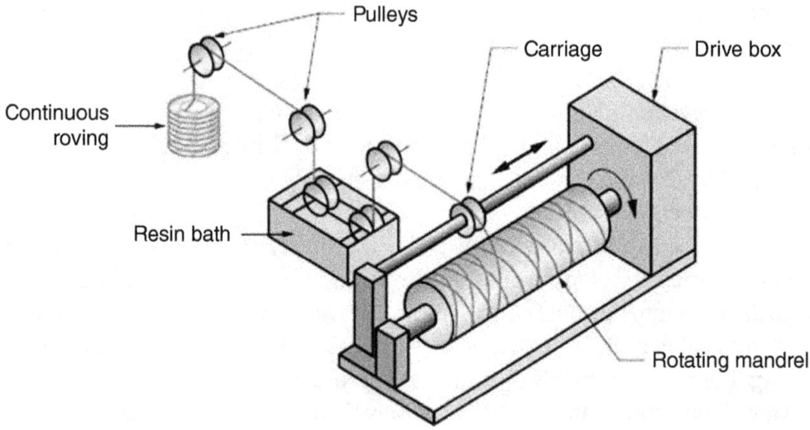

Figure 12.33 Filament winding molding system. Source: Filament winding (2020).

References

Czel, G. et al. (2017). Design and characterization of high performance, pseudo-ductile all-carbon/epoxy unidirectional hybrid composites. *Compos. Part B* 111 (15): 348–456.

Dominik, B., Schuster, J. et al. (2006). Flow rate control during vacuum-assisted resin transfer molding (VARTM) processing. *Compos. Sci. Technol.* 66 (13): 2265–2271.

Furqan, A. et al. (2014). A review: natural fiber composites selection in view of mechanical, light weight, and economic properties. *Macromol. Mater. Eng.* 300: 10–24. https://doi.org/10.1002/mame.201400089.

Gacitua, W. et al. (2005). Polymer nanocomposites: synthetic and natural fillers a review. *Madera.Ciemcia y technologia* 7 (3): 159–178.

Glass, F. et al. (2000). *Ullmann's Encyclopedia of Industrial Chemistry, Fibers, 5. Synthetic Inorganic*. Weinheim, Germany: Wiley-VCH Verlag GmbH & Co. KGaA. ISBN 3527306730. doi:https://doi.org/10.1002/14356007.a11_001.

Hallad, S.A. (2018). Kevlar reinforced polymer matrix composite for structural applications. *IOP Conf. Ser. Mater. Eng.* 376 (9): 1–13.

Han, B. et al. (2017). Nano-core effect in nano-engineered cementitious composites. *Compos. A: Appl. Sci. Manuf.* 95: 100–109.

Hercules (2011). Determining the Fatigue Life of Composite Aircraft Structures Using Life and Load-Enhancement Factors. Final Report. William J. Hughes Technical Center at actlibrary.tc.faa.gov.

Iredale, R. et al. (2017). Modern advances in bismalemide resin technology: A 21st century perspective on the chemistry of addition polyimides. *Prog. Polym. Sci.* 69: 1–21.

Macosko, Chris, *RIM Fundamentals of Reaction Injection Molding*, Hanser Gardner Publications (December 1, 2000) **ISBN-10:** 1569900558.

Romano, A. et al. (2018). On the development of self-controlled bio-based panels for building's thermal management. *Conference: ECCM18*, Athens. https://www.researchgate.net/publication/325539191_On_the_development_of_self-controlled_bio-based_panels_for_building's_thermal_management (accessed February 2022).

Sutcliffe, M.P.F. (1998). Design of sandwich structures. Ph.D. Dissertation. Cambridge University Engineering Department.

Websites

Carbon fiber (2020). http://www.madehow.com/Volume-4/Carbon-Fiber.html (accessed September 2020).

Carbon fiber process (2020). https://en.wikipedia.org/wiki/Carbon_fibers (accessed September 2020).

Composite laminate (2020). https://www.asminternational.org/documents/10192/1849770/05287G_Sample_Chapter.pdf (accessed September 2020).

Compression molding (2020). https://www.substech.com/dokuwiki/doku.php?id=compression_molding_of_polymers (accessed September 2020).

Compression press (2020). https://www.freemanschwabe.com/schwabe-compression-molding-presses.html (accessed September 2020).

Cored composite (2022). https://aerospaceengineeringblog.com/sandwich-panel/ (accessed February 2022).

Fatigue (2020). http://nexusprojectes.com/en/durability (accessed September 2020).

Fiberglass (2020). https://www.redorbit.com/reference/fiberglass/Redorbit.com (2014-06-20). Retrieved on 2016-06-02 (accessed September 2020).

Fibers (2020). www.matweb.com (accessed September 2020).

Filament winding (2020). https://www.pinterest.com/pin/702631979334405930/ (accessed September 2020).

Foam cores strength (2020). https://www.boatdesign.net/articles/foam-core-properties/index.htm (accessed September 2020).

Glass (2020). https://dev-in-saint-gobain-glass-com.content.saint-gobain.io/glass-manufacturing (accessed September 2020).

GMT (2020). https://www.azom.com/article.aspx?ArticleID=318 (accessed September 2020).

GMT oven (2020). https://www.katom.com/054-25011346.html?gclid=EAIaIQobChMIvaOt8buA7AIVVT2tBh1GEwO7EAQYCiABEgKQEfD_BwE (accessed September 2020).

Matweb (2020). http://www.matweb.com/ (accessed September 2020).

Nanocor (2022). http://www.nanocor.com/tech_papers/plastic_nanocomposites.asp (accessed February 2022).
Natural fibers (2022). https://www.aplustopper.com/types-of-natural-fibres-uses/ (accessed February 2022).
Prepreg. Wikipedia (2022). https://en.wikipedia.org/wiki/Pre-preg (accessed February 2022).
RTM (2020). https://www.indiamart.com/proddetail/resin-transfer-moulding-10495972962.html (accessed September 2020).
RTM process (2020). http://songwoltech.com/en/molding/ (accessed September 2020).
SCRIMP (2020. https://www.compositesworld.com/articles/seemann-composites-inc-acquires-materials-sciences-corporation (accessed September 2020).
SMC (2020). http://greenhulk.net/forums/showthread.php?t=22098&p=1096906&viewfull=1 (accessed September 2020).
SRIM (2020). https://reactioninjectionmolding.com/rim-process/ (accessed September 2020).
Stress strain fibers (2020). https://www.chegg.com/homework-help/using-fig-127-estimate-amount-energy-required-fail-fiber-ran-chapter-12-problem-87p-solution-9780071311144-exc (accessed September 2020).

13

Natural Fiber Polymer Composites

13.1 Natural Fibers

Natural fibers such as flax, hemp, and jute can be used as reinforcement for thermoset/plastic composites. The source of natural fibers can be from animals, plants, or minerals. The sources of natural fiber are shown in Figure 13.1. The animal fibers can be wool or hair from sheep, camel, or rabbit. The plant fibers can be cotton, Kapok, Akon, Flax, Hemp, Jute, Ramie, Agave, Banana, and Coco Bromelia. The mineral fibers can be from asbestos. Flax, hemp, and Jute are the most common natural fibers. Also, rice straw and almond shells can be used as fillers for plastic parts.

The global natural fiber composites market size was valued at USD 4.46 billion in 2016 (Global natural fiber market 2021). It should increase by about 12% from 2016 to 2024. Automotive and other industries are interested in lightweight products that are sustainable. They can be produced with lower greenhouse gases and lower pollution than petroleum-based plastics. However, moisture sensitivity is a concern of the natural fibers.

Natural fibers are biobased materials manufactured using materials such as wood, cotton, flax, kenaf, and hemp. All these materials are less harmful to the environment and sustainable. Raw materials used to manufacture natural fiber composites are environment friendly and sustainable. These fibers can replace synthetic fibers in many industries in the future.

Green products made with natural fibers can be a boon for the automotive, home furnishings, and other markets. Composites made from natural fibers can be more sustainable than petroleum-based composites and have a lower mass of the component since the natural fibers are lighter than glass fibers. This can help lower the total energy consumption in making the product.

Sustainable Plastics: Environmental Assessments of Biobased, Biodegradable, and Recycled Plastics, Second Edition. Joseph P. Greene.

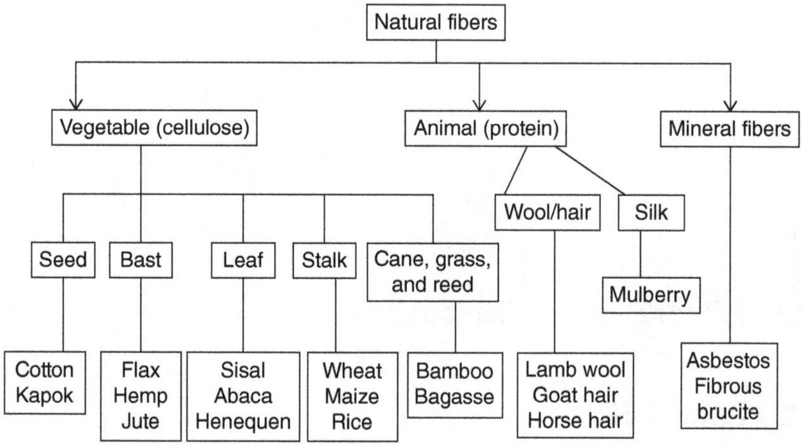

Figure 13.1 Sources of natural fibers. Source: Natural fibers (2020).

13.2 Raw Material Information

Some of the major raw materials include wood, flax, kenaf, cotton, and hemp. Wood dominated the market and represented 59.3% of the overall revenue in 2015 (Global natural fiber market 2021). The trend should continue until 2024. Wood has advantages of high strength and stiffness which can fuel its utilization in the future.

Flax was one of the most widely used materials in 2015, with a market share of 13.0% (Global natural fiber market 2021). Flax is CO_2 neutral, vibration damper, and renewable as compared to glass and carbon fibers. Flax has many advantages including high tensile strength, ultraviolet rays blocking properties, vibration absorbent, and high-water retention. This can help it be one of the most used raw materials in the industry.

Cotton belongs to the seed fiber category, which is commonly used for textiles all over the world (Global natural fiber market 2021). Cotton is one of the weakest fibers due to its moisture absorption property. It can absorb moisture up to 20% of its dry weight. The use of cotton can grow in textile and sporting goods industries.

Use of kenaf has been growing in various industries such as construction, oil & chemical absorbent, food packaging, and automotive (Global natural fiber market 2021). The material has good tensile properties and lightweight. It is also highly sustainable and can be completely recycled.

Composites that are made from hemp can replace glass fiber in many applications, and they are biodegradable (Global natural fiber market 2021). Hemp is desirable due to its being made from natural materials.

Natural fibers can be used in a variety of industries shown in Figure 13.2.

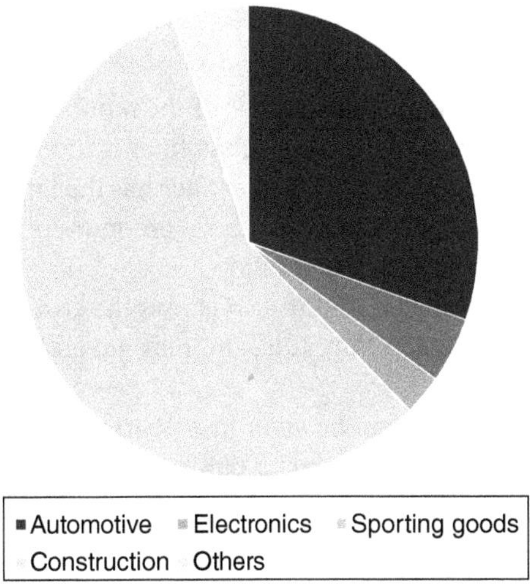

Figure 13.2 Global natural fiber composites market revenue, by application in 2016.

Figure 13.2 shows that the construction industry can use the most natural fibers in the future followed by the automotive industry. Several automotive companies are researching the use of natural fibers for interior components. The market projection for natural fibers from 2013 to 2024 is shown in Figure 13.3. Figure 13.3 shows that all fibers should experience growth in applications in the future. Wood fiber should see the biggest growth in the near future.

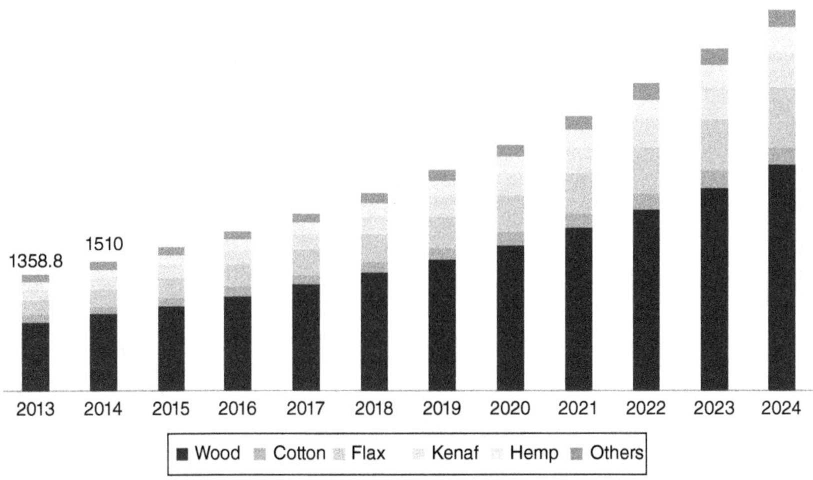

Figure 13.3 Growth of natural fibers in the United States of America from 2013 to 2024.

13.3 Fiber Properties

The aspect ratio (L/D) of the fiber is an important characteristic of the reinforcing capability of the fiber. As a comparison, the aspect ratio of glass fiber is around 100. The aspect ratio of natural fibers is listed in Table 13.1. Jute fiber has the largest aspect ratio followed by hemp and flax fiber. Straw and shells are more of a filler rather than a reinforcement due to the low as aspect ratio.

The cost of hemp fiber is less than $1 per pound. Figure 13.4 shows the cost of many fibrous materials for polymer composites. Hemp, Jute, and Flax are all less expensive than E-glass and carbon fiber.

Jute is a long, soft, shiny vegetable fiber that can be spun into coarse, strong threads. It can be blended with epoxy and polyester to make a polymer composite.

Table 13.1 Aspect ratio of natural fibers.

Fiber	*L* (mm)	*D* (mm)	Aspect ratio (*L*/*D*)
Hemp	0.4	0.01	40
Flax	0.3	0.02	15
Jute	2	0.01	200
Straw (rice, wheat, grain)	0.1	0.1	1
Shells (walnuts other nuts)	0.1	0.1	1

Source: Peltola et al. (2011).

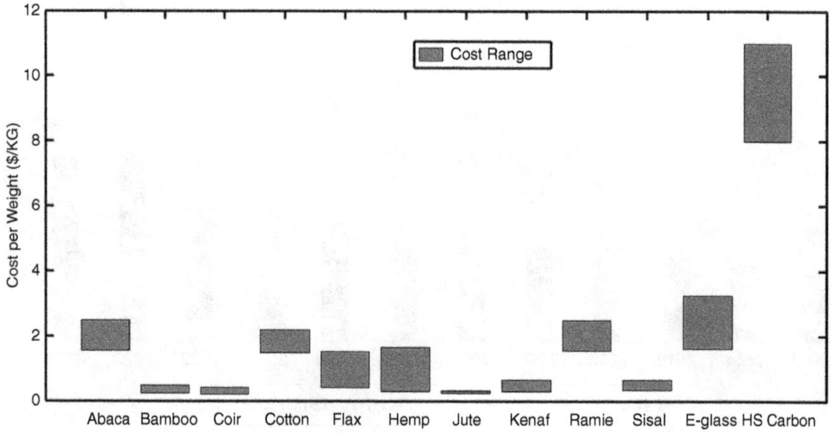

Figure 13.4 Cost per mass of reinforcing fibers. Source: Furqan et al. (2014).

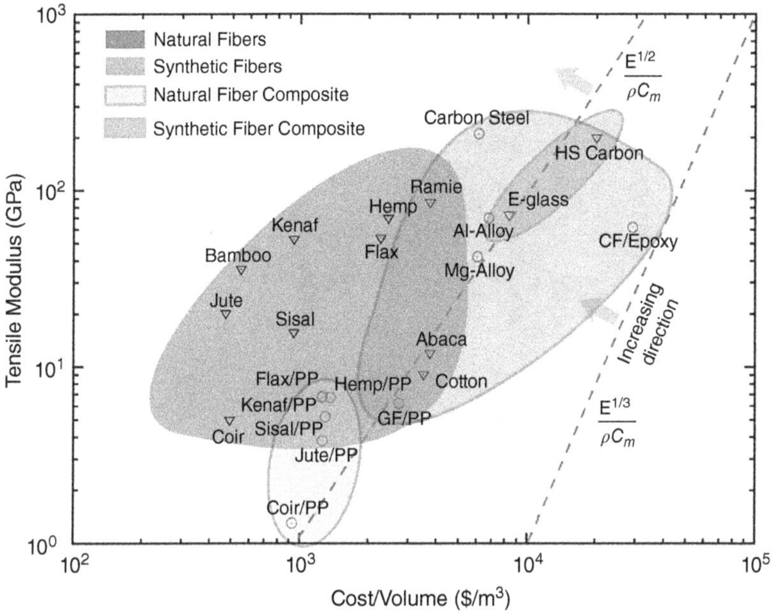

Figure 13.5 Tensile moduli and cost of reinforcing fibers. Source: Furqan et al. (2014).

Figure 13.4 shows the cost of reinforcing fibers. Hemp, Jute, and Flax have lower modulus and lower costs per volume than E-glass and carbon fiber.

Figure 13.5 illustrates the tensile moduli and density of reinforcing fibers. Hemp, Jute, and Flax have lower modulus and lower density than E-glass and carbon fiber.

Natural fibers like hemp and flax have lower energy use in the production of the fibers than polyethylene terephthalate (PET), thermoplastic starch (TPS), polypropylene (PP), acrylic, glass fibers, and nylon fibers. The natural fibers also have lower global warming potential than plastics, such as polystyrene, PET, and polyurethane, as shown in Figure 13.6.

13.4 Automotive Use of Natural Fibers

Automotive companies are looking for ways to produce a more sustainable car. Low-cost natural fibers can be an attractive source of meeting their goals. Hemp fiber is 61% cellulose and was used by Henry Ford in 1941 to build car doors with cellulose plastic (Henry Ford fiber use 2021). It can be used with biodegradable starch-based plastics, TPS. Hemp is already being made into the compressed door panel and dashboards for carmakers such as Ford, GM, Chrysler, Saturn, BMW, Honda, and

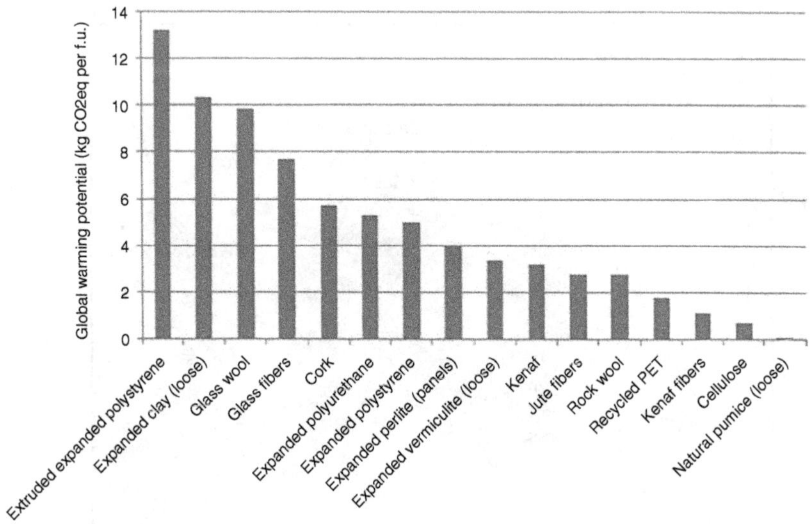

Figure 13.6 Global warming, kg CO_{2eq} per potential of thermal insulation material (Cradle to Gate Approach, CTGA). (Natural fibers global warming). Source: Asdrubali (2021).

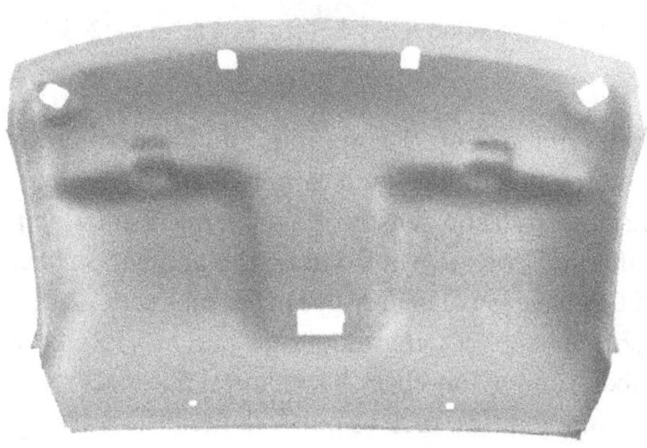

Figure 13.7 Automotive headliner made from ABS for the 1999 Ford F-150. Source: Auto headliner (2021).

Mercedes. The interior of the car has low-cost parts made from plastic and inexpensive additives. The headliner, shown in Figure 13.7, is a prime example of a low-cost part with an opportunity to use more sustainable materials. Research is needed to provide low-cost materials made from natural fibers and biodegradable plastics.

13.5 Processing of Natural Fibers

Almond shells and rice straw can be blended with polypropylene plastic to make an interior part for an automotive company. The propylene can be made from recycled materials. The recycled polypropylene plastic needs to be blended with natural fibers. The materials can be compounded with a Leitztriz twin-screw extruder. Compounding is necessary to combine the recycled PP plastic into plastic pellets with almond shells.

The first step in the process is to chop the almond shells or rice straw to less than 300 microns. That is needed to prevent the rice straw or almond shells to plug the vents in the twin-screw extruder. Figure 13.8 shows examples of some choppers. Then, the almond shells or rice straw should be sifted with a sieve to produce the small particles that are less than 300 microns as shown in Figure 13.9. The small particle powder of almond shells or rice straw can be blended with polypropylene to make pellets of the right concentrations of natural fiber. After blending the polypropylene and natural fibers into pellets, the parts can be injection molded into the interior part. This will be explained further in the following.

Almonds shells and rice straw begin as large particles shown in Figures 13.10 and 13.11.

The almond shells are taken from an almond mill that removes the almonds from the shells. The shells are an organic waste product. In 2020, the state of California produces over 1.645 billion pounds of waste almond shells per year. (Almond waste 2021). The almond shells and rice straw can be ground down into a powder as shown in Figure 13.12.

The almonds shells can be ground down into a powder as well as shown in Figure 13.13.

The almond powder was heated to 175 °C to see if it would burn. Luckily, it did not burn until 200 °C. This is important when running the twin-screw extruder

Figure 13.8 Fiber chopper example.

Figure 13.9 Sieve example.

Figure 13.10 Raw almond shells.

and the injection molding machine to not burn the rice straw or almond shells. Figure 13.13 shows the results after heating the almond shell powder to 175 °C. It did not burn and did not leave a burning smell.

The twin-screw extruder, American Leitzritz Model ZSE-18HP twin-screw extruder system, with 40:1 L/D is shown in Figure 13.14. In the feed zone, the particles are conveyed away from the hopper and fed into the heated barrel. In the transition zone, particles are melted and the melt homogenized, completing a process that started at the end of the feed zone. This section is designed to enhance

Figure 13.11 Raw rice straw.

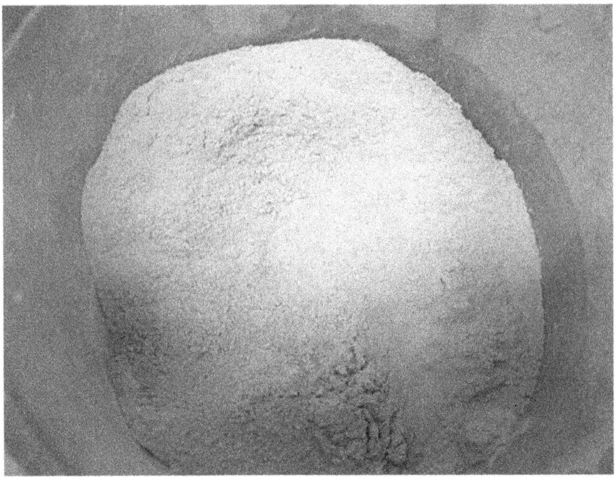

Figure 13.12 Rice straw powder.

Figure 13.13 Almond shell powder.

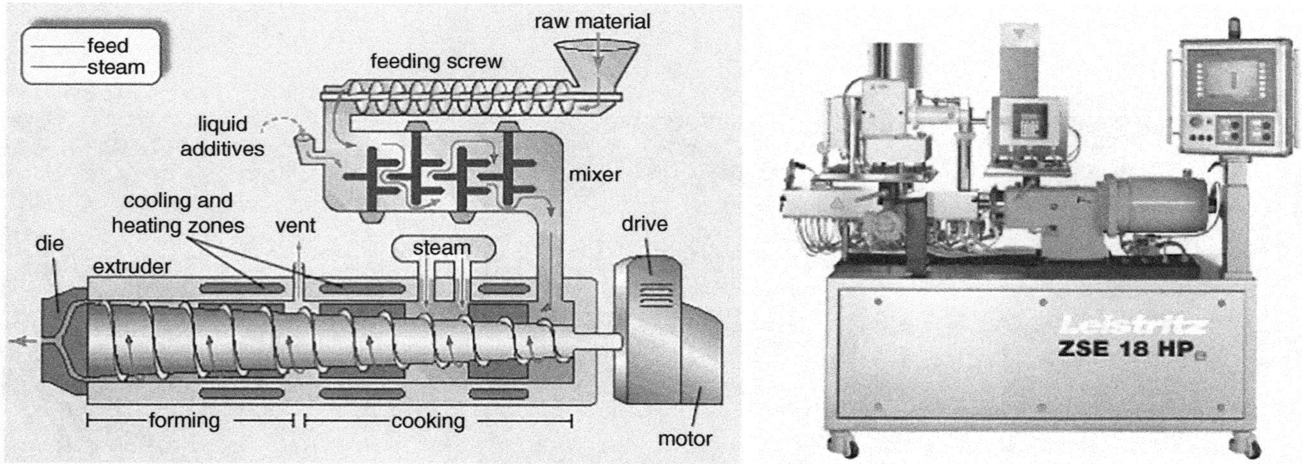

Figure 13.14 American Leitzritz Model ZSE-18HP. Source: American Leitxritz (2021).

Figure 13.15 Plastic pellets of almond shell powder blended with recycled polypropylene plastic and extruded in a twin-screw extruder. These pellets can be injection molded in an injection molding machine to make parts.

the friction and contact with the barrel. Finally, in the metering zone, the screw section is designed to act efficiently as a pump by generating pressure in the now homogenously molten mass of plastic.

Figure 13.15 shows the product of combining almond shell powder with recycled polypropylene plastic.

The interior part would need to be machined into an aluminum or steel mold with computer numerical control (CNC). The injection mold can be made out of aluminum for prototype purposes and with steel for production purposes. The interior part can be designed with SolidWorks then machined into a mold with all of the components of vents, gates, and ejector pins. SurfCAM® can be used to export the digital model to the machine code for a VF3 Haas vertical mill. The Haas VF3 is shown in Figure 13.16

The CNC mill will cut the shape of the two mold halves and inserts to take the shape of the automotive interior part. The design of the automotive interior part can be adapted to cause it to break upon the impact of shot or pellet. The design of the automotive interior part will be adjusted based upon flow analysis from the SolidWorks Plastic Flow program. The flow of plastic into a mold is an essential component of effective mold and tooling design. The design of the plastic part is crucial to its success during molding.

Solidworks® Plastics simulates how melted plastic flows during the injection molding process to predict flow distribution, pressure distribution, temperature distribution, and warpage analysis. It also provides troubleshooting steps and practical design advice to help diagnose and avoid potential molding problems.

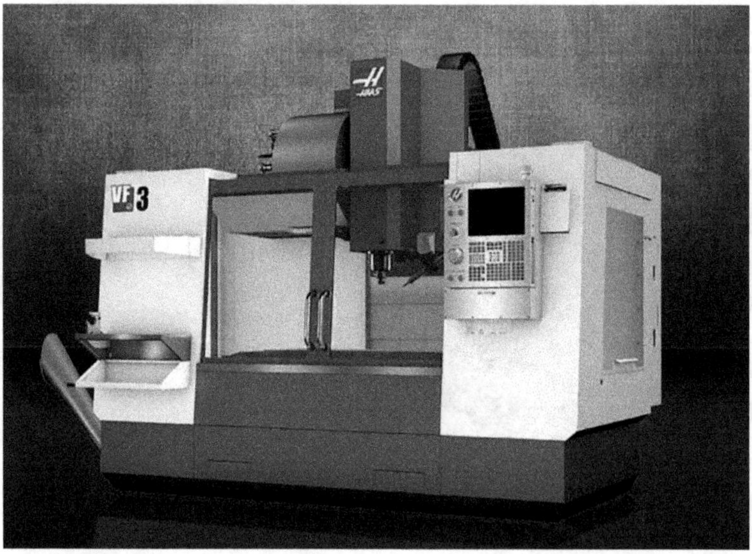

Figure 13.16 Haas VF3 vertical mill. Source: Haas mill (2021).

Solidworks® Plastics can design the shape of the mold and help design venting, gates, and ejector pins locations (Solidworks® Plastics 2021). The information will be used to CNC the shape of the vents and gates in the mold. The mold is usually not very complicated and the use of SolidWorks flow should be very useful.

Then, the interior parts can be made with an injection molding machine to shape the plastic into the interior part. Figure 13.17 lists the picture of the Arburg 28-ton injection molding machine at California State University (CSU), Chico.

Injection molding is a process where solid plastic is melted, injected into a mold, and then cooled back to a solid. Plastic injection molding equipment, typically, comprises an extruder and a compressions press. The plastic pellets are placed in the hopper and fed through the extruder in three main heated zones. The temperature profile in the three zones is dependent on the plastic type. The first zone is called the feed zone. The second zone is called the compaction zone. The third zone is called the metering zone.

Students at CSU, Chico compounded the polypropylene plastic with almond shells or rice straw and molded the parts. Figures 13.18 and 13.19 show one student working with the natural fibers and polypropylene with the molding machines.

The parts can be injection molded as shown in Figure 13.20. The part is a prototype sporting pigeon made from poly lactic acid (PLA) biodegradable plastic and rice straw powder. The biodegradable plastic part can replace the clay pigeons. The sporting pigeon is more sustainable than the clay pigeon.

Figure 13.17 Arburg 25 ton injection molding machine at California State University, Chico.

Figure 13.18 Students working on the twin-screw machine adding almond shells. Source: Permission granted from Mr. Marshall Ellis.

Figure 13.19 Students working on the injection molding machine. Source: Permission granted from Mr. Marshall Ellis.

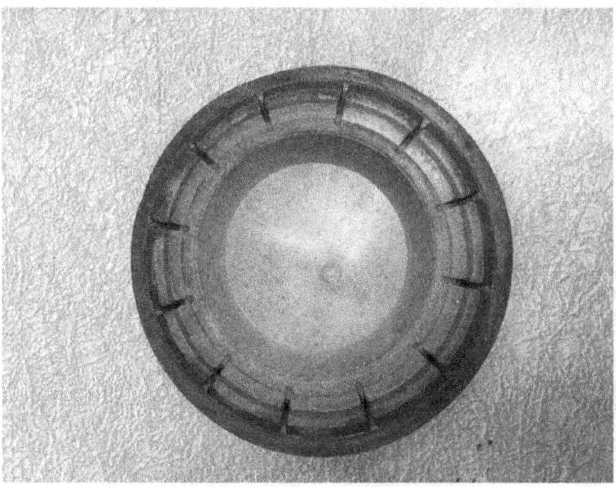

Figure 13.20 Sporting pigeon made from PLA plastic and rice straw powder.

13.6 Test Results of Natural Fibers

After parts are made, then they need to be tested. Tensile bar and impact bar parts can be injection molded with the natural fibers and polypropylene and then tested for impact and tensile properties. Figure 13.21 illustrates a tensile machine.

Previous work at CSU, Chico reported that the strength of the plastic increased with the addition of almond shells. Figure 13.22 shows the tensile testing of PP with 20% almond shells on a tensile test machine.

The testing produced the results that showed that almond shells at 40 and 20% increased the tensile load. Figure 13.23 shows the results. PP and almond shells at 40% were 630% stronger than neat PP and that PP and almond shells at 20% were 170% was stronger than neat PP. The other parts in the graph were made with PP and different combinations of rice straw, rice hulls, and hemp fiber.

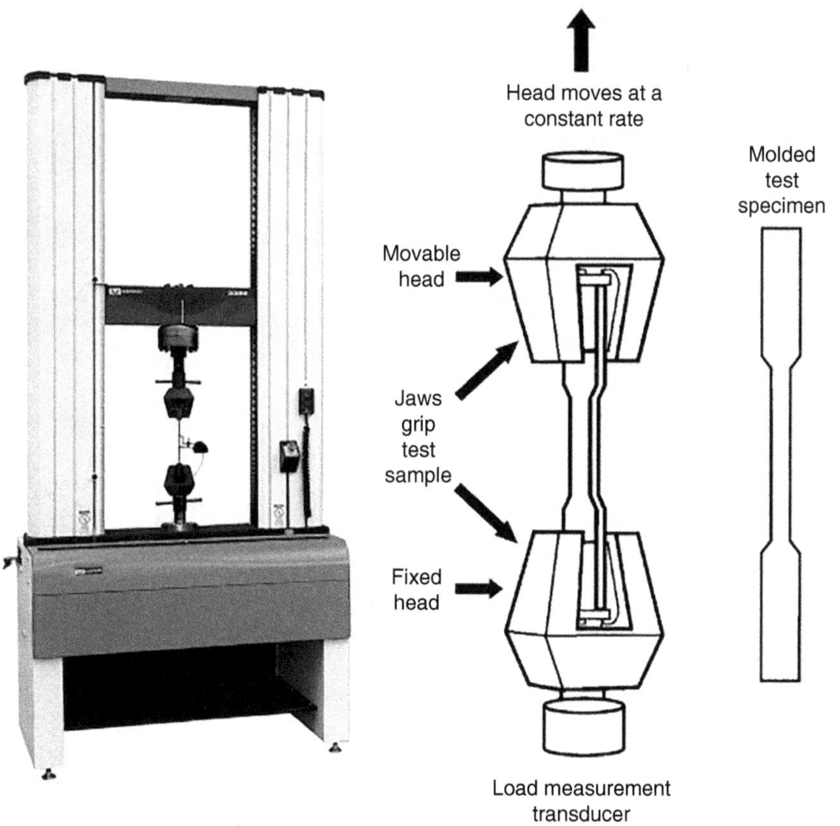

Figure 13.21 Tensile test machine. Source: Tensile test machine (2021).

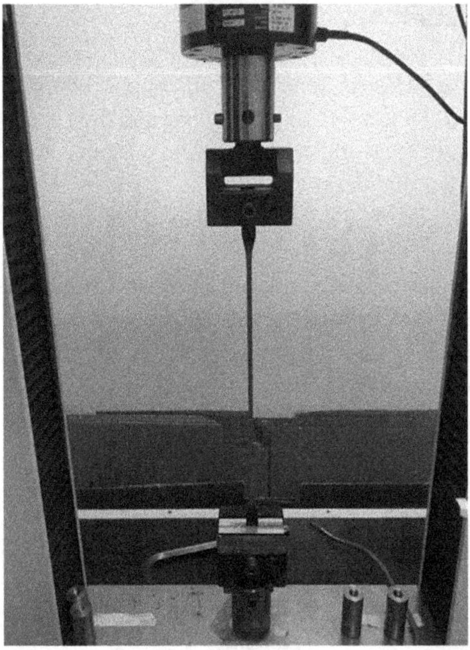

Figure 13.22 Tensile testing of PP and almond shells.

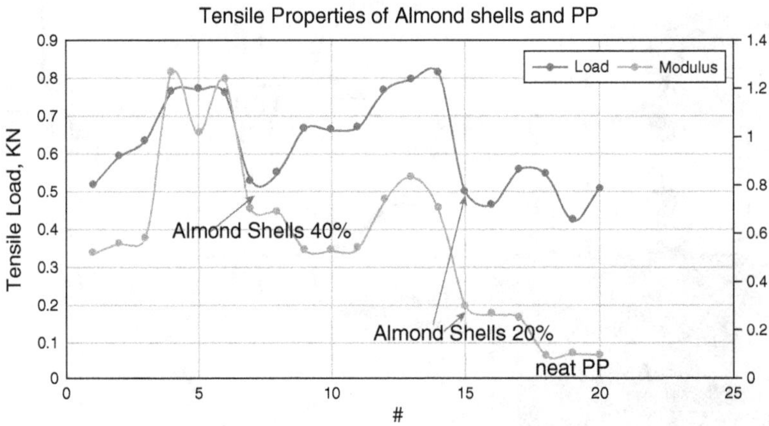

Figure 13.23 Tensile results for almond shells and PP.

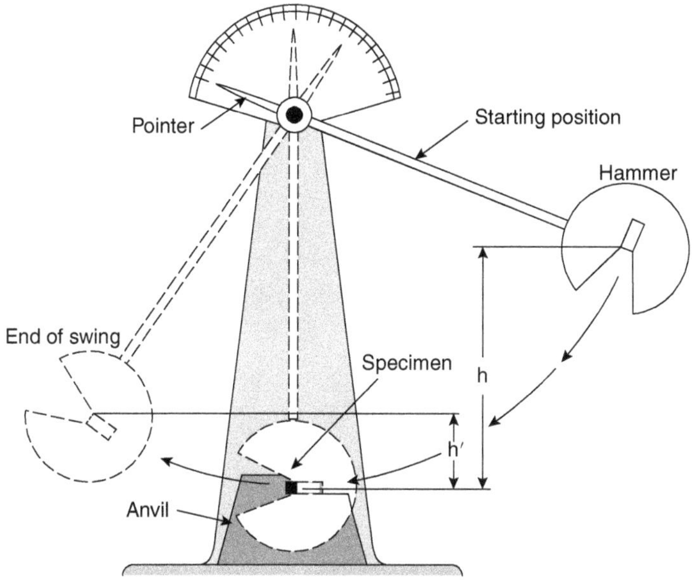

Figure 13.24 Izod impact test. Source: Impact test (2021).

The impact tests are based on an Izod impact test machine. Figure 13.24 illustrates the Izod impact test. In this test, the sample is placed upright. Then a weighted pendulum swings down and impacts the sample. The energy absorbed by the swinging pendulum is listed on a dial on the machine.

Previous work at CSU, Chico reported that the PP and almond shells at 40and at 20% increased the impact strength. Figure 13.25 shows the results. PP and almond shells at 40% were lower than neat PP and that PP and almond shells at 20% higher impact strength than neat PP. PP and almond shells at 40% were 20% lower than neat PP and PP and almond shells at 20% were 140% higher than neat PP. The other parts in the graph were made with PP and different combinations of rice straw, rice hulls, and hemp fiber.

Once the testing is completed for the tensile and impact specimens, the optimum processing conditions can be found to produce high-strength parts for the automotive interior parts. Further research is needed to develop a production process with new and innovative fiber materials and a cost-competitive manufacturing process.

Other researchers have studied the influence of natural fibers on mechanical properties of plastics. One researcher looked at the mechanical and thermal properties of polypropylene and almond shells. (El Mechtali et al. 2015).

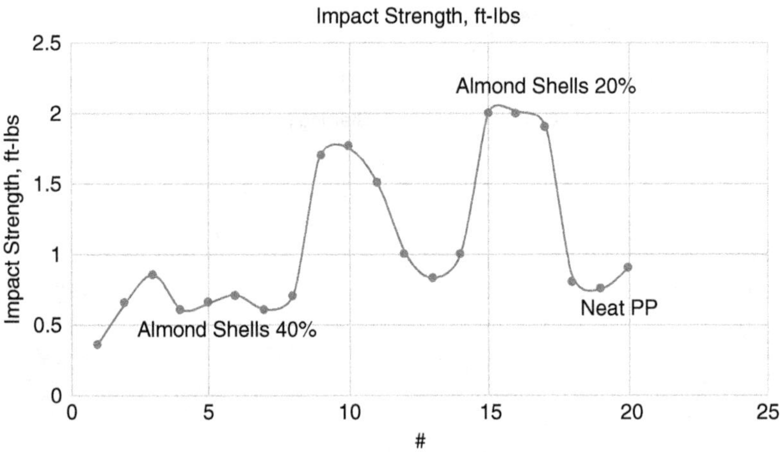

Figure 13.25 Izod impact test results for almond shells.

They found that chemical modifications of almond shell particles affected the mechanical and viscoelastic properties of almond shell/PP composites. The almond shells increased the Young's modulus by 14%, the yield strain by 31%, and the thermal stability. Another research found that almond shell flour and nanoclay particles with polypropylene plastic decreased the tensile and flexural strength by about 37.7 %. The impact strength of composites significantly decreased with the addition of 30, 35, and 40% almond shell flour with polypropylene plastic (Lashgari et al. (2013).

Almond shells were shown to increase the tensile properties of polypropylene (Essabir et al. (2013). They also added maleic anhydride to increase bonding between the almond shells and the polypropylene up to 30% shells. Thirty percent almond shells increased the Young's modulus between 35 and 56.2%. The 30% shells reinforced polypropylene also had an increase in thermal decomposition temperature versus virgin polypropylene.

Jute and hemp fibers have been shown to increase the mechanical properties of plastic. One researcher looked at adding hemp fiber to recycled high-density polyethylene (Lu and Oza 2013). They found that hemp fiber with 40% by volume concentrations and high density polyethylene (HDPE) increased the tensile strength to 60.2 MPa and flexural strength to 44.6 MPa. The hemp–reinforced HDPE composites with 30% of fiber loading demonstrated the best impact strength of 51.1 KJ/m^2. Mechanical properties of natural fibers are shown in Figure 13.26.

Fiber	Tensile strength (MPa)	Young's modulus (GPa)	Density (g cm^{-3})	Refs
Cotton	330–585	4.5–12.6	1.5–1.54	119
Flax	345–1035	27.6–45.0	1.43–1.52	119
Hemp	690–1000	50.0	1.47–1.50	119
Jute	393–800	13–26.5	1.3–1.45	82
Silk	650–750	16	1.3–1.38	82
Kenaf	930	53.0	1.5	119
Ramie	400–1000	61.5	1.5–1.6	119
Sisal	511–635	9.4–15.8	1.16–1.5	119
Banana	500–700	7–20	1.4	120
Softwood	100–170	10–50	1.4	120
Hardwood	90–180	10–70	1.4	120
E-glass	1800	69.0–73.0	2.5	119
HM carbon	2400	380	1.95	121,122
HS carbon	3400	230	1.75	121,122
Kevlar 49	3000	130	1.45	121,122

HM: high modulus, HS: high strength.

Figure 13.26 Mechanical properties of natural fibers Source: Natural fibers properties (2021).

References

Almond waste (2021). https://www.almonds.com/sites/default/files/2020-12/2020%20Almond%20Almanac.pdf (accessed May 2021).

American Leitxritz (2021). https://extruders.leistritz.com/en/extruders-systems/overview (accessed May 2021).

Asdrubali, F. (2021). Sustainable materials and metamaterials for acoustical applications. https://ceas-asc-workshop-2019.ing.uniroma3.it/wp-content/uploads/2019/10/CEAS19_ASDRUBALI.pdf (accessed May 2021).

Auto headliner (2021). https://www.ocautocarpets.com/wn-product/abs-plastic-headliner-uncovered-56324/1999/ (accessed May 2021).

El Mechtali, F., Esssabir, H., Nekhlaoui, S. et al. (2015). Mechanical and thermal properties of polypropylene reinforced with almond shells particles: Impact of chemical treatments. *Journal of Bionic Engineering* **12**: 483–494.

Essabir, H., Nekhlaoui, S., Malha, M. et al. (2013). Bio-composites based on polypropylene reinforced with almond shells particles: mechanical and thermal properties. *Materials and Design* 51: 225–230.

Furqan et al. (2014). A review: Natural Fiber Composite Selection in View of Mechanical, Light Weight, and Economic Properties, Macro Molecular Materials and Engineering, September 2014, https://doi.org/10.1002/mame.201400089 (accessed May 2022).

Global natural fiber market (2021). https://www.grandviewresearch.com/industry-analysis/natural-fiber-composites-market (accessed May 2021).

Haas mill (2021). https://cncmachines.com/haas-mini-mill-2011/l/3726?utm_source=google&utm_medium=cpc&utm_campaign=google-shopping-ads%20&utm_term=&utm_content=google-shopping-low&gclid=Cj0KCQjwp86EBhD7ARIsAFkgakhwihnSBMKPPoMmEmfuTjpzmoEwTcHQ_BQys-9wFE49ZPHZ37EAPBIaAvzWEALw_wcB (accessed May 2021).

Henry Ford fiber use (2021). https://www.thehenryford.org/collections-and-research/digital-resources/popular-topics/soy-bean-car/ (accessed May 2021).

Impact test (2021). https://web.itu.edu.tr/~arana/charpyfatigue.pdf (accessed May 2021).

Lashgari, A., Farsi, M., Eshghi, A. et al. (2013). A study on some properties of polypropylene based nanocomposites made using almond shell flour and organoclay. *Asian Journal of Chemistry* 25 (2): 1043–1049.

Lu, N. and Oza, S. (2013). A comparative study of the mechanical properties of hemp fiber with virgin and recycled high density polyethylene matrix. *Composites Part B: Engineering*, Lu, Na, and Oza, Shubhashini, "A comparative study of the mechanical properties of hemp fiber with virgin and recycled high density polyethylene matrix," Composites Part B: Engineering 45 (1): 1651–1656.

Natural fibers properties (2021). https://www.researchgate.net/figure/Tensile-strength-of-different-natural-fibers_tbl2_305113608 (accessed May 2021).

Peltola, Heidi, Bo Madsen, Roberts Joffe et al., Experimental study of fiber length and orientation in injection molded natural fiber/starch acetate composites, *Advances in Materials Science and Engineering* Volume 2011, Article ID 891940, pages 1–7 (2011).

Solidworks® Plastics (2021). https://www.solidworks.com/product/solidworks-plastics#:~:text=SOLIDWORKS%C2%AE%20Plastics%20delivers%20easy,part%20manufacturability%20while%20you%20design (accessed May 2021).

Tensile test machine (2021). https://www.sciencedirect.com/topics/materials-science/tensile-testing (accessed May 2021).

14

Design Aspects in Automotive Plastics

The chapter will include design aspects for injection molding of plastics, including shrinkage, warpage, tooling, costs, and design for manufacturing (DFM) principles, and how they relate to design of plastic parts. Standard practices for parts that are injection molded, compression molded, or extruder will be discussed. The costs associated with injection molding and compression molded parts will be discussed. Common examples of design standards for automotive components will be discussed. Common parts used in automotive applications are presented by Wilhelm (1993).

14.1 Introduction

Plastics are made with unique design principles. Quality of the plastic part is most affected by the design of the product. Manufacturing cost is mostly affected by the design of the product. A plastic part can be designed properly based on the proper knowledge of plastic materials and processing of plastics. All three are related. The design of plastic parts is influenced by the following:

- Shrinkage
- Part tolerances
- Draft angles
- Inserts
- Undercuts and slides
- Ribs
- Runners
- Bosses
- Radii

Sustainable Plastics: Environmental Assessments of Biobased, Biodegradable, and Recycled Plastics, Second Edition. Joseph P. Greene.

- Sinks
- Fasteners
- Shrinkage

The design of plastic parts is influenced by part design and process design. Typically, a prototype part is made first to validate the designing and processing. This results in pilot run to prove design part and process works. The part is usually tested for key performance specifications, and the process is run at near production levels. Then, the design is changed to incorporate improvements to the design part and process. Finally, production will follow with tool build and production run of the part. The production design step needs coordination with the following:

- Production
- Quality Control
- Purchasing
- Maintenance
- Marketing
- Finance
- Management

14.2 Design Process

A typical development process for an automotive part can take four to six years as shown in Figure 14.1.

- Phase 0. Prototype stage
  - Prove part concept works
  - Test prototype parts to specs
  - Design production process
  - Validate part
  - Design frozen
  - Request funding for full program

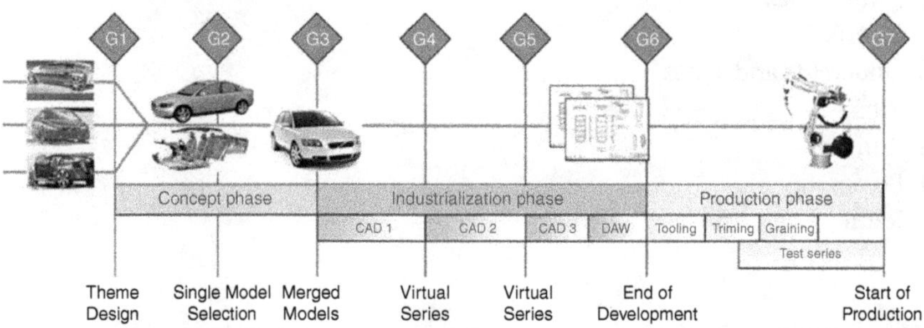

Figure 14.1 Automotive production process. Source: Wickman and Soderberg (2010).

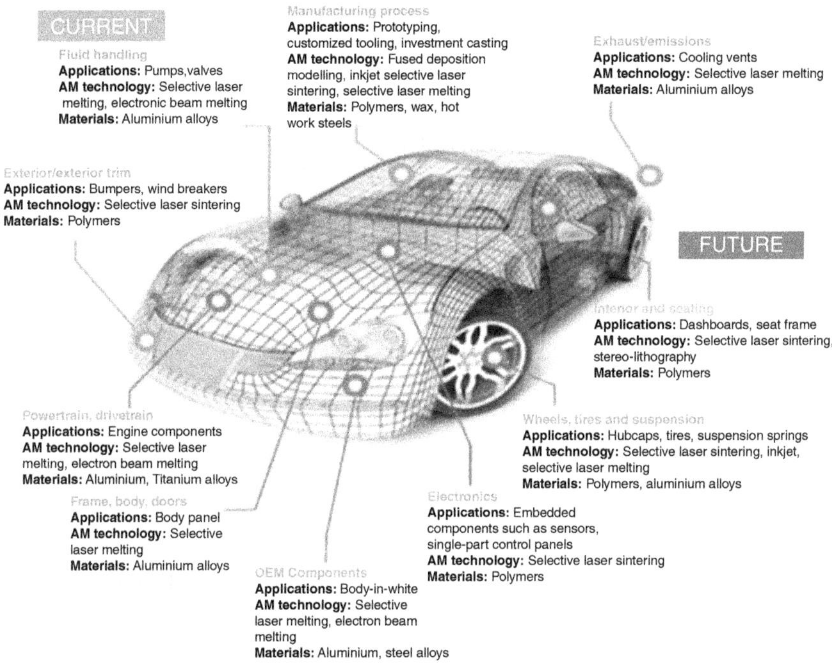

Figure 14.2 Parts in an automobile. Source: Mekonnen and Savas (2014).

- Phase 1. Production tools built
  - Design and manufacture production tooling
- Phase 2. Plant development
  - Set up production line and stations
- Phase 3. Production
  - Build parts with production tools

The typical parts in an automobile are shown in Figure 14.2. Many of the parts use plastic materials. Typically, plastic parts are designed by the engineering team and given to manufacturing to build the production parts.

14.3 Manufacturing Checklist for Quality

The manufacturing engineer must be sure that the parts can be produced. Typically, the engineer will have a checklist to improve the manufacturing process of the plastic part. An example of checklist is given as follows:

1) The part must include a shrinkage value for the mold. The mold should be expanded by a factor of the shrinkage. Shrinkage of plastics are listed in Table 14.3. For example, a part made with polypropylene (PP) should expand

the mold in all dimensions – 2% to account for the 2% shrinkage. (Shrinkage plastics 2020). Thus, a part of 200 mm × 100 mm × 3 mm thick should have the mold machined to 204 mm × 102 mm × 3.06 mm. Then, the molded part will shrink to 200 mm × 100 mm × 3 mm.

2) The part must include draft angles between 1 degree and 3 degrees with an optimum value of 2 degrees.
3) The part should be of the same thickness between 2 and 3 mm.
4) The part should not have any holes near the edges.
5) The part should not have any holes near structural areas.
6) The part should have generous radii.
7) The mold should be made with common parts.
8) The mold should be made out of P-20 steel if production is more than 100 000 parts.
9) The mold should be chrome plated if there is glass-filled plastic.
10) The mold can be made out of aluminum if the part production is less than 100 000.

Problem 14.1 How much should a mold be expanded for a HDPE part with dimensions, 250 mm × 200 mm × 3 mm?

1. Answer: Need to include shrinkage to expand the mold by 1.5 to 4%. If 2% is used then the mold dimensions should be expanded by 1.02%. Thus, a part of 250 mm × 200 mm × 3 mm thick should have the mold machined to 255 mm × 204 mm × 3.06 mm. Then, the molded part will shrink to 250 mm × 200 mm × 3 mm.

14.4 Plastic Materials for Automotive Use

Figure 14.3 shows the parts that can be made with plastic and polymer composite materials. The exterior body panels, door panels, and bumper beams can be made with sheet molding compound (SMC). The floor luggage compartment, trim luggage compartment, and sun roof panel can be made with glass-filled PP or glass mat thermoplastic (GMT) with PP and glass fibers.

The automotive industry relies heavily on proper materials selection. The proper choice of materials for a vehicle is a very important factor for automotive design. There are many materials that can be used in the automotive body, engine, interior, and powertrain. Plastics and composites can be used in all of the major components of the automobile. Plastics offer the following advantages versus steel for the vehicle components:

✓ Lightweight: This criterion enables reduction of greenhouse gas and improved fuel efficiency.
✓ Economic effectiveness: The cost for plastic and composite parts is more on the part basis but much less when tooling is included.

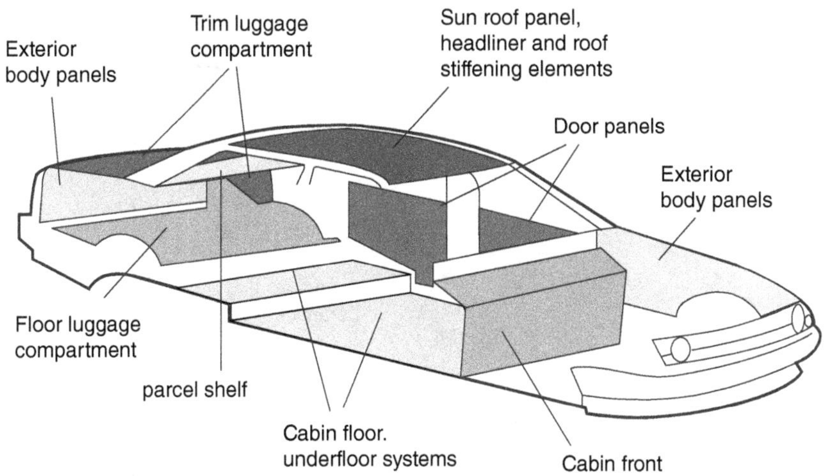

Figure 14.3 Automotive materials applications. Source: Todor et al. (2016).

✓ Safety: Plastics and composites have higher impact energy absorption than metals.
✓ Recyclability and sustainability: Plastics and composite materials can provide recycling options for the automotive products and provide a part with lower carbon footprint, lower waste, and lower pollution than metal components.

The weight reduction in the automotive industry can be obtained in three ways:

∢ Replacing metal materials of high specific weight with lower density materials, e.g., plastic and composite materials. Plastics and composite parts have higher strength-to-weight ratio and higher stiffness-to-weight ratio than metals parts. Figure 14.4 shows the modulus and density for several automotive materials.
∢ Optimizing the design of load carrying elements and exterior attachments through part consolidation to reduce their weight without any loss in rigidity or functionality.
∢ Optimizing the production process.

The design process, at GM and Ford, starts with 3D CAD model development includes the following:

- Phase 0. Prototype.
  - Prototype parts with 3D models.
    - Models are sent to potential part suppliers to quote production job based on part design.
    - Prototype tooling and parts are made based on prototype 3D model.
- Phase 1.
  - Preproduction parts made with modified design based on manufacturing requests.
    - Engineering change orders (ECOs) are made to update drawing and 3D model.

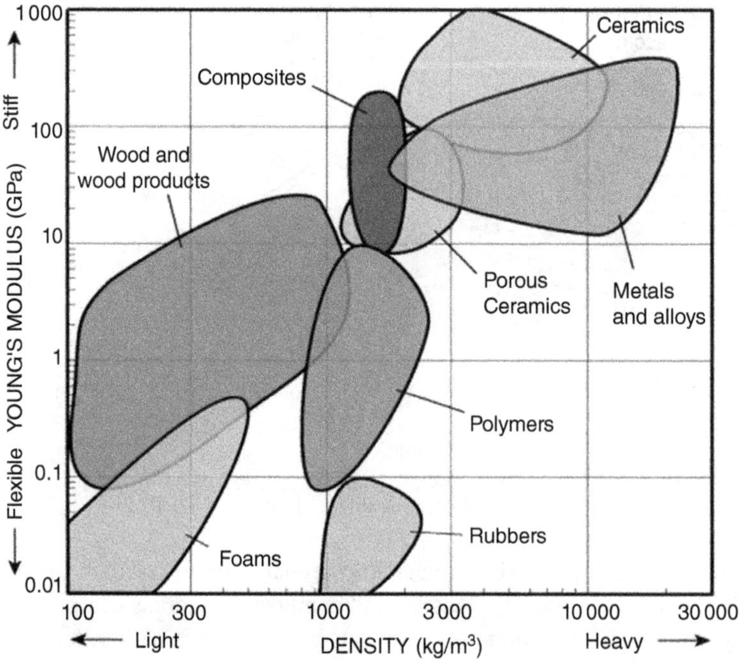

Figure 14.4 Tensile modulus versus density for several automotive materials. Source: Material modulus (2020).

- Phase 2.
  - Production layout may affect part design with more ECOs.
- Phase 3.
  - Production begins with part design modified based on manufacturing needs.

14.5 Plastic Guidelines for Injection Molding

When designing parts for injection molding, the manufacturing process is an important consideration. Injection molding is a process in which solid thermoplastic resin pellets are melted, injected into a mold, and then cooled back to a solid state in a new form. During both the injection and cooling stages of the manufacturing process, there are several factors that may affect the quality of the final product and the repeatability of the manufacturing process. The following recommendations are some of the most fundamental guidelines when designing parts for injection molding:

- Wall thickness should be less than 5 mm.
- Keep wall thickness constant throughout part.
- Avoid gating near an area with large variations in wall thickness.

- Parts will fill from thick regions to thin regions that can cause race tracking of the plastic that can cause dry areas.
- Inside radius should be at least 50% of nominal wall thickness.
- Outside radius should be nominal wall thickness plus the inside radius that can be 150% of nominal wall.
- Avoid sharp corners in bosses and in the part.
- Maximum rib thickness should be 0.5 × 0.75 of nominal wall thickness.
- Ribs should have a draft angle of at least ½ inch per side.
- Maximum rib height should not be greater than three times the nominal wall thickness.
- Draft angle should be at least ½ inch per side (RTP Design guidelines 2020).

The advantages and disadvantages of plastics are shown in Table 14.1.

Material selection is based on properties that can be found at www.matweb.com. Figures 14.5 and 14.6 show the MatWeb website and some example properties. Material selection is based on properties that can be found at www.prospector.com. Figures 14.7 and 14.8 show the Prospector website and some example properties.

Table 14.1 Advantages and disadvantages of plastics.

Advantages
• Ease of molding with fast cycle times
• Low density
• Low material cost due to fast cycle times
• Little or no corrosion
• High strength-to-weight ratio
• High stiffness-to-weight ratio
• Low thermal conductivity
• High electrical resistance
Disadvantages
• Low compression strength
• Low bearing strength
• Low shear strength
• High temperature limitations
• Warpage due to differential shrinkage
• Weathering in some environmental conditions
• Low crazing in some plastics
• Low fatigue strength
• Shrinkage after molding, especially with Polyoxymethylene (Delrin)

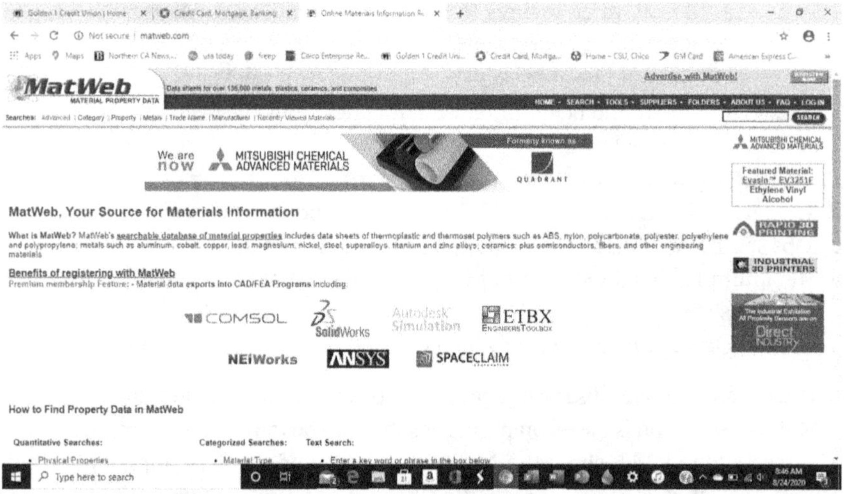

Figure 14.5 Matweb website. Source: Matweb (2020).

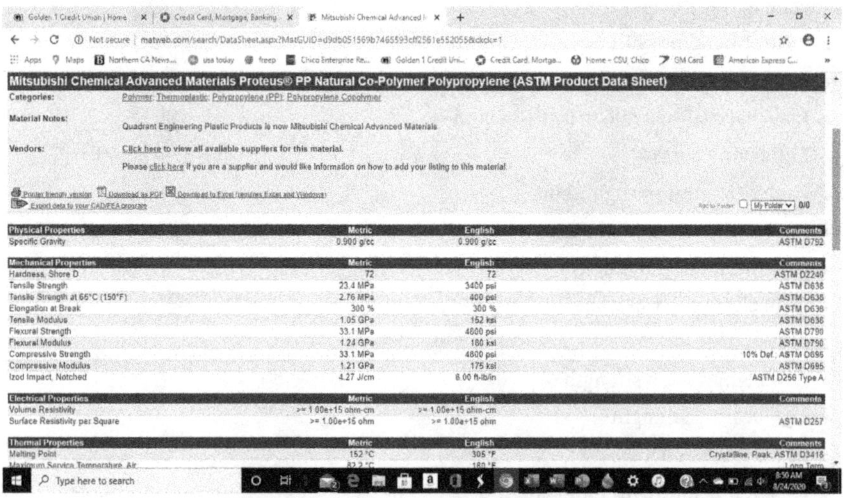

Figure 14.6 MatWeb mechanical properties of PP. Source: MatWeb (2020).

Figure 14.7 Prospector website. Source: Prospector (2020).

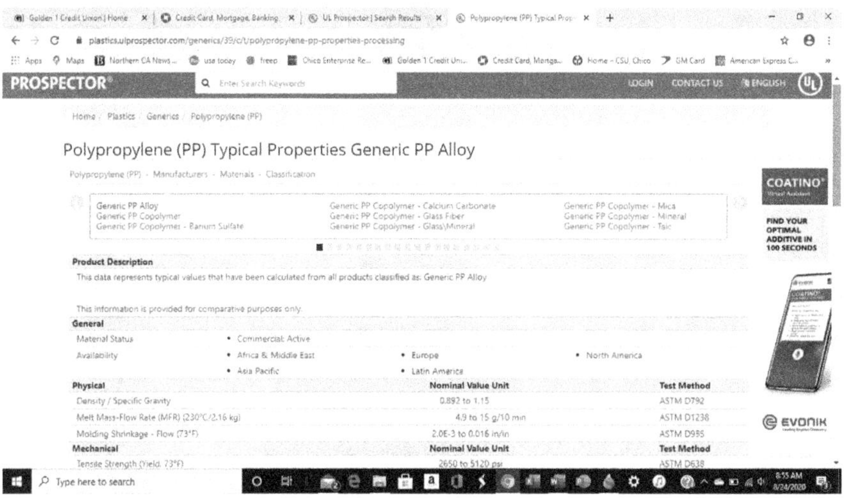

Figure 14.8 UL prospector mechanical properties of PP. Source: Prospector (2020).

14.6 Plastic Prototypes and 3D Printing

Prototypes can be made to simulate production part. Prototypes must simulate a part with fit, form, and function. The fit is that the prototype part must have the same shape as the production part and fit in the assembly like the production part would. The form of the prototype part is that it must fit in the same way as the production part. The function is that the prototype part must have similar mechanical properties or test performance as the production part. The choices for a prototype part are that it can be made with the following:

- 3D printing of ABS, PC, PETG, or PLA.
- Composite mock-up with fiberglass and polyester.
  - Need a mold to be produced from epoxy, aluminum, and kirksite.
- Part cast with polyurethane, acrylic, and other castable resin.
  - Need a mold to be produced from EPDM, rubber, and aluminum.

Rapid prototyping is a process of building objects during the design phase to have a 3D object to check for fit, form, and function. The rapid prototype will help you visualize the dimensions of the part and see if 3D object is desired. The object is made from the solid model geometry in a stereolithography file. Stereolithography represents the 3D object as a shell of the part that is broken into triangles. Rapid prototype process can bring a design on the computer to 3D shape in a matter of minutes to hours. The process involves CAD system to digitizer to rapid part as shown in Figure 14.9.

Additive manufacturing is another term for 3D printing. It refers to a series of processes that allows the creation of parts and components in an additive instead of a subtractive manner. Car components are created layer by layer in a plastic extrusion process rather than machining the parts out of metal. Additive

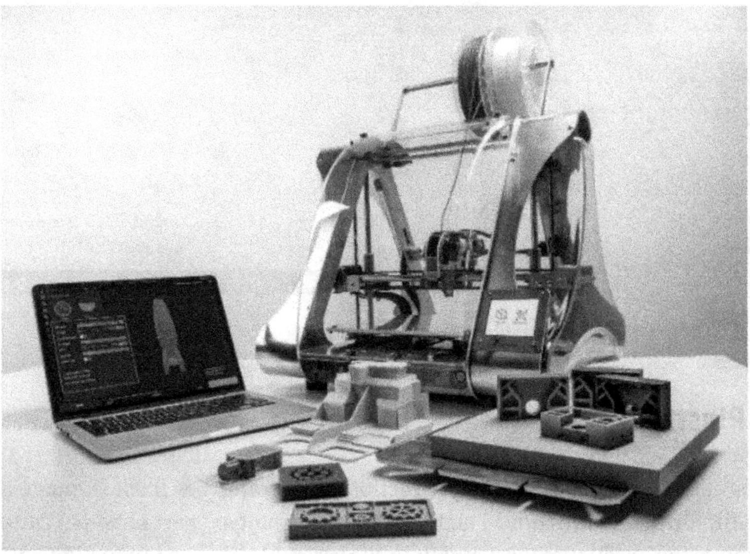

Figure 14.9 3D printing for the automotive industry. Source: 3-D printing (2020).

manufacturing would be used in the prototyping phase to make parts that are similar to the production part. The parts can be assessed for fit and for to see how the parts look and connect together. The quality of the plastic part though is low and does not have the mechanical properties of the injection molded part. So, the function could not be assessed. Additive manufacturing offers many benefits, including

- More design freedom
- Parts consolidation
- Low mass materials
- Parts consolidation
- Increased customization

14.7 SolidWorks Flow Simulation

Injection molding can be used to make plastics parts for the cars. SolidWorks simulator is useful for plastic parts.

Design of part can be aided with simulation of the injection molding process. Moldflow is a useful tool that is a part of Autodesk/AutoCad. SolidWorks simulator is useful for plastic parts/designing plastic part simulation of

- mold filling distribution,
- pressure flow distribution,
- temperature distribution, and
- warpage distribution.

Design tool of the part includes the following:

- Venting
- Gates and runners
- Slides

14.8 Design for Manufacturing (DFM) with Plastics

Designing for manufacturing is an essential component of design for plastic parts. This includes the following:

- Use of standards
- Use of common components
- Design to specifications and tolerances
- Use of manufacturing guidelines in the early stages of design that maximize quality of manufactured part

- Minimize the use of materials
- Minimize the use of floor space in plant
- Locate all necessary components near functional operation
- Use of automated machining for minimal errors

14.9 Shrinkage in Plastics

Control of shrinkage is another essential component of the plastics design. Note that shrinkage for plastics are listed in Tables 14.2, 14.3, and 14.4. The effects of

Table 14.2 Shrinkage and moisture growth for plastics.

Material	Thermal expansion (mm/mm $K^{-1} \times 10^6$)	Mould shrinkage (%)	Water absorp. (%)
ABS (rigid)	80	0.3–0.8	0.3
Acetal	80	2.0–3.5	0.2
Cellulose acetate	100	0.3–0.7	2–6
Fluorinated ethylene propylene	90	3.0–6.0	0
Nylon 6,6	120	1.5–2.0	1.5
Nylon 6	100	1.0–1.5	1.6
Nylon 11	150	1.2	0.4
Nylon 12	104	1.0	0.3
Polybutyleneterephthalate	90	1.5–2.0	0.2
Polycarbonate	70	0.6–0.8	0.16
Polyethylene (LD)	170	2.0–3.5	0.02
Polyethylene (HD)	120	2.0–3.5	0.01
Polymethylmethacrylate	85	0.1–0.8	0.35
Polypropylene	110	1.5–2.5	0.01
Polyphenylene oxide (modified)	55	0.5–0.7	0.1
Polystyrene (GP)	70	0.2–0.6	0.2
Polystyrene (rubber modified)	120	0.2–0.8	0.2
Polyethersulphone	55	0.6–0.8	0.15
Polyvinyl chloride (rigid)	55	0.1–0.5	0.05
Styrene acrylonitrile	70	0.2–0.5	0.3
Steel	11–13		

Source: Plastic mold (2020).

Table 14.3 Shrinkage for semi-crystalline plastic materials.

Typical shrinkage values for semi-crystalline materials Based on a 1/8″ (3.2 mm) section using ASTM D 955	
Material	**Mold shrinkage (in./in.)**
Unfilled Polypropylene (PP)	0.010–0.025
Talc filled Polypropylene (PP)	0.007–0.015
High Density Polyethylene (HDPE)	0.015–0.040
Unfilled Nylon 6	0.005–0.015
Unfilled Nylon 6/6	0.008–0.015
Acetal (POM)	0.020–0.025

Source: RTP Design guidelines (2020).

shrinkage are that the parts are designed with shrinkage included early design and before tool build. The shrink rates for common materials are given in Table 14.2.

Shrinkage is different for semi-crystalline plastic materials and amorphous plastics.

The commodity plastics have typically higher shrinkage than the engineering plastics. The effects of shrinkage are sink marks shown in Figure 14.10.

Table 14.4 Summary of shrinkage for plastic materials.

Material	**Max. shrinkage (%)**
• Acetal	2.5
• Acrylic	0.8
• ABS	0.8
• Nylon	1.5
• PC	0.7
• PE	5.0
• PP	2.5
• PS	0.6
• PVC rigid	0.5
• PVC flexible	5.0

Source: RTP design guidelines (2020).

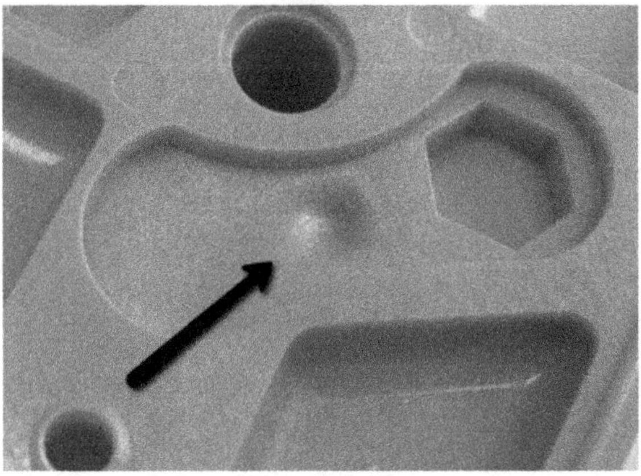

Figure 14.10 Sink marks in plastic molded parts. Source: Sink marks (2020).

14.10 Design Guidelines

Several types of gates are available for injection molding. These can include edge, fan, submarine, flash, tunnel, ring, diaphragm, disk, or sprue gate as shown in Figure 14.12. Key design considerations for a gate are as follows:

- Not on a show surface
- Not near a structural member or hole or fastener
- Minimize flow length
- Minimize the number of weld lines
- Gate thick to thin

The gate is usually thinner than the part thickness in injection molding. Typically, the wall thickness for injection molding is 2–3 mm. It is important to have constant wall thickness in the part. If thicker sections are required, then it is important to have transitions from thick to thin regions. The thicker regions can have gentle and no sharp transitions as shown in Figure 14.12. The design guidelines for injection molding are as follows:

- Holes
  - Holes are possible with slides but can cause weld lines
  - Minimum spacing between two holes or a hole and a sidewall should be 1D
  - Should be located 3D or more from the edge of a part to min stresses
  - Through hole is preferred to a blind hole because core pin that produces hole can be supported at both ends and is less likely to bend
  - Holes in bottom of part are better than holes in side, which requires retractable core pins

- Blind holes should not be more than 2D deep
- Use steps to increase the depth of a deep blind hole
- For through holes, cutout sections in the part can shorten the length of a small-diameter pin
- Use overlapping and offset mold cavity projections instead of core pins to produce holes parallel to the die parting line (perpendicular to the mold-movement direction).

The sprue must not freeze before any other cross-section. This is necessary to permit sufficient transmission of holding pressure. The sprue must demold easily and reliably. Common runners are shown in Figure 14.11 and shown below.

The runners include the following:

- Full-Round Runner
  - The best in terms of a maximum volume-to-surface ratio, which minimizes pressure drop and heat loss
  - Tooling cost is generally higher because both halves of the mold must be machined so that the two semi-circular sections are aligned when the mold is closed
- Trapezoidal Runner
  - Works well and permits the runner to be designed and cut on one side of the mold. Commonly used in three-plate molds and at the parting line in molds, where the full-round runner interferes with mold sliding action
  - Hydraulic diameter and flow resistance
  - To compare runners of different shapes, you can use the hydraulic diameter, which is an index of flow resistance

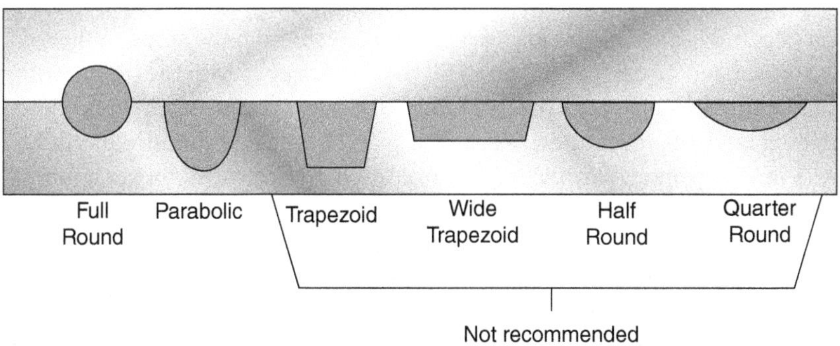

Figure 14.11 Injection molding runners. Source: Runners (2020).

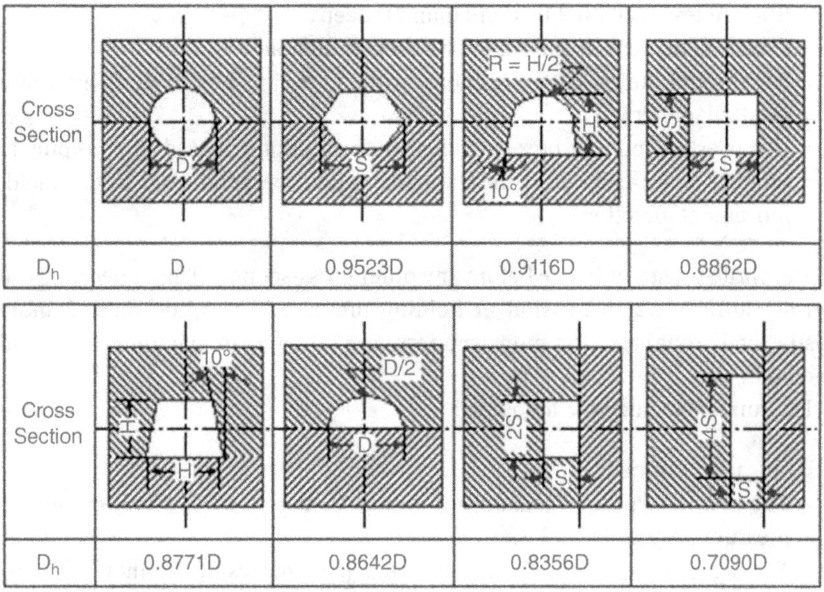

Figure 14.12 Equivalent hydraulic diameters for runners. Source: Hydraulic diameter (2020).

- The higher the hydraulic diameter, the lower the flow resistance Hydraulic diameter can be defined as:
- Equivalent hydraulic diameters, shown in Figure 14.12.
  - $D_k = 4A/P$,
  - Where,
    - D_k = hydraulic diameter
    - A = cross-section area
    - P = perimeter

Runners provide a distribution system for the resin from the sprue to the cavities. Runners must be large enough to fill the cavities even with the resin cooling along the runner walls. Materials with high viscosity (low MFR) runners should be large enough to prevent gate freeze-off, e.g., PC has a large viscosity and should have a larger runner than nylon that has a low viscosity. Runners should be designed to minimize shear, e.g., acetal can decompose if subjected to excessive shear. Length of runner should be minimized. When secondary runners are used in multi-cavity molds, the flow should be streamlined. Streamlining the runners minimizes the shear stress on the plastic.

When the same part is made in a multicavity design, the runners should be balanced and fill at the same time with similar pressures. When cavities are producing different parts, balanced runner design is more complicated and flow analysis should be used. Runner balancing is an essential for a balanced filling pattern with a reasonable pressure drop. Payoffs of good runner design. A runner system that has been designed correctly will

- achieve the optimal number of cavities,
- deliver melt to the cavities,
- balance filling of multiple cavities,
- balance filling of multigate cavities,
- minimize scrap,
- eject easily,
- maximize efficiency in energy consumption, and
- control the filling/packing/cycle time.

There are several types of runner systems available for plastics, as shown in Figure 14.13. It is important not to flow directly into the mold but instead have a right angle turn at the entrance to the mold. This prevents stretching of the plastic.

Runner balancing is also important for plastics with larger runners needed for larger parts and smaller runners for smaller parts in a family-type mold. This will help the part will fill evenly at the proper time. This is shown in Figure 14.14.

Gate design is important parameter for good quality injection molded parts. The gate can be single or multiple gates. The single gate is usually desirable because multiple gates have weld lines. The gate thickness is usually two-thirds the part thickness. The gate thickness controls packing time. Typically, a larger gate is preferred for better appearance, low residual stress, and better dimensional stability. The gate location is positioned away from load-bearing areas. The position of the gate is away from the thin section areas or regions of sudden thickness change to avoid hesitation and sink marks. The gate types are manually trimmed or automatically trimmed gates. The manually trimmed gates

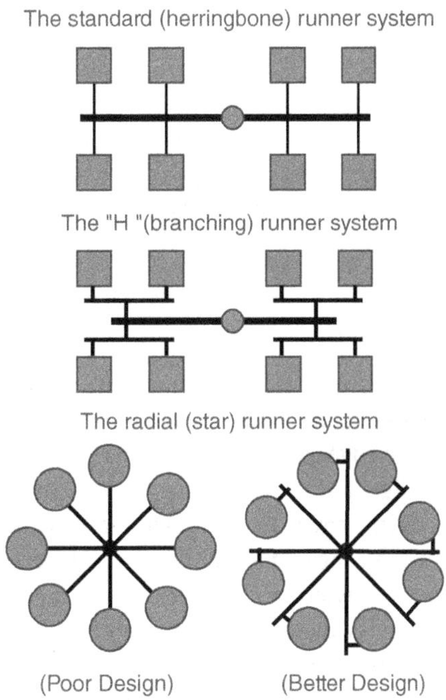

Figure 14.13 Runner types for injection molding. Source: Herringbone runners (2020).

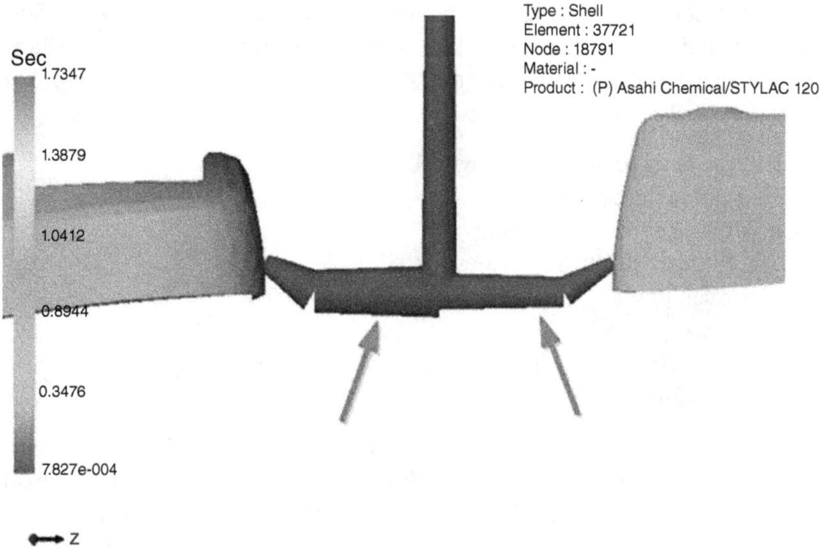

Figure 14.14 Runner balancing for plastics. Source: Runner balancing (2020).

are cheaper and requires an operator to separate parts from runners during a secondary operation. The types include sprue, tab, edge, overlap, fan, disk, ring, film, diaphragm, and spider gates. The automatically trimmed gates incorporate features in the tool to break or shear the gate. They should be used to

- avoid gate removal as a secondary operation,
- maintain consistent cycle times for all shots, and
- minimize gate scars.

The types include sprue, fan, tunnel, ring, tab, pin, submarine, and others. Types of gates are shown in Figure 14.15.

The edge gate is a small rectangular opening at end of runner channel. The cost of edge gate is low. The submarine gate starts from the edge of the runner and goes into cavity edge at an angle. Gate is sheared off at part ejection. This can create high shear stress in the part. The cost is higher than edge gate. The tab gate is formed by connecting the runner directly into the cavity with no reduction in runner cross-section. It is used for large parts. The fan gate is made by reducing only the thickness and not the diameter of the runner channel as it goes into the cavity. It can be used for parts of intermediates size and when reinforcements can't flow though edge gate. The ring gate is used to make hollow cylinder parts. It covers the entire top of the cylindrical part so the resin flow is downward into the walls of the part. It is used to minimize weld lines.

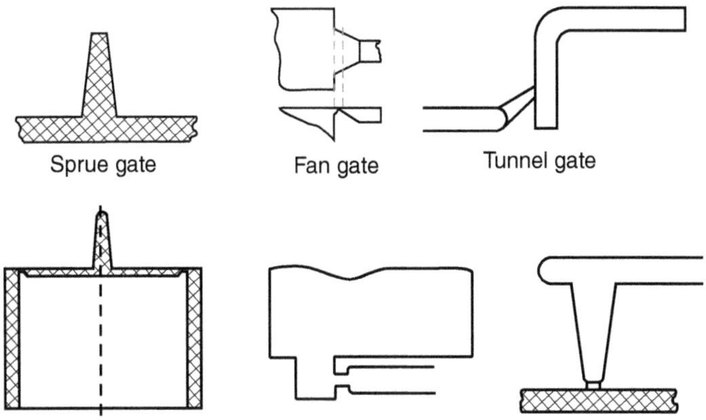

Figure 14.15 Types of gates for injection molding. Source: Types of gates (2020).

The gate location should be at the thickest area of the part, preferably at a spot where the function and appearance of the part are not impaired and should be central so that flow lengths are equal to each extremity of the part. The gate should be symmetric to avoid warpage. It should be vented properly to prevent air traps. The gate should be large enough to avoid jetting. It should be positioned so that the weld and meld lines are minimized.

The gate length should be as short as possible to reduce an excessive pressure drop across the gate. The gate length ranges from 1 to 1.5 mm (0.04–0.06 inches). The gate thickness is normally 50–80% of the gated wall section thickness. Pin and submarine gates range from 0.25 to 2.0 mm (0.01–0.08 inches). The freeze-off time at the gate is the maximum effective cavity packing time. Fiber-filled materials require larger gates to minimize breakage of the fibers.

- Gate is the end of the runner and the entry path into the cavity.
  - Shape of gate strongly affects the ease with which the part is removed from the runner system.
  - Gate is the most restricted point in injection path and first to freeze-off.
  - Designs
- Gate Design Overview
  - Single versus multiple gates
    - Single gate is usually desirable because multiple gates have weld lines.
- Gate dimension
  - The gate thickness is usually two-thirds the part thickness.
  - The gate thickness controls packing time.
  - Chose a larger gate if you're aiming for appearance, low residual stress, and better dimensional stability.

Structural integrity is the goal of every design. The major component of designing for structural integrity, in many cases, is to design the structure to be stiff enough to withstand expected loads. Increasing the thickness to achieve this is self-defeating, since it will

- increase part weight and cost proportional to the increase in thickness,
- increase molding cycle time required to cool the larger mass of material, and
- increase the probability of sink marks.

Well-designed ribs can overcome these disadvantages with only a marginal increase in part weight (Ribs 2020). Typical uses for ribs are as follows:

- Covers, cabinets, and body components with long, wide surfaces that must have good appearance with low weight.
- Rollers and guides for paper handling, where the surface must be cylindrical.
- Gear bodies, where the design calls for wide bearing surfaces on the center shaft and on the gear teeth.

Frames and supports can improve structural integrity of the part. The major component of designing for structural integrity, in many cases, is to design the structure to be stiff enough to withstand expected loads. Increasing the thickness to achieve this is self-defeating, since it will

- increase part weight and cost proportional to the increase in thickness,
- increase molding cycle time required to cool the larger mass of material, and
- increase the probability of sink marks.

Well-designed ribs can overcome these disadvantages with only a marginal increase in part weight. Keep part thickness as thin and uniform as possible.

- This will shorten the cycle time, improve dimensional stability, and eliminate surface defects.
- If greater stiffness is required, reduce the spacing between ribs, which enables you to add more ribs.

Rib geometry is an important aspect of good quality injection molded parts. Rib thickness, height, and draft angle are related: excessive thickness will produce sinks on the opposite surface whereas small thickness and too great a draft will thin the rib tip too much for acceptable filling. Ribs should be tapered (drafted) at one degree per side. Less draft can be used, to one-half degree per side, if the steel that forms the sides of the rib is carefully polished. The draft will increase the rib thickness from the tip to the root, by about 0.175 mm per centimeter of rib height, for each degree of draft angle. The maximum recommended rib thickness, at the root, is 0.8 times the thickness of the base to which it is attached. The typical root thickness ranges from 0.5–0.8 times the base thickness.

Ribs aligned in the direction of the mold opening are the least expensive design option to tool.

- As illustrated in Figure 14.16, a boss should not be placed next to a parallel wall; instead, offset the boss and use gussets to strengthen it.
- Gussets can be used to support bosses that are away from the walls. The same design rules that apply for ribs also apply for gussets.

As shown in Figure 14.17, ribs can take the shape of corrugations. The advantage is that the wall thickness will be uniform and the draft angle can be placed on the opposite side of the mold, thereby avoiding the problem of the thinning rib tip.

Honeycomb ribbing attached to a flat surface provides excellent resistance to bending in all directions. A hexagonal array of interconnected ribs will be more effective than a square array, with the same volume of material in the ribs. Bosses are protruding pads that are used to provide mounting surface or reinforcements around holes. They use the same guidelines as for ribs.

Undercuts are required for sliding cores, split molds, or stripping plate. Shallow undercuts may be strippable from mold without need for core pulls. When you have holes or bosses on the side wall, it is difficult to demold the part. The molded

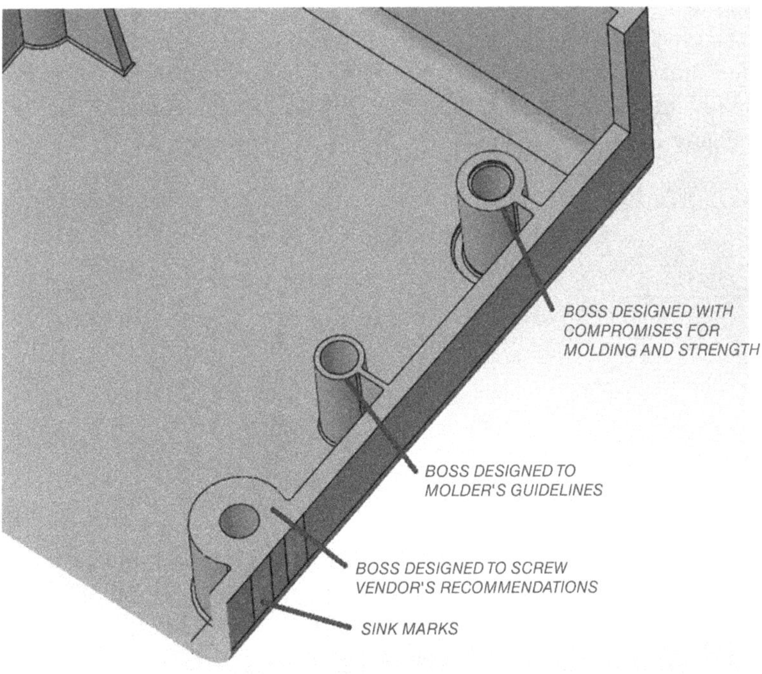

Figure 14.16 Bosses for injection molding. Source: Bosses (2020).

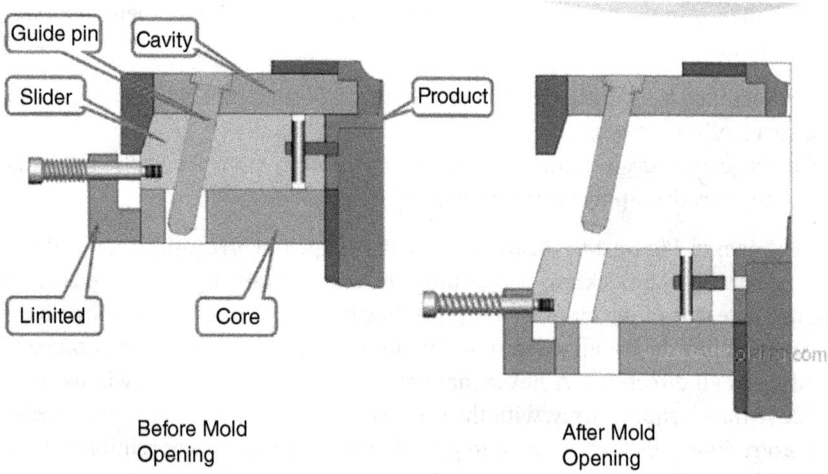

Figure 14.17 Undercuts with injection molding. Source: Undercuts (2020).

part has to move laterally. This is called undercuts. The undercut is referred to as a slider or lifter. An example of an undercut for common materials is shown in Figure 14.17.

Molded screw threads should be avoided if possible. They use a core that is rotated after molding is complete to unscrew part. Put axis of the screw at the parting line of the mold. Make threads few, shallow, and of rounded form as shown in Figure 14.18.

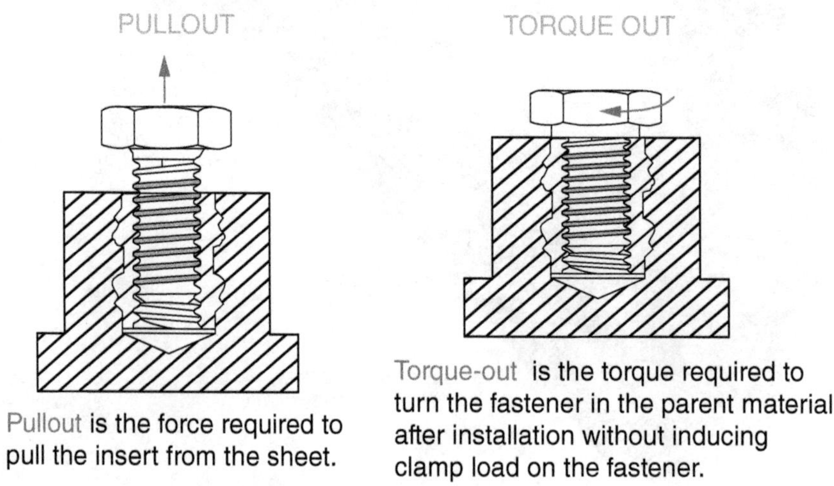

Figure 14.18 Screw threads. Source: Screw threads (2020).

Inserts are useful and practical to provide reinforcement where stresses exceed the strength of the plastic. Sharp corners should be avoided. Recommended designs are shown in Figure 14.18. Lettering and surface decorations can be injection molded. Lettering in the part should be raised. Lettering should be perpendicular to the parting line of the mold; otherwise, there will be an undercut.

Draft angles are needed to assist in demolding. Common draft angles for materials are shown in Table 14.5. Draft angles are needed to mold part. The rule of thumb is an increase in dimension per degree of draft. Minimum draft angle is 0.125 degree and typical draft angle is 2 degrees.

Design parts with a minimum of ½° per side draft in order to accommodate easier ejection from the mold. Typically, injection molded parts have 1° to 2° draft angle. 3° is usually the maximum for draft angle. Sharp corners should be avoided as they add stress to polymer. Fillets and radii should be generous as possible. Desirable minimum is 0.5 mm and preferable minimum is 1.0 mm.

Surface polish and textures are possible with plastic parts. High gloss finishes are feasible but can accentuate sinks and blemishes. Dull, matte, or textured surfaces are preferred. Flat surfaces are feasible but prone to show irregularities than gently curved surfaces that are preferred. Molding parting line is the line created when the two mold halves come together. It should be straight as the two mold halves come together in one plane. Offset dies may help avoid appearance defects, as shown in Figure 14.19.

Table 14.5 Draft angles of plastic materials.

Plastic type	**Draft angle**
Polycarbonate (PC)	1–2°
Polystyrene (PS)	>0.5°
Polyarylsulfone (PSU)	1–2°
Polyarylethersulfone (PES)	1–2°
Liquid crystal polymers (LCP)	>0.5°
Polybutyleneterephthalate (PBT)	1–1.5°
Polyethyleneterephthalate (PET)	1–1.5°
Acrylonitrile/Polybutadiene/Styrene (ABS)	>0.5°
Polymethylmethacrylate (PMMA)	>1–2°
Polyethylene (PE)	>0.7–0.8°
Polypropylene (PP)	>0.7°

Source: Draft angles (2020).

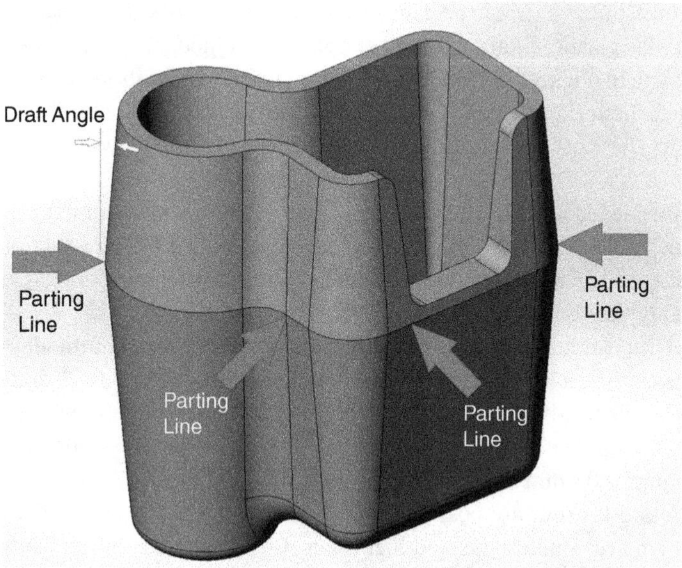

Figure 14.19 Parting lines in injection molding. Source: Parting line (2020).

Dimensional factors and tolerance recommendations are useful for a good plastic part design. Injection molding can have tight tolerances for injection molded parts. The factors that affect tolerances are as follows:

- Material shrinkage
- CLTE
- Variations in process settings
- Runners and cooling channels are located in incorrect location
- Two different materials coming together

Rule of thumb is different plastics can have different tolerance capabilities. Low-shrink plastics can have tight tolerances. Glass- and mineral-filled plastics can have tight tolerances.

The shrinkage can be reduced and the plastic warpage can be controlled by properly designing the part, mold, and process, as well as through careful material selection. Maintain a wall thickness of less than 5 mm because thick walls can lead to long cycle times and poor mechanical properties.

- Avoid large variations in wall thicknesses in order to simplify flow pattern and minimize variations in shrinkage that can lead to warpage. Avoid abrupt changes in wall thickness, as this can create stress concentration areas that may reduce a part's impact strength. Wall thickness changes should have transition zones that reduce the possibility of stress concentrations, sinks, voids, and warp.

- Avoid gating near an area with a large variation in wall thickness because hesitation and race tracking can create non-uniform flow and shrinkage. The following design rules provide some guidelines for developing low-shrinkage, warp-free parts.
- Wall thickness: Avoid non-uniform wall thickness or design a transition length of three times the thickness of the thinner region as shown in Figure 14.20.

The following design rules provide some guidelines for developing low-shrinkage, warp-free parts:

- Thick sections: Alter the design to replace thick sections that cause significant shrinkage and lead to sink marks or internal voids.
- A thin, uniform wall with ribs provides for uniform shrinkage, strength to weight ratio, and cost effectiveness.

Due to the shrinkage of the plastic, sink marks, and voids can be created if the part is too thick, as shown in Figure 14.21. The part should be redesigned with constant wall thickness to minimize sinks and voids in the part.

The following design rules provide some guidelines for developing low-shrinkage, warp-free parts.

Balanced Filling

- Design the delivery system to provide a balanced filling pattern with a constant melt-front velocity
- Packing Pressure
- While high packing pressure can help reduce the shrinkage, it also potentially increases the level of residual stress and the clamp force requirement.
- For a better process design, use a proper packing pressure level, allow sufficient packing time, and remove the pressure after the gate freezes off.
- The packing pressure should be able to deliver additional material to compensate for the volumetric shrinkage in the part.

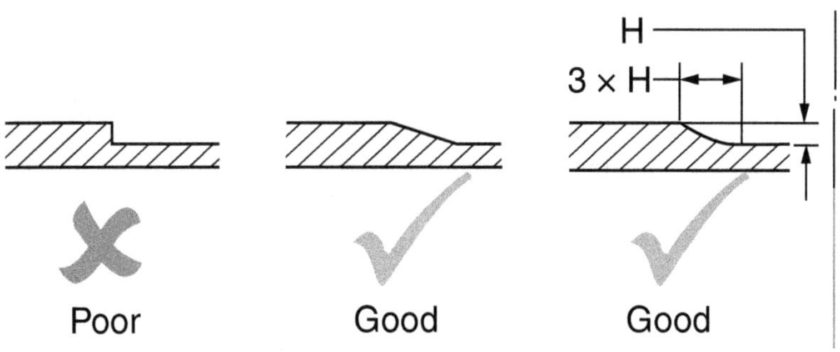

Figure 14.20 Part thickness transition guidelines. Source: Transition (2020).

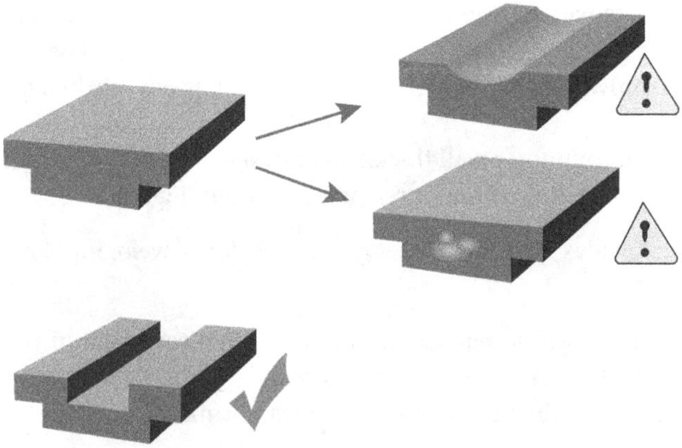

Figure 14.21 Sink marks and voids in injection molding. Source: Sink marks and voids (2020).

Balanced Cooling System

- Design the cooling system to apply uniform, balanced cooling, both across the thickness and throughout the part.

Balanced Residual Stress

- Reduce residual stress and molecular or fiber orientation by increasing the melt temperature, mold-wall temperature, fill time, and cavity thickness, or by decreasing the packing pressure and flow path.

14.11 Undercuts

Undercuts are those portions of the part that cannot be pulled in the line of draw. If an *undercut* was to be machined into a mold without a mechanism to relieve it, the part would be destroyed upon mold opening or ejection (Undercut design 2020).

Slides are usually used for relieving external undercuts or to allow for zero-degree draft on part exteriors. Their motion can be driven by one of several mechanisms. Any slide mechanism must also employ a locking device to hold the slide against the ravages of molding pressures in excess of 10 000 psi. *Undercuts* can usually be divided into two categories, internal and external. *Slides* are used to pull external undercuts and lifters to pull internal undercuts. Referred to in prudish circles as a *cam-pin* or *angle pin*, this is the most common and versatile of all slide mechanisms and is shown in Figure 14.22. It is driven by the opening of the press. The action of the angled pin withdrawing from the angled hole drives the slide back. Springs hold the slide in the retracted position. Upon press closing the pecker pin returns to drive the waiting slide back almost to molding position.

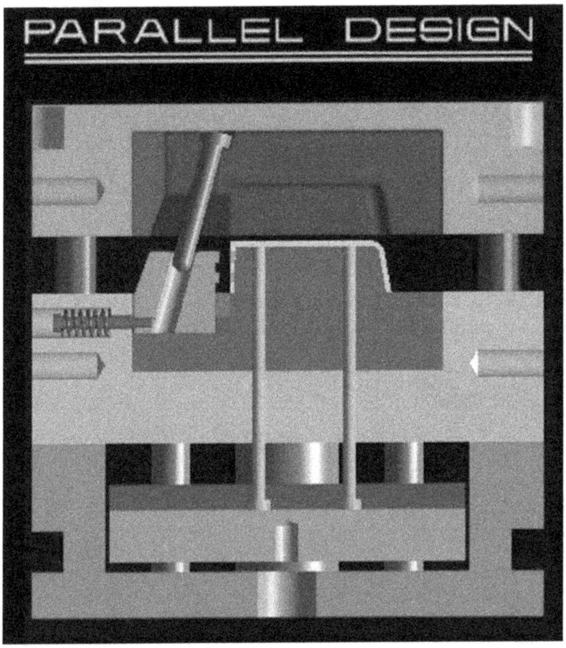

Figure 14.22 Cam-pin or angle pin. Source: Undercut design (2020).

Finally, then the lock seats against the back of the slide, driving it to its home position and clamping it there for the duration of the next shot.

Lifters are used primarily for relieving internal undercuts or for zero-draft internal faces. Their motion is driven by the molding press pushing on the ejector plates. A simple side action lifter withdraws at 90 degrees to the line of draw, simultaneous with ejection. The angle of the lifter passing through the core frequently confuses the novice into thinking that it pulls the feature at an angle. Relative to the core the lifter indeed moves at an angle. Relative to the part, however, a lifter moves directly sideways because the part is being ejected forward at the same rate as the lifter. *Slides* move in die direction to slide a piece of tooling away from part that is used for holes, slots, or openings. *Lifters* move in die direction to lift the part and free an undercut or "in-turn" flange.

14.12 Mold Stack Design

Molds are designed as two- or three-plate molds, depending on the type of component to be manufactured. Mold construction depends on the shape of the component, which determines the parting line selection, runner and gate selection and component ejection system selection. Polymer flow analysis is used to determine the location of runners, gates, and injection. The mold base size depends on component size and number of cavities to be planned per mold. Stack molds are a series of molding faces "stacked" together to create multiple faces or levels for molding. Each level or face is a parting line and produces molded product. The benefit of stack molding is to increase the output of a given molding machine and operation. There are five steps to proper mold stack design (Stack mold 2020).

Step 1: Venting

- Rule 1: Vent to suit anticipated flow of plastic at all points remote from gate.
- Rule 2: Vent at closed end of all pockets, any area not in the direct plastic stream. Ejector pins, slides, and the parting line provide natural vent locations.
- Rule 3: Vents should be large as possible, without causing flash.
- Rule 4: Vents are needed in areas where flow changes direction. Ribs and bosses are particularly troublesome for trapped air and need to be vented with the insert or slide used.
- Rule 5: Vent should be moving and open with every time opens, e.g., ejector pins. Good practice to pressurize stationary vent pins or slots to flow air.
- Rule 6: Vent in land area between cavity and core may be blocked by clamp forces deforming metal.
- Rule 7: Every vent must vent to the outside. If not, then use of vent grooves and channels must be provided.

Step 2: Ejection

- Select and detail method, i.e., stripper, ejector, pins, sleeves, and air poppets.

Step 3: Cooling

- Mold is a heat exchanger
- Want even distribution of the heat in the mold during cooling
- Select appropriate number, diameter, and location of cooling channels

Step 4: Alignment

- Mold cavity and core close together
  - Use mold shoes and leader pins
  - Use tapers, wedges, or pins for cavities and cores

Step 5: Balancing of Runners

- Balanced flow into the cavities is a prerequisite for a quality part. This can be achieved by changing the runner size and length.
- Whenever possible, a naturally balanced runner system should be used to balance the flow of material into the cavities. If a naturally balanced runner is not possible, then the runner system should be artificially balanced, as shown in Figure 14.14.

14.13 Mold Costs

Costs of mold are dependent upon the following:

- Size of part
  - For example, 7 ft long × 2 feet wide P-20 steel tool costs ~\$300 000 (bumper)
  - For example, 10 inches × 8 inches mud tool costs ~\$15 000
- Number of cavities
- Number of slides and lifters
- Tolerance of dimension
- Finish and gloss on mold
- How much is the tool polished
- Time for machining and plunge EDM
- Mold material type:
  - P-20 Pre-hard steel. Rc 29-36. The most common tool steel used, and it is a favorite among mold makers for its durability and machining qualities.
  - H-13 Air hard steel. Rc 46-54. Great for high production cavities and cores, slide bodies, lifters, and gibs.
  - S-7 Air hard steel. Rc 54-56. Used for many of the same applications as H-13; however, somewhat less stable and more prone to cracking. Great for long wearing gibs and guides.
  - M-2 High speed steel. Rc 60-62. Used where tough rigidity is required such as tall thin core pins or blades.
  - 420 SS stainless steel. Rc 49-53. Best for achieving high polish finishes. Also for corrosive polymers such as PVC.
  - Beryllium Copper. Primarily used for inserting small areas in tools that may be prone to overheating. The thermal characteristics of beryllium copper make it the best choice for wicking heat out these areas.
  - 7075 T-6 Aluminum. Generally used for short run or prototype tooling; however, the hardness, wear resistance, and ability to take a good polish have caused many a short run mold to outlast its projected life.
  - Lamina. "Lamina" is a name brand material that is used to wear plates. This material is a bronze over steel laminate that imparts long life to moving parts with a minimum of lubrication. In addition to wear plates, it is commonly used for large locking surfaces.
- Productivity (number of parts per minute)
- Number of cavities
- Quality of cooling
- Cycle time (seconds per part)
- Speed and timing of ejection; Strength and durability of the mold
- Number of cavities

- Mold cost is determined by cost of mold/total number of parts
  - Example,
    - Container mold is built for $50 000 and is expected to produce 5 million parts. Machine costs are $100 per hour.
    - Cycle time is 10 seconds per part in single cavity tool.
    - Mold cost is $50 000/5 000 000 = $0.0100 per part.
    - Dual cavity costs $80K
    - Mold cost is $80 000/5 000 000 = $0.0160 per part.
    - Mold cost is $30 000 more expensive than single cavity mold.
- Machine cost is cost per number of parts per hour.
  - Single cavity = 1 part per 10 seconds or 360 parts per hour = $100/360 = $0.2778 per part
  - Dual cavity = 2 parts per 10 seconds or 720 parts per hour = $100/720 = $0.1289 per part.
  - Savings = $0.1289/part * 5 million parts = $644 500 over life of tool

Multicavity molds require fewer molding machines and thus less floor space.

Cycle time can increase with multicavity molds. Recovery time: time required for the extruder to prepare the plastic for the next cycle. More cavities to fill typically require more time, especially, for large shots. Injection time: time required to inject resin. Volume of resin required is higher though not exceed machine rating. Pressure drop is affected by number of cavities. Wherein, more cavities equal longer flow length and higher pressure. Other plant operates related to multicavity molds. Mold stoppages are due to sloppy molding conditions. Housekeeping is needed for preventative maintenance. Power supply disruptions can cause mold stoppage. Quality of cooling can cause disruptions if mold gets too hot. There is a need for efficient heat transfer to remove heat from plastic and keep cooling time low and cycle time low.

The cost of the molds are also dependent upon the following:

- Speed and Timing of Ejection.
  - Ejection time is lost time, similar to mold open and close
  - Try to eject during mold opening time
  - Free fall ejection requires more stroke and time than Husky "guide shuts," which are more expensive
- Strength and durability of mold.
- Productivity depends upon the quality of the mold.
- Quality of the part starts with the quality of the mold.
- Properly maintained and built mold should last 2–3 million parts (less with glass filled materials).

The cost of injection molds can be reduced with the use of Master Unit Dies (MUD) molds. MUD dies can cost about $5000 for a 6 × 9 × ½ inch MUD base. An

Figure 14.23 MUD die example. Source: MUD (2020).

injection mold of that size can cost up to $30 000. They are the frames that the A and B side molds are inserted into. The A side is the side near in injection unit. Typically, the A side will have the part design machined into it. The B side is the back side of the mold and typically has the ejector pins. Like the A side, the part is machined into the B side. The use of the MUD dies will enable less machining so that the tool is cheaper (DME design guideline 2020). MUD dies enable quick change of dies. Several dies can be built to use the same MUD die. When a product is needed, the A and B sides of the new mold are inserted into the MUD base and then production can start. The smaller production run allows the customers to provide customized products through and optimized supply chain. The low investment cost and operating cost are combined for savings in inventory. Figure 14.23 shows an example of an MUD die.

References

Mekonnen, A.F. and Savas, M.A. (2014). Materials used in automotive manufacture and material selection using ashby charts. *International Journal of Materials Engineering* 40–54. http://article.sapub.org/10.5923.j.ijme.20180803.02.html.

Todor, M.-P., Kiss, I., and University Politehnica Timisoara, Faculty of Engineering Hunedoara, Romania (2016). Systematic approach on materials selection in the automotive industry for making vehicles lighter, safer and more fuel–efficient. *Applied Engineering Letters* 1 (4): 91–97. e-ISSN: 2466-4847).

Wickman, C. and Soderberg, R. (2010). Retrospective analysis and evaluation of non-nominal visualization as means for evaluation of perceived quality in the automotive industry. *Proceedings of NordDesign 2010 International Conference on Methods and Tools for Product and Production Development* (August 2010), Göteborg, Sweden.

Wilhelm, M. (1993). Materials used in automobile manufacture – current state and perspectives. *J. Phys. IV France* **03**. *C7-31-C7-40*, DOI: https://doi.org/10.1051/jp4:1993703.

Websites

3-D printing (2020). https://www.3dprintingtoday.com/benefits-of-3d-printing-in-the-automotive-industry (accessed September 2020).

Bosses (2020). https://studiored.com/injection-molded-part-design-part-3-bosses/ (accessed September 2020).

DME design guideline (2020). https://www.dme.net/mud/ (accessed September 2020).

Draft angles (2020). https://www.slideshare.net/SrinivasTorlakonda/plastics-pptsrinivas (accessed September 2020).

Herringbone runners (2020). http://www.zhilianmould.com/blog/planning-the-runner-system-layout/ (accessed September 2020).

Hydraulic diameter (2020). http://www.zhilianmould.com/blog/designing-runner-cross-sections/ (accessed September 2020).

Material modulus (2020). http://www-materials.eng.cam.ac.uk/mpsite/interactive_charts/stiffness-density/NS6Chart.html (accessed September 2020).

Matweb (2020). www.matweb.com (accessed September 2020).

MUD (2020). https://store.milacron.com/shop-parts?isCategoryRedirect=true&productCategoryId=18050&productCatalogId=94&orgId=2 (accessed September 2020).

Parting line (2020). https://www.engineersrule.com/plastic-parts-design-101/ (accessed September 2020).

Plastic mold (2020). https://www.myplasticmold.com/mold-shrinkage.html (accessed September 2020).

Prospector (2020). www.prospector.com (accessed September 2020).

Ribs (2020). https://www.xcentricmold.com/2018/09/26/ribs-can-enhance-your-injection-molded-part/ (accessed September 2020).

RTP Design guidelines (2020). swww.rtpcompany.com/wp-content/uploads/RTP-Part-Design-Guidelines-for-Injection-Molded-Thermoplastics.pdf (accessed September 2020).

Runner balancing (2020). https://www.design-point.com/resource/blog/9-22-15-runner-balancing-in-plastics/ (accessed September 2020).

Runners (2020). https://www.beaumontinc.com/injection-molding-glossary/runner/ (accessed September 2020).

Screw threads (2020). https://www.pemnet.com/fastening-products/four-ways-to-tackle-threaded-inserts-for-plastics/ (accessed September 2020).

Shrinkage for plastics (2022). https://www.plastikcity.co.uk/useful-stuff/material-shrinkage-rates (accessed September 2022).

Shrinkage plastics (2020). https://www.plastikcity.co.uk/useful-stuff/material-shrinkage-rates (accessed September 2020).

Sink marks (2020). https://hawkridgesys.com/blog/solidworks-plastics-sink-marks (accessed September 2020).

Sink marks and voids (2020). http://www.viewmold.com/sources/Injection_Molding_Defects/Molding_sink_mark_and_void.html (accessed September 2020).

Stack mold (2020). https://www.myplasticmold.com/stack-mold (accessed September 2020).

Transition (2020). https://www.pinterest.co.kr/pin/224265256426784729/?autologin=true (accessed September 2020).

Types of gates (2020). https://www.okmolding.com/how-to-choose-the-correct-mold-steel-and-gate-type/ (accessed September 2020).

Undercut design (2020). http://www.paralleldesign.com/moldability_101/index.htm (accessed September 2020).

Undercuts (2020). https://www.injectionmould.org/2019/04/19/undercuts/ (accessed September 2020).

15

Future of Sustainable Plastics

The future of sustainable plastics can be described as excellent growth, especially for biobased plastics. In 2010, bioplastics comprised less than 1% of the 181 million metric tons of synthetic plastics (Nampoothiri et al. 2010). Biobased polymer production capacity increased from 3.5 million tons in 2011 to 12 million tons in 2020. Bioplastics are expected to comprise 3% of the global polymer production in 2020 (Nova 2013). A market boom is expected for biobased and sustainable plastics in the next few years with production capacities to increase by more than 30% by 2035 (Future market for bio-based plastics 2021). Other sources claim that sustainable plastics will comprise more than 15% of plastic production by 2030 (Biobased plastic capacity 2021).

North America is expected to had modest gains from 159 000 tons in 2011 to 202 000 tons in 2016 (Environmental Leader Calculations 2012). The largest gains are expected for South America and Asia due to better access to agricultural feedstock and a favorable political framework (Nova 2013).

The growth in bioplastics can be supported by the development of biobased plastics from non-food-based agricultural sources, the development of durable products in addition to biodegradable products, and the development of new biobased chemicals for thermoplastic and thermoset polymers.

15.1 Sustainable Biobased Plastics Made from Renewable Sources

Global production of polyhydroxyalkanoates (PHA) and poly lactic acid (PLA) is expected to quadruple in capacity from 2011 to 2020. PHA is expected to be produced in 14 companies at 16 locations in 2020. PLA is expected to be produced in 27 companies at 32 locations in 2020 (Nova 2013).

Sustainable Plastics: Environmental Assessments of Biobased, Biodegradable, and Recycled Plastics, Second Edition. Joseph P. Greene.

PHA and PLA are both produced from sugars that are generated from starchy agricultural sources, or biomass. Currently, corn starch or sugarcane can be used to produce the biopolymers through bacterial synthesis. In 2009, approximately, 190 billion bushels of corn were produced in the United States. Of that as shown in Figure 15.1, approximately 44% of the corn is used as livestock feed, approximately 22% of the corn is used for effective ethanol production, approximately 13% of the corn is exported, approximately 11% of the corn is used for other purposes, and approximately 9% of the corn is used for human food, seed, and industrial use (National Corn Growers 2013). In the world, approximately, 60% of the harvested agricultural biomass is used for animal feed, 32% for human food, 4% for material use, and 4% for energy use (Carus and Dammer 2013).

Future sources of sugars can come from plant oils, lignocellulose, natural gas, and agricultural waste products. In 2008, 10 billion tons of biomass were harvested worldwide. In 2008, 60% of the worldwide agricultural land was used to produce animal feed, 32% of the agricultural land was used to produce agricultural products for food, 4% was used for the creation of materials, and 4% was used for the creation of energy. The materials use included raw materials for the production of biobased plastics and chemicals (Nova Food 2013).

Table 15.1 lists the production yields and products from the fermentation of sugars from agricultural sources. Combustion of bagasse fibers can provide sufficient power and heat requirements for a typical sugar mill (Audsley and Wilkinson 2014).

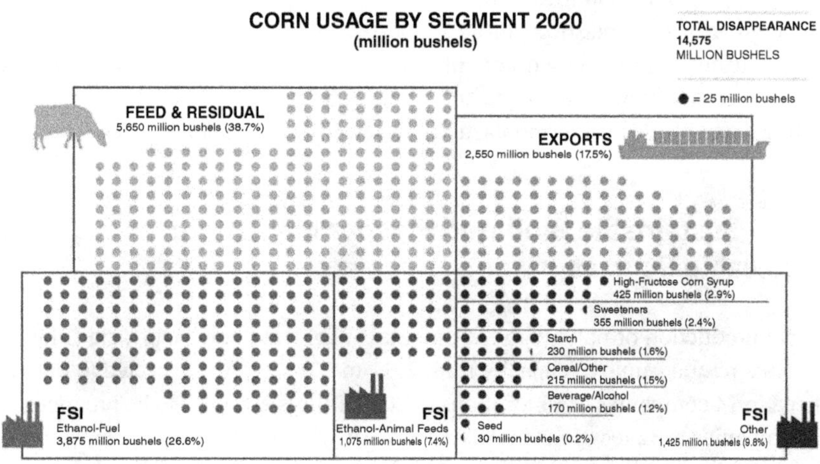

Figure 15.1 Corn utilization in the United States for 2020. Source: USDA (2021).

Table 15.1 Production yields and products from the fermentation of sugars from agricultural sources.

	Yield	DM	ME	CP	GHG, kg CO_2e per			
Crop	**t/ha**	**g/kg**	**MJ/kg DM**	**g/kg DM**	**kg**	**GJ ME**	**kg CP**	**£ value**
Winter bread wheat	7.7	860	13.6	130	0.51	0.044	4.56	3.00
Winter teed wheat	8.1	860	13.6	116	0.46	0.039	4.61	2.83
Winter barley	6.5	860	13.2	123	0.42	0.037	3.97	2.57
Spring barley	5.7	860	13.2	116	0.38	0.033	3.81	2.38
Winter oilseed rape	3.2	930	23.1	212	1.05	0.049	5.33	3.42
Sugar beet	63.0	220	13.2	68	0.04	0.015	2.87	1.25
Main-crop potatoes	52.0	200	13.3	93	0.14	0.053	7.53	2.57
Second-early potatoes	48.0	200	13.3	93	0.10	0.038	5.38	2.90
Field beans	3.4	860	13.3	298	0.51	0.045	1.99	1.98
Soya beans	2.4	860	14.5	415	0.70	0.056	1.96	2.13
Maize grain	7.2	860	13.8	102	0.38	0.032	4.33	2.43
Forage maize (DM)	11.2	280	11.0	101	0.30	0.027	2.97	1.91

DM = dry matter, ME = metabolisable energy, CP = crude protein. Concentrations of DM, ME, CP from Thomas, 2004.
Source: Audsley and Wilkinson (2014).

Table 15.1 shows that sugarcane and beets produce the most amounts of annual sugar and agricultural residues per hectare. Corn produces the most green house gas (GHG) emission and nitrous oxide (NO2) emission per hectare. PLA can be made with starchy biomass and other natural waste products. The production of PLA from starchy substrates can yield a range of lactic acid production.

Table 15.2 provides a list of starch and lignocellulosic substrates for the production of lactic acid (John et al. 2008).

Table 15.2 shows that cellulosic materials are a viable carbon source for bioplastic production. Sugarcane bagasse can be used to produce lactic acid (John et al. 2005). PLA biorefinery system is economically viable with reduced GHG emissions with the use of agricultural residues.

Table 15.2 illustrates that cassava bagasse and whole wheat produce equivalent sugars and lactic acid as corn starch. Cellulose produces less sugars and lactic acid than corn starch.

The costs of the biorefinery depend greatly on the production levels of the manufacturing operation. PLA production from short rotation wood leads to a positive gain of land use per year in Europe (Dornburg et al. 2006).

Table 15.2 Starchy and cellulosic substrate materials for production of lactic acid.

Substrate cone (g/l)	Starch content (g/l)	Fermentation time (d)	Substrate (starch) degraded (g/l)	Lactic acid produced (g/l)	Lactic acid yield (g/g)[a]
Sorghum flour					
20	13.6	2	12.6	11.6	0.92
60	40.8	4	33.0	29.4	0.89
100	68.0	6	35.3	28.7	0.81
CD at 5%	—	—	1.7	3.5	0.03
Cassava flour					
20	16.5	2	15.2	14.0	0.92
60	49.4	4	38.1	33.6	0.88
100	82.0	6	37.0	30.3	0.82
CD at 5%	—	—	1.6	3.7	0.03
Wheat flour					
20	13.8	2	12.6	11.7	0.93
60	41.4	4	33.2	29.9	0.90
100	69.0	6	35.8	29.0	0.81
CD at 5%	—	—	2.6	3.3	0.01
Rice flour					
20	15.6	2	14.3	13.0	0.91
60	46.8	4	35.9	30.9	0.86
100	78.0	6	36.9	29.5	0.80
CD at 5%	—	—	2.0	2.6	0.01
Barley flour					
20	13.8	2	12.6	11.3	0.90
60	41.4	4	31.8	27.3	0.86
100	69.0	7	31.5	25.5	0.81
CD at 5%	—	—	2.6	3.2	0.01

CD at 5% — confidence difference was arrived at 5% level of significance for comparison of any combination or two mean values.
[a] g/g — gram lactic acid per gram substrate (starch) utilized.
Source: John et al. (2005).

PHA can be produced with sugarcane bagasse, waste cheese whey, starchy waste water, and from renewable oil-palm biomass (Yu 2001; Albuquerque et al. 2007; Obruca et al. 2010; Hassan et al. 2013).

A review of the PHA production method with carbon-rich wastes provides promising economic promise for reduced environmental impacts with less land use and food crop use (Nikodinovic-Runic et al. 2013).

15.2 Sustainable Traditional Plastics Made from Renewable Sources

Traditional plastics can be made from agricultural products, including poly ethylene terephthalate (PET), poly butylene terephthalate (PBT), nylon 6, nylon 10, and acrylics. The largest interest has occurred with the biobased PET bottle. Currently, Coca-Cola introduced a biobased PET bottle, for example, Plant Bottle, with 30% biobased and 70% petroleum based (After Dasani Test 2009). PET bottles are made typically from 30% monoethylene glycol (MEG) and 70% terephthalic acid. The MEG can be made with biobased sources and the terephthalic acid is made from petroleum sources. The production capacity of the 30% bio-PET bottle is 452 000 tons. Coca-Cola plans to launch a 100% PET bottle in the future (The Race to 100% bio-PET 2012).

Cradle-to-gate nonrenewable energy use from producing traditional polymers with agricultural product sources is listed in Table 15.3. The energy to produce polyethylene is lowest for sugarcane than for corn, sugar beet, or wheat. The residues from sugarcane processing into sugar are converted into energy resulting in a negative energy use for the production of polyethylene from sugarcane. In all cases, the production of polyethylene from agricultural sources requires 60 to 70% less energy than from fossil fuel sources (Bos et al. 2012).

Production of polyethylene from sugarcane or sugar beets requires half of the land use than from corn (Bos et al. 2012).

Table 15.3 Nonrenewable energy use for the production of one ton of biobased polyethylene.

Biobased material	Energy (GJ per ton)	Land use (ha per ton)
Corn (maize)	29	0.6
Sugar beet	27	0.3
Sugarcane	−18	0.35
Wheat	25	0.68
Fossil fuel (as a reference)	78	—

15.3 Growth in Biobased Plastics with Development of Durable Goods

Biobased plastics can be made into durable goods to complement the biodegradable plastic products. In 2009, durable biopolymers accounted for 7% of the estimated global production capacity of 318 000 tons.

In 2010, durable biobased plastics accounted for 40% of the global capacity of 724 000 tons. The significant contributor to the growth of durable bio-based plastics was Braskem Company, Brazil, which produces sugarcane-based high density polyethylene (HDPE). In 2015, the projected global biobased plastic production capacity is expected to reach 1 700 000 tons with durable biopolymers accounting for 60% of the market (Market Update Bioplastics 2012).

In 2015, the growth in bio-based plastics is expected to be led by bio-polyethylene and bio-PET. Table 15.4 lists the anticipated production volumes of biobased polymers (Market Update Bioplastics 2012).

Traditional plastics can be produced with starch, lignocellulose, plant oils, and sugars. Glucose can be produced from starch, lignocellulose, and sugars. Glucose can be fermented to produce thermoplastic polymers. Glycerol and fatty acids can

Table 15.4 Biobased polymer production capacity for 2015.

Biobased polymer	Capacity (tons)	Percentage of market
Bio-PE	450 000	26
Bio-PET	290 000	17
PLA	216 000	13
PHA	147 100	9
Biopolyesters	143 500	8
Starch blends	124 800	7
Bio-PVC	120 000	7
Bio-PA	75 000	5
Regenerated cellulose	36 000	2
PLA blends	35 000	2
Bio-PP	30 000	2
Bio-PC	20 000	1
Others	22 300	1
Total	1 709 700	100

Table 15.5 Biobased polymer production from renewable sources and intermediates.

Biobased polymer	Intermediate	Intermediate
Bio-PET and bio-PBT	Ethanol, isobutanol	Ethylene, MEG, *p*-Xylene, terephthalic acid
Bio-HDPE and bio-LDPE	Ethanol	Ethylene
Bio-PVC	Ethanol	Ethylene, vinyl chloride
Bio-PP	Ethanol	Ethylene, propylene
Bio-PMMA	Ethanol	Methyl methacrylate
Bio-nylon	Adipic acid	Hexamethylene diamine (HMDA)
Bio-PBS	Succina:e	1,4-butanediol
Bio-PC	Sorbitol	Isosorbide
Bio-SBR	Isobutanol	
Bio-epoxy (thermoset)	Glycerol	Epichlorohydrin
Bio-nylon (thermoset)	Glycerol	Diacids
Bio-polyesters (thermoset)	Glycerol	Polyols, diacids
Bio-PU (thermoset)	Glycerol, triglycerides	Polyol, isocyanates

be produced from plant oils and then produce thermoset polymers. Table 15.5 lists the renewable source and intermediates in the production of traditional thermoplastics and thermosets (Nova 2013). Note that PE represents polyethylene, PVC represents poly vinyl chloride, PP represents polypropylene, PC represents polycarbonate, PMMA represents poly methyl methacrylate, PU represents polyurethane, LDPE represents low density polyethylene and SBR represents styrene butadiene rubber.

Polyurethane polymer can be made with agricultural sources for the polyol and isocyanates (More et al. 2013). Epoxy polymer can be made from grapeseed oil and epoxidized linseed oil (Stemmelen et al. 2011).

15.4 Growth in Biobased Plastics for Pharmaceuticals and Medical Applications

Biobased plastics can be used for coatings for drug delivery, bioabsorbable, and other medical devices. Biopolymers are made from nontoxic materials that are compatible with human tissues. PLA, poly(glycolic acid) (PGA), and polycaprolactone (PCL) are commonly used in biomedical devices (Chen et al. 2004).

The biopolymers are degraded with simple hydrolysis of the ester bonds without the use of enzymes that prevent inflammation. The biodegraded bioproducts are eliminated from the body through normal cellular activity and urine. PLA can be used as a bioabsorbable polymer for resorbable plates and screws (Lasprilla et al. 2012). PLA can provide a biocompatible alternative to titanium plates or other metallic implants that do not erode the bone structures (Dearnaley and Arps 2005).

PHA biopolymers are promising materials for biomedical applications because they are biodegradable, biocompatible, and made from nontoxic materials. The physical and material properties of PHA can be modified with different functional groups formed on the side of the polymer chain. PHAs can be used in applications for tissue engineering, drug delivery, and scaffolding for arteries (Hazer et al. 2012; Chen et al. 2010; Zinn et al. 2001).

15.5 Summary

The future of sustainable plastics is expected to have excellent growth, especially for biobased plastics. The growth in bioplastics can be supported by the development of biobased plastics from non-food-based agricultural sources, the development of durable products in addition to biodegradable products, and the development of new biobased chemicals for thermoplastic and thermoset polymers.

Bioplastics in the future can be produced from plant oils, lignocellulose, and agricultural waste products. Sugarcane and beets produce the most amounts of annual sugar and agricultural residues per hectare.

Corn produces the most GHG emission and NO2 emission per hectare. For PLA production, cassava bagasse and whole wheat produce equivalent sugars and lactic acid as corn starch. Cellulose produces less sugars and lactic acid than corn starch. In a PLA biorefinery system, agricultural residues can be used to produce PLA with economic viability and reduced GHG emissions.

Traditional plastics can be made from agricultural products. PET can be made currently with up to 30% from biobased sources. PET in the near future can be made with 100% plant-based materials. Polyethylene is currently made from sugarcane. Polypropylene will soon be made from sugarcane.

Biobased plastics can be used for coatings for drug delivery, bioabsorbable, and other medical devices. Biopolymers are made from nontoxic materials that are compatible with human tissues. PLA, PGA, and PCL are commonly used in biomedical devices.

References

After Dasani Test (2009). Coke begins global rollout of plant bottle. *Environmental Leader, Environmental and Energy Management News*. http://www.environmentalleader.com/2009/11/16/after-dasani-test-coke-beginsglobal-rollout-of-plantbottle/ (last accessed June 2013).

Albuquerque, M., Eiroa, M., Torres, C. et al. (2007). Strategies for the development of a side stream process for polyhydroxyalkanoate (PHA) production from sugar cane molasses. *J. Biotechnol.* 130: 411–421.

Audsley, E. and Wilkinson, M. (2014). What is the potential for reducing national greenhouse gas emission from crop and livestock production systems? *J. Clean. Prod.* 73: 263–268.

Bio-based plastic capacity (2021). https://www.prnewswire.com/news-releases/sustainable-plastics-will-make-up-more-than-15-of-production-by-2030-301274538.html (accessed June 2021).

Bos, H., Meesters, K., Conijn, S. et al. (2012). Accounting for the constrained availability of land: a comparison of bio-based ethanol, polyethylene, and PLA with regard to non-renewable energy use and land use. *Biofuels Bioprod. Biorefin.* 6 (2): 146–158.

Carus, M. and Dammer, L. (2013). Food or non-food. Which agricultural feedstocks are best for industrial uses? *Bioplastics Magazine* 8: 42.

Chen, J., Zheng, Z., Chen, J., and Chen, G. (2004). Plyhydroxyalkanoate synthases PHAC1 and PHAC2 from pseudomonas stutzeri 1317 had difference substrate specificities. *FEMS Microbiol. Lett.* 234 (2): 231–237.

Chen, F., Zhang, M., and Wu, Z. (2010). Toward delivery of multiple growth factors in tissue engineering. *Biomaterials* 31 (24): 6279–6308.

Dearnaley, G. and Arps, J. (2005). Biomedical applications of diamond-like carbon (DLC) coatins: A review. *Surf. Coat. Technol.* 200: 2518–2524.

Dornburg, V., Faaij, A., Patel, M., and Turkenburg, W. (2006). Economics and GHG emission reduction of a PLA bio-refinery system – combining bottom-up analysis with price elasticity effects. *Resour. Conserv. Recycl.* 46: 377–409.

Environmental Leader Calculations (2012). Environmental leader calculations based on European bioplastics/Institute for bioplastics and biocomposites, European bioplastics, labelling, logos, charts. http://en.european-bioplastics.org/press/press-pictures/labelling-logos-charts/ (accessed June 2013).

Future market for bio-based plastics (2021). https://www.businesswire.com/news/home/20210107005460/en/Global-Market-for-Biobased-and-Sustainable-Plastics-2020-2030---ResearchAndMarkets.com (accessed June 2021).

Hassan, M., Yee, L.-N., Yee, P. et al. (2013). Sustainable production of polyhydroxyalkanoates from renewable oil-palm biomass. *Biomass Bioenergy* 50: 1–9.

Hazer, D., Kilicay, E., and Hazer, B. (2012). Poly(3-hydroxyalkanoate)s: diversification and biomedical applications: a state of art review. *Mater. Sci. Eng.* 32 (4): 637–647.

John, R., Nampoothiri, K., Nair, A., and Pandey, A. (2005). L (+)-lactic acid production using Lactobacillus casei in solid-state fermentation. *Biotechnol. Lett.* 27: 1685–1688.

John, R., Gangadharan, D., and Nampoothiri, K. (2008). Genome shuffling of Lactobacillus delbrueckii mutant and Bacillus amyloliquefaciens through protoplasmic fusion for L-lactic acid production from starchy wastes. *Bioresour. Technol.* 99: 8008–8015.

Lasprilla, A., Martinez, G., Lunelli, B. et al. (2012). Poly-lactic acid synthesis for application in biomedical devoces – a review. *Biotechnol. Adv.* 30: 321–328.

Market Update Bioplastics (2012). Why green plastics are here to stay. Compounding World. http://www.compoundingworld.com/Issue.aspx?IssueId=113.

More, A., Lebarbe, T., Maisonneuve, L. et al. (2013). Novel fatty acids based di-isocyanates towards the synthesis of thermoplastic polyurethanes. *Eur. Polym. J.* 49: 823–833.

Nampoothiri, K., Nair, N., and John, N. (2010). An overview of the recent developments in polylactide (PLA) research. *Bioresour. Technol.* 101: 8493–8501.

National Corn Growers (2013). U.S. Corn growers meet the need for fuel and food. National Corn Growers Association. http://www.ncga.com (accessed October 2013).

Nikodinovic-Runic, J., Guzik, M., Kenny, S. et al. (2013). Chapter four–carbon-rich wastes as feedstocks for biodegradable polymer (polyhydroxyalkanoate) production using bacteria. *Adv. Appl. Microbiol.* 84: 139–200.

Nova (2013). Market study on bio-based polymers in the world. Nova-Institute, Press Releases. http://www.bio-based.eu/market_study/pressrelease (accessed June 2014).

Nova Food (2013). Food or non-food: Which agricultural feedstocks are best for industrial uses?. Nova – Institut GmbH. http://www.nova-institut.de/download/paper2.

Obruca, S., Marova, I., Snajdar, O. et al. (2010). Production of poly(3-hydroxybutyrate-co-3-hydroxyvalerate) by Cupriavidus necator from waste rapeseed oil using propanol as a precursor of 3-hydroxyvalerate. *Biotechnol. Lett.* 32: 1925–1932.

Stemmelen, M., Pessel, F., Lapinte, V. et al. (2011). A fully bio-based epoxy resin from vegetable oils: from the synthesis of the precursors by thiol-ene reaction to the study of the final material. *J. Polym. Sci. A Polym. Chem.* 49 (11): 2434–2444.

The Race to 100% BioPET (2012). The race to 100% BioPET. The Plastics Engineering Blog, Society of Plastics Engineering. http://www.plasticsengineeringblog.com/2012/08/13/the-race-to-100-bio-pet/ (accessed September 2022).

USDA (2021). USDA, ERS Feed outlook, Jan. 15, 2021; ProExporter Network, Projected Crop Year Ending Aug. 31, 2021.

Yu, J. (2001). Production of PHA from starchy wastewater via organic acids. *J. Biotechnol.* 86: 105–112.

Zinn, M., Witholt, B., and Egli, T. (2001). Occurance, synthesis, and medical application of bacterial polyhydroxyalkanoate. *Adv. Drug Deliv. Rev.* 53 (1): 5–21.

Index

Sustainable Plastics: Environmental Assessments of Biobased, Biodegradable, and Recycled Plastics, Second Edition. Joseph P. Greene.

d

e

f

j

k

l

r

S

CPSIA information can be obtained
at www.ICGtesting.com
Printed in the USA
BVHW050821210123
656311BV00034B/29

9 781119 882060